Notebooks, English Virtuosi, and
Early Modern Science

Notebooks, English Virtuosi, and Early Modern Science

Richard Yeo

The University of Chicago Press

Chicago and London

Richard Yeo is adjunct professor in the School of Humanities, Griffith University, Australia, and a fellow of the Australian Academy of the Humanities. He is the author and editor of numerous books, including *Defining Science* and *Encyclopaedic Visions*.

The University of Chicago Press, Chicago 60637
The University of Chicago Press, Ltd., London
© 2014 by The University of Chicago
All rights reserved. Published 2014.
Printed in the United States of America

23 22 21 20 19 18 17 16 15 14   1 2 3 4 5

ISBN-13: 978-0-226-10656-4 (cloth)
ISBN-13: 978-0-226-10673-1 (e-book)
DOI: 10.7208/chicago/9780226106731.001.0001

Library of Congress Cataloging-in-Publication Data

Yeo, Richard R., 1948– author.
    Notebooks, English virtuosi, and early modern science / Richard Yeo.
        pages   cm
        Includes bibliographical references and index.
    ISBN 978-0-226-10656-4 (cloth : alk. paper)
    ISBN 978-0-226-10673-1 (e-book)
    1. Science—England—History. 2. Scientists—England. 3. Hartlib, Samuel, –1662.
    4. Beale, John, 1603–1683? 5. Boyle, Robert, 1627–1691. 6. Locke, John, 1632–1704.
    7. Hooke, Robert, 1635–1703. I. Title.
    Q127.G4Y47 2014
    509.2′242—dc23

                                                            2013043063

♾ This paper meets the requirements of ANSI/NISO Z39.48-1992 (Permanence of Paper).

For Mary Louise

[Contents]

[Editorial Notes]

Early modern England used the Julian calendar and did not adopt the Gregorian calendar (used throughout continental Europe) until 1752. Some sources, such as letters between England and the Continent, include two dates, which indicate the ten-day difference between the two calendars. References are cited using short titles; their full details appear in the bibliography. For those sources not included in the bibliography, full details are given in the notes. For quotations from primary sources, all contractions have been expanded, and original spelling has been maintained. Editorial inference is included in square brackets.

In chapter 7, the three English versions of John Locke's "New Method" will be referred to thus: BL, Add. MS 28728, fols 54–63, as "New Method," English draft; the English translation in *Posthumous Works* (1706), as "New Method," *Posthumous Works*; the English translation published by Greenwood (1706), as "New Method," Greenwood.

The following abbreviations have been used throughout the text:

BL	British Library, London
Bodl	Bodleian Library, The University of Oxford
BP	Robert Boyle Papers, The Royal Society of London
CUL	Cambridge University Library

EEBO	Early English Books Online. http://eebo.chadwyck.com/home
HP	*The Hartlib Papers*
ODNB	*Oxford Dictionary of National Biography*, 61 vols., edited by H. C. G. Matthew and Brian Harrison (Oxford: Oxford University Press, 2004–12). http://www.oxforddnb.com
OED	*Oxford English Dictionary*, edited by J. A. Simpson and E. S. C. Weiner, 20 vols, 2nd ed. (Oxford: Oxford University Press, 1989). http://www.oed.com
OFB	*The Oxford Francis Bacon*, 15 vols., Chair of Editorial Advisory Board, Sir Brian Vickers (Oxford: Clarendon Press, 1996–).
RS	The Royal Society of London
RS CP	The Royal Society of London, Classified Papers.

[Preface]

Notes can record passages from books, register observations of the world, and capture passing thoughts. They can be taken and kept in various places — in manuscripts and printed books, on loose slips of paper, on index cards, in specially bound notebooks of various dimensions from pocket-size to large folios; and they can be gathered according to different methods — by date, by topic, or by roughly arranged numbered points in lists of facts, queries, and wishes. Notes can be straightforward drafts of future compositions or, as footnotes, they can reveal the underpinnings of a published text.[1] Due in part to a combination of contingency and deliberation, notebooks can become talismans: consider Leonardo da Vinci's small folded pages of mirror writing; René Descartes' lost notebook, painfully reimagined by Gottfried Wilhelm Leibniz; Isaac Newton's youthful pocket-book, heavily occupied with inventories of his sins; Charles Darwin's "red notebook" containing his first graphic depiction of the branching tree of evolution by natural selection; the famous Moleskine notebooks favored by artists and travelers in the nineteenth century; or the sketch-book of fantastic machines treasured by the young Hugo Cabret in the novel by Brian Selznic.[2] Tantalizingly, notes can preserve a tiny part of a lost whole, or act as pithy condensations of ideas never fully committed to paper. Notes work to jog the memory of those who make them, or serve as records for others, including future generations.

In this book I show how some leading English figures in early modern science kept notebooks and thought about note-taking and the collection and retrieval of information. Robert Boyle, John Aubrey, William Petty, John Evelyn, John Locke, Martin Lister, John Ray, and Robert Hooke, in association with their European network of friends and correspondents, drew upon Renaissance humanist practices of excerpting from texts to build storehouses of proverbs, maxims, quotations and other material in personal notebooks. These were called "commonplace" books because they grouped like with like — in a common place, under subject or topic headings. The complication here is that this method of "commonplacing" belonged to a traditional textual mode of pedagogy and scholarship often disparaged by the scientific "moderns" as out-dated and unsuitable for the investigation of the natural world. For example, in the early 1600s Galileo Galilei wrote to Johannes Kepler about the stupidity of scholars who persisted in believing that nature could be understood by studying books:

> This kind of man thinks that philosophy is a sort of book like the *Aeneid* and *Odyssey*, and that truth is to be found not in the world or in nature but in the collation of texts (I use their terminology). I wish I could spend a good long time laughing with you.[3]

Here Galileo voiced, somewhat in jest, what would soon be construed as an inherent tension between philological scholarship and the empirical (including experimental) study of the world.

The English philosopher and Lord Chancellor Francis Bacon also denigrated overreliance on books, identifying it as a major factor in the stagnant condition of natural knowledge in his time. His message found a receptive audience among members of the Royal Society of London (founded in 1660) who, in polemics against their critics, constructed an opposition between books used by previous generations of natural philosophers and their own observations and experimental methods. At the same time, the people I discuss, mostly Fellows of the Society and admirers of Bacon, understood that his castigation of "everything philological" coexisted with a call for the gathering of information in natural histories. This project required extensive and unrelenting note-taking.[4] Today we take for granted the gradual stockpiling of data in the form of detailed notes carefully made in laboratories, on field excursions, or automatically registered by machines. In the seventeenth century there was a conviction that, to a large extent, copious knowledge could be reliably stored and manipulated in memory. However, during the Scientific Revolution a contrary view was emerging: namely, that the advancement of natural knowledge entailed a reconfiguration of the balance between memory and other ways of storing information. It was accepted that the empirical sciences demanded large quantities of detailed information that needed to be recorded with precision, and kept as durable records to be shared and communicated. By reflecting on the best use of memory, notebooks, and other records in the collection and analysis of empirical information, the English *virtuosi* were prominent contributors to such a reassessment.[5] They took abundant notes, both as a way of dealing with the proliferation of printed books, and as a means of assembling and securing information that books did not supply. They were able to reach into the past beyond both Bacon and the humanists to find recognition of the need for long-term inquiry, and its challenges, in the ancient Hippocratic medical tradition. In doing so, they made note-taking and information management a crucial part of the modern scientific ethos.

This book contributes to ongoing scholarship in the history of science, the history of the book, and the cultural history of information in early modern Europe. First, it has implications for a version of "the two culture" question in the early modern period, as implied in the formulations of Galileo and Bacon. Important work has already been done on the close relations between humanists and scientific figures, thereby cautioning against any severe oppositions of "bookish" and empirical, or solitary and collective, inquiry.[6] Indeed, on occasions, when alluding to their own extensive reading, or the company they kept, men such as Boyle and Locke identified themselves as "bookish."[7] What was in dispute, in their view, was not the value of books but the use made of them. So Locke criticized "learned bookish Men, devoted to some Sect" who were incapable of rational conversation with others.[8] He also ridiculed those who depended on commonplace books stocked with arguments "pro and con" on questions they had not properly considered: they were able, he said, "only to talk copiously on either side, without being steady and settled in their own judgments."[9] My exploration of the note-taking of the English virtuosi offers new perspectives on the ways in which individuals inhabited what later would become distinct intellectual worlds—of philological and empirical scientific inquiry.[10]

Second, the book adds to discussion of the complex relationship between print and manuscript cultures. Since Elizabeth Eisenstein's strong claims about the nexus between printing and the advances of the Scientific Revolution, there has been a reassertion of the continuing function of manuscript exchanges in early modern intellectual life.[11] Harold Love demonstrated how such exchanges operated in political and musical coteries, and more recent work has indicated a similar process at work in intellectual and scientific networks.[12] Recent studies of medical, chemical, botanical, experimental, and antiquarian circles in early modern Europe show that information in print was converted through note-taking and correspondence into what Love called scribal communication. Such notes were regarded as more flexible than print, better to think with, and well suited to quick communication among small self-defining groups. This was not nostalgia for manuscript culture; rather, it reflected a judgment that the fixity offered by print was most beneficial toward the end, not the start, of scientific inquiry. Certainly, there was some use of printed questionnaires for wider distribution, but since these usually included an invitation to suggest additional queries, more manuscripts were generated.[13]

Third, I am concerned with the interplay between individual memory and externalized records in the storing and processing of information. This issue is a vibrant topic in contemporary cognitive psychology. I consider a historical dimension by analyzing the notebooks and reflections of the scientific virtuosi who confronted the vast mass of information required in empirical inquiry. An appreciation of their thinking requires close attention to distinctions between memory, recollection, and recorded information. It is tempting to delineate a shift during the late 1600s from a celebration of memory, especially the value of memory training, to greater and more regular dependence on written and printed records. However, it is difficult at this stage of our historical understanding to make definitive statements with confidence. My work is based on the notes, notebooks, and letters now extant in the papers of some significant figures of the seventeenth-century English intellectual scene. One inference from this material is that the choice between reliance on memory, with or without the aid of notes, and recourse to external records was negotiated in different ways, and with different outcomes, by individuals and groups confronting various intellectual problems. Although I show that Bacon and his followers in the Royal Society warned that memory was unreliable in managing details (or as they said, "particulars"), some of their responses involved new ways of arranging material in order to aid memory, and hence thinking. Furthermore, although these people wanted good archives to form the bedrock of scientific inquiry, there remained a conviction that the deep memory and lifetime experience of individuals were also important factors in stimulating intellectual advances.

In *The Art of Travel* (1855), Francis Galton, a Victorian polymath and pioneer of psychometric studies, offered practical tips on how to take notes and keep notebooks. He discussed the best kinds of paper, pencil, and ink, as well as various kinds of pocket-books and the transfer of information across them. He stressed that notes must be made while the memory of events was fresh, and in a manner that made them potentially useful to others: "It is very important that what is written should be intelligible to a stranger after a long lapse of time."[14] Most of this advice would have been familiar in the seventeenth century, so we must confront this apparent need to rediscover past lessons about tools and techniques for gathering and storing information. The English virtuosi drew on the practices of traditional pedagogy that they often attacked; they also made what they considered to be innovations, albeit not always realizing that some scholars before them had made similar moves. This was often because their sense of what was standard in earlier note-taking relied on published tips and precepts—especially those of humanist and Jesuit authors—rather than on close acquaintance with the private notes of individuals. For example, they were not aware that Martin Crusius (1526–1607), professor of Greek at Tübingen, filled his copy of the works of Homer with annotations linked to a massive diary and other loose notes and letters, all displaying intricate ways of copying, dating, cross-referencing, and analyzing material from an impressive array of sources and informants.[15] In this respect, he was typical of a generation of European scholars who created lasting techniques for condensing, organizing, and retrieving printed information.[16]

As we shall see, in the 1640s the Prussian émigré Samuel Hartlib investigated similar practices in his efforts to coordinate information within scholarly and scientific circles in London and Europe. He was therefore excited about the appearance of a novel device, designed (but not produced) by Thomas Harrison for indexing notes on loose slips, an invention that subsequently lapsed into obscurity. Subsequently, the twentieth-century German sociologist Niklas Luhmann (1927–98) pursued a sophisticated method for numbering, filing, and linking notes on slips of papers in a set of boxes (*Zettelkästen*), probably without awareness of this early modern predecessor.[17] It seems that Galton's reissuing of old advice (albeit to a new audience of explorers and adventurers) is a feature of the history of note-taking: at various times the skills and tricks already fashioned in one context need to be reinvented in another.[18] Thus in discussing early modern note-takers, I am not contending that they created wholly novel methods or techniques; rather, I am interested in how they developed an established practice to meet what they perceived as new challenges associated with the progress of the new sciences.

The English virtuosi regarded note-taking as an essential component of empirical inquiry in the sciences. In their view, this followed from the observational requirements of both natural history and experimental chemistry and physiology. It therefore applied to Evelyn's work on horticulture, to Boyle's pioneering research on the weight of the air and on human blood, to Locke's interest in weather phenomena and the aetiology of diseases, and to Hooke's contentions about fossils and earthquakes. These interests were conceived as contributions to what Bacon defined as "natural history."[19] For some, such pursuits were driven by a curiosity for the rare, monstrous, and marvelous, but this passion does not fit the leading virtuosi without significant qualifications.[20] They agreed that the focus of the new science could not be restricted to what Aristotle understood as "common" experience; they accepted Bacon's insistence that it must also incorporate "*Deviating Instances*, i.e. nature's mistakes, vagaries and monsters, where nature strays and turns aside from her ordinary course."[21] They also took seriously his demand for more detailed information concerning what appeared to be banal and everyday phenomena — as part of the task of establishing a baseline for generalizations. Boyle announced that this project called for a certain kind of person — one with an intellectual disposition open to all kinds of information. He specified a "Docility" (or evenness) of temperament required "to be a *Virtuoso*."[22] Such a posture eschewed both rhetorical ornamentation and theoretical presuppositions in favour of first-hand observations and experimental reports.[23]

What followed from this was a realization that careful note-taking was the way to secure all information. First, although curious objects and artificially produced phenomena might make a lasting mental impression, diverse testimony, observation, and experiment had to be recorded in order to be properly assessed and compared, often over long periods of time. Second, in the absence of an overarching theory or system, specific details, including circumstantial ones of time and place, could not be entrusted to memory. Third, it was recognized that pieces of data were more easily moved around and combined on paper than in the mind.[24] In adhering to these three caveats, the intellectual stance of the English virtuosi became closely connected with taking notes and thinking about the process of distributing information across memory and external records.

For those involved in Baconian inquiries, the satisfaction of discovery and understanding might not be guaranteed within a single human lifetime. The English virtuosi therefore wanted to ensure that personal notebooks contributed to a collaborative enterprise and a lasting scientific archive. One difficulty here was that personal notes came under the control of their owners, who improvised on modes of entering and arranging to suit their habits and preoccupations; indeed, individual choices such as these gave notes their power to remind and prompt. However, to be effective, long-term scientific inquiry required agreement on standard methods of collecting, displaying, and communicating material—a consensus that, ideally, held over generations. Furthermore, empirical information acquired through various networks, such as those of commerce, trade, religious missions, and the Republic of Letters, had to be filtered and vetted.[25] The early Royal Society relied on such networks and was therefore compelled to monitor the criteria by which credit, plausibility, and doubt were attributed.[26] Leading members, often in correspondence with Henry Oldenburg, their industrious secretary, wanted to reduce the contingencies involved, and they decided that one fundamental variable was the form and function of notes. By considering this issue as part of the process by which information was gathered, recorded, and transmitted, they initiated the steps that eventually made routine collection of information a powerful method. By the 1840s it was possible for *Punch* to satirize what was, by that time, an uncontentious feature of scientific research: "Mr Softinwitz, who had been appointed to keep a register of the shocks of earthquakes in Regent-street during the past year, handed in four reams of blank foolscap as a result of his observations."[27]

Chapter 1 introduces the English virtuosi and their attitudes toward reading, learning, and science; the kinds of notes and notebooks used in early modern Europe, and the ways in which the relationships between memory, notes, and thinking were conceived. Chapter 2 discusses the humanist and Jesuit traditions of note-taking and memory training, and includes an examination of the notebooks of John Evelyn and Robert Southwell. Chapter 3 addresses the new imperative for empirical information as forecast by Bacon and embraced by Fellows of the Royal Society; it argues that thinking about notes clarified some of the challenges and opportunities of the nascent empirical sciences. The next four chapters deal with key individuals, delineating their views on the methods and purposes of note-taking, deep personal memory, systematic display of information, and retrieval from written records. Each chapter taps a rich personal archive, beginning in chapter 4 with the massive surviving set of papers collected by the London intelligencer Samuel Hartlib. These include his own diaries and the letters and documents generated by the network of correspondence he orchestrated, including material from John Beale, William Petty, John Pell, and Thomas Harrison. Chapter 5 considers Boyle's views on memory and note-taking in the light of his relationship with Hartlib, Beale, and other members of this circle. Chapter 6 draws on a section of Boyle's papers to describe his mode of note-taking and his ideas about the relationships between notes, memory, and the communication of scientific information. Chapter 7 explains the method and rationale of Locke's lifetime of assiduous note-taking, with illustrations from his personal notebooks; and it raises the problem of making this method work for large-scale inquiries. Chapter 8 examines the challenge of collective note-taking in the service of collaborative projects, such as those of John Ray and Martin Lister, and the importance of Oldenburg as the Royal Society's information manager. The main figure in this chapter is Robert Hooke, whose views on note-taking are closely interwoven with his vision of an institutional archive. Chapter 9 provides a conclusion.

[1]

Introduction

Is it not evident, in these last hundred years (when the study of philosophy has been the business of all the Virtuosi in Christendom), that almost a new Nature has been revealed to us?

John Dryden, *Of Dramatic Poesy* (1668)

: : :

The most famous diarist of the seventeenth century, Samuel Pepys, gave a vivid eyewitness account of the Great Fire of London, which began about midnight on Sunday, September 2, 1666. We have his description of its "most horrid malicious bloody flame, not like the fine flame of an ordinary fire." Some days later Pepys felt compelled to record that at last he was able to return to "my Journall-book to enter for five days past."[1] This entry displays the devotion to a daily task, interrupted only by catastrophe and now made good, relying on memory.[2] Thanks to similar punctilious note-taking, we have another, less direct, notice of the fire from John Locke, about fifty miles away in Oxford.

Locke's record is interesting because he did not know what he was seeing. On the morning of Tuesday, September 4, he made his daily weather observations from his rooms in Christ Church. He had started to keep a "Register" of the weather on June 24 of that year and continued to do so—with various interruptions—until June 30, 1683.[3] At 9 a.m. on that day in September 1666, he entered his usual set of figures about temperature, air pressure, humidity, and the strength and direction of the wind. At 1 p.m. he made another series of entries, adding "dim redish sun shine" in the column marked "Weather."[4] At 8 p.m. he made a third set.[5] Either at that time or later, he inserted a note in the far right of the page reserved for comments: "This day the sun beams were dim'd & of an unusuall colour heare at Oxford which I observed at 12 a clock & all that afternoon & others in the morning, which was occasiond by the smoak of London burning."[6] Locke now realized the significance of his earlier observational report (see fig. 1.1).

Here we have two notebooks kept by English virtuosi. Both evince the methodical note-keeping habits urged by Renaissance pedagogues and Puritan moralists. Yet examination of Locke's notebook also highlights an instructive complication. Locke entered his daily weather observations in the carefully ruled pages he prepared for this purpose. This "Register" occupies the final pages of one of his large folio-sized commonplace books, arranged by subject, not by chronology as in the case of all diaries. This placement may seem to indicate a distinction in kind, separating the weather diary from the topical arrangement of the copious notes from his reading of scientific, especially medical, books. However, by making the weather observations under the heading of "Aer 66" (top center of the first page) he signaled the use of the method of commonplacing. The "Head" (heading) indicated that this information belonged to a topic — namely, the nature and phenomena of the "air," one that occurs elsewhere in this and other notebooks (see fig. 1.2).[7] Locke regarded his observations and measurements as a contribution to a long-term collaborative project on the natural history of the air that had implications for the genesis and transmission of diseases. The "Register" was later published in Robert Boyle's *The General History of the Air* (1692), a book prepared for the press by Locke in his role as one of Boyle's literary executors.[8] This work collated information collected by various observers in response to Boyle's queries, probably first issued in the 1660s. The challenge was to coordinate notes made by individuals, often under different principles — such as topical or chronological — so that they might be of collective and public value.

Scholarship over the last few decades has contended that the rise of early modern science was deeply enmeshed within a range of humanist, legal, and social ways of inquiring. As an intellectual pursuit cultivated in new institutions, empirical scientific inquiry drew to some extent upon the methods of older disciplines, such as medicine and law, and the ethos of gentlemanly codes of civil conduct.[9] I think note-taking needs to be added to this set of

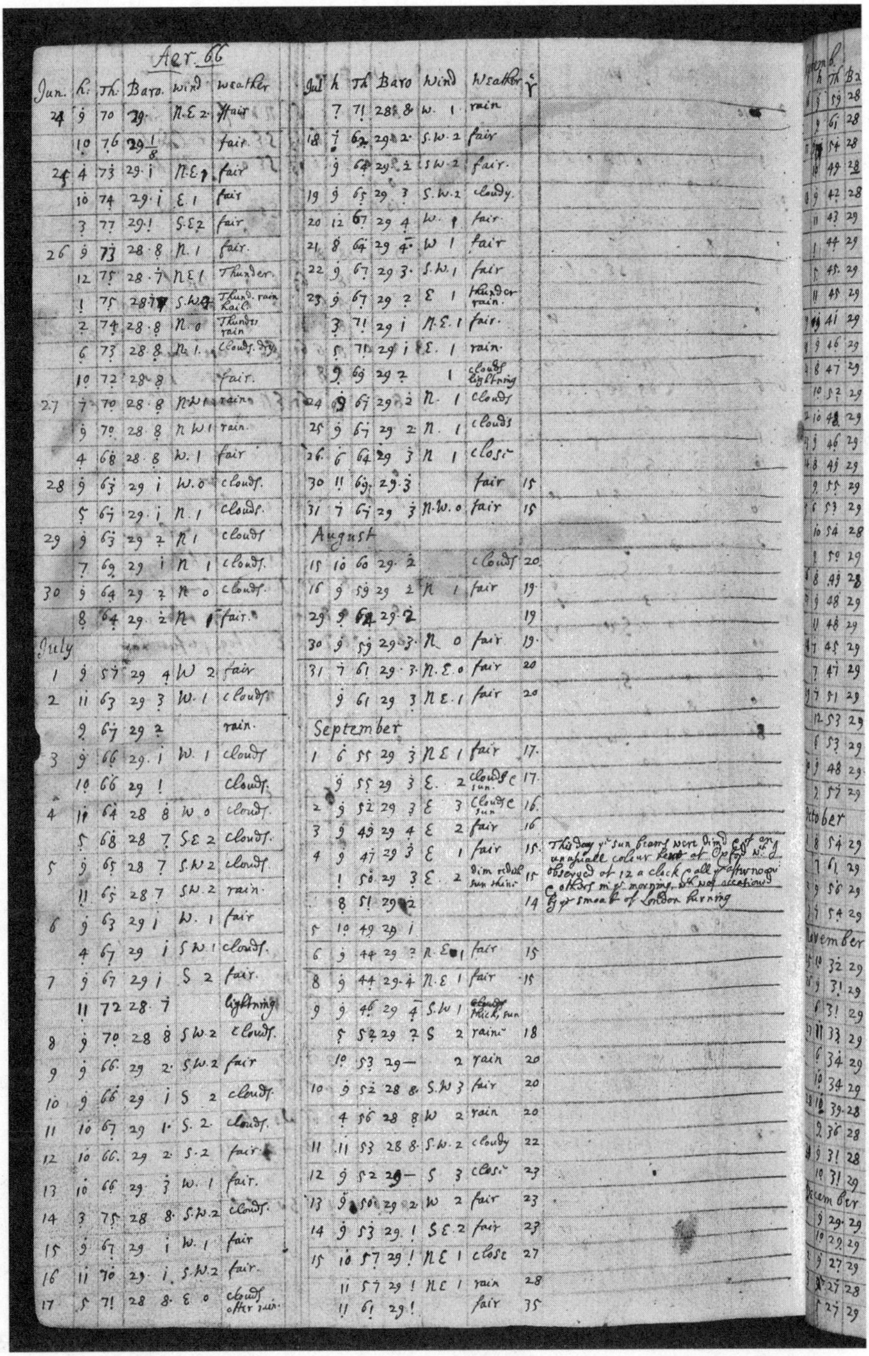

Figure 1.1. Locke's weather Register in Oxford: on September 4, 1666, in the right-hand column, he notes the effects of the Great Fire of London. MS Locke d. 9, p. 530. By permission of the Bodleian Libraries, the University of Oxford.

preoccupations and techniques.[10] In *Notebooks, English Virtuosi, and Early Modern Science* I consider this practice as a way of exploring attitudes toward both memory and information in the early modern period (say, 1550 to 1700), focusing mainly on the second half of the seventeenth century. In particular, the case of scientific knowledge (especially Baconian natural history) is revealing because the virtuosi who sought new empirical information used notebooks in both traditional and novel ways, as aids to memory and as records of information that might potentially have value for those who did not collect the original material. These English figures were aware of the note-taking techniques promoted by Renaissance humanists and inculcated in grammar schools and universities throughout Europe. They were not the first to extend this practice from the selection of textual passages to the recording of observations of the social and natural worlds.[11] However, they may be distinctive in their reflections on note-taking and its relationship with the demands of empirical inquiry. As members of the early Royal Society of London, these virtuosi had to confront the personal problem of dealing with empirical information that exceeded the capacity of memory, and the institutional one of collecting private notes in a scientific archive. I aim to bring these people together, comparing their assumptions and methods, and considering them in the context of European concerns about the management of information and knowledge from Francis Bacon and René Descartes to Gottfried Wilhelm Leibniz.

Any discussion of "science" in the seventeenth century requires a discussion of terminology. It is now well recognized that this word did not exclusively denote the study of the natural world, as it came to do (especially in English) by the mid-nineteenth century. In the early modern period, the term "science" referred to bodies of systematic knowledge, including grammar, geometry, and theology. When it was applied to knowledge of nature it was usually limited to those disciplines that claimed, or sought, certain or axiomatic knowledge. In her study of Elizabethan inventors, surgeons, apothecaries, and mathematicians, Deborah Harkness has suggested that they began to use the word "science" in ways that anticipate the later, more restricted, "modern" usage.[12]

But even if this is so, the older concept of "science" was still alive among some of Bacon's ardent followers who cultivated natural knowledge. Thus in 1648, Samuel Hartlib, a mentor to several members of the Royal Society, entertained at least quasi-demonstrative aspirations: "By Science wee conceive a Certain Body of Notions set in order to inable the Mind of Man to discerne the Principles of all Thinges; whereof there are certain and constant causes existent in nature; and to Demonstrat the production of the Effects which naturally follow thereupon."[13] With this criterion as a benchmark, most of the new experimental inquiries fell short, although natural philosophy and mixed mathematics, understood as the study of causes anchored in principles, came closer than natural history, if regarded solely as an exercise in description and classification. As late as 1690, Locke suspected "that natural Philosophy is not capable of being made a Science."[14] He certainly did not mean that this subject was unworthy of study, but that its achievements were likely to be less stable than the demonstrative knowledge expected of disciplines, such as mathematics and ethics, which he called "sciences." Similar caution must be taken with words we use in speaking about the people who sought knowledge of nature. The word "scientist" was coined by William Whewell in 1833 and not widely used until after 1900. In writing about the seventeenth century, I therefore use the actors' terms: natural historians, natural philosophers, physicians, experimental chemists, and virtuosi.

In addition, some of the people I discuss referred to themselves as "moderns." Thus Boyle frequently spoke of the learned or ingenious "moderns," usually in contrast with those adhering to older views on medicine, chemistry, and natural philosophy deriving from Hippocrates, Aristotle, or Galen. In an early reflection (c. 1650), Boyle wrote that once he had become "inclin'd to the Study of Naturall Philosophy" he set out "to instruct my selfe in the Aristotelian Doctrine," but soon became disenchanted by it, in part because "I observ'd many things in my Travells which were wholly unintelligible from Aristotles Theory." Later, in his *Memoirs* (1684) about the study of human blood, he judged that "the curiosity of the Moderns" had delivered far more than "the Ancients."[15] In tune with this attitude, the poet John Dryden declared in 1668 that in his time "more errors of the school have been detected, more useful experiments in philosophy have been made, more noble secrets in optics, medicine, anatomy, astronomy, discovered, than in all those credulous and doting ages from Aristotle to us . . . [and] that nothing spreads more fast than science, when rightly and generally cultivated."[16] Here we do see an attempt to make the "new philosophy" synonymous with "science."

Virtuosi

The first two terms of my title—"notebooks" and "English virtuosi"—also need explanation. I shall start with the people themselves. The first reputed use of the word *virtuoso* in English occurs in Henry Peacham's *The Compleat Gentleman* (1634), in his chapter "Of Antiquities." Referring to collectors of "Statues, Inscriptions, and Coynes," Peacham added that "such as are skilled in them, are by the *Italians* called *Virtuosi.*" It soon became common in English, used either with or without observance of its Italian plural form, and often without italics or other indication of it being a foreign word. In his *Glossographia* (1656), Thomas Blount defined a "virtuoso" simply as "a learned or ingenious person, one that is well qualified." In 1673 the naturalist John Ray continued this usage but allowed it to extend to connoisseurship as well as skilled performance. Speaking of the Italians encountered on his travels, he remarked: "Though all of them cannot paint or play on the music, yet do they all affect skill and judgement in both: And this knowledge is enough to denominate a man a *virtuoso.*"[17]

By the 1660s the leading members of the newly established Royal Society applied the label "virtuosi" to themselves.[18] When Pepys was offered an introduction to this group in early 1662, he referred to it as "the college of Virtuosoes."[19] In his unpublished biographical sketches (later called *Brief Lives*), John Aubrey used this term to identify some of the leading physicians, mathematicians, and natural philosophers of his day. Nominating Oxford as the home of some "ingeniose scholars" such as Ralph Bathurst, John Wilkins, Seth Ward, and Thomas Willis, he reported that "experimentall philosophy first budded here and was first cultivated by these vertuosi in that darke time."[20] Samuel Hartlib, seeking to associate his own earlier hopes for a Baconian academy with the new society in London (soon to be called "Royal"), remarked in December 1660 that "there is a meeting every week of the prime virtuosi," and that "His Maj[esty]. is sayd to profess himself one of those virtuosi." A month later he wrote of "Mr. Boyle" being "one of the virtuosi."[21] Boyle himself used the word regularly when referring to those interested in experimental philosophy, and awarded it as a badge of honor to Locke after the latter's less-than-successful attempt to take barometric measurements in a mine in the Mendip Hills in Somerset in 1666.[22] He occasionally spoke of a "*Virtuosa*" when describing a learned woman.[23] When John Evelyn listed the range of visitors keen to meet Boyle, he mentioned "Princes, Ambassadors, Forrainers, Scholars, Travellers & Virtuosi," thereby giving the last category a separate recognition.[24]

During the seventeenth century, one could admire virtuosi for their skill in a particular subject, or for the breadth of their curiosity. In *English Scientific Virtuosi* (1979), Barbara Shapiro and Robert Frank stressed the overlap of interests between antiquarians and physicians in the early Royal Society, and the operation of an intellectual framework that did not demarcate historical and natural observations. Abraham Hill, one of the first secretaries of the Society, acknowledged such a range of interests when describing the lawyer Sir John Hoskyns as one who "understands painting and sculpture exceedingly well, and is a virtuoso in most other branches, particularly gardening." Aubrey's examples also show that the virtuosi were involved in a broad range of inquiries that included medicine, botany, horticulture, physiology, chemistry, experimental philosophy, and archaeological research on historical monuments.[25] In 1668 the Italian visitor Lorenzo Magalotti noticed the diverse interests of the members of the Royal Society, and Evelyn, as the very model of an English virtuoso, embraced this wider suite of subjects.[26]

It was possible, however, to put a negative slant on this broad range of interests, regarding it as indicative of a promiscuous approach. To some extent, the historian Walter Houghton did this in his pioneering study that identified virtuosi as not only those collecting coins, medals, and paintings, but also those boasting various scientific interests. However, he viewed these multiple pursuits as responsible for the "dilution and distortion of the scientific mind . . . to the spirit of virtuosity."[27] In a sense, this echoed some of the contemporary attacks from those whom Thomas Sprat, in *The History of the Royal Society* (1667), called "these terrible men."[28] Almost a decade later, in his comedy *The Virtuoso* (1676), Thomas Shadwell had his lead character, Sir Nicholas Gimcrack, pursuing projects close to those of the Society. These are satirized as abstract speculation marked by nonutility: "We virtuosos never find out anything of use." In fact, Gimcrack is refused entry by "the College" (a reference to Gresham College, London, the first location of the Society's meetings), but such a twist only heightened the satirical effect.[29] Robert Hooke, the curator of the Society, returned from the play and vented his anger in his diary: "Damned Doggs, *Vindica me Deus*. People almost pointed."[30] Similarly, the physician and political statistician William Petty expressed private dismay at the manner in which worthy and difficult inquiries were portrayed.[31] We can detect the impact of Shadwell's work in William Wotton's *Reflections upon Ancient and Modern Learning* (1694). He commented that "nothing wounds so much as Jest," and so ridicule of the Royal Society might discourage participation in scientific studies, especially if it got about that "every Man whom they call a *Virtuoso*, must needs be a Sir Nicolas Gimcrack."[32]

Continuing aftershocks of this caricature are discernible in Ephraim Chambers' *Cyclopaedia* (1728), which carried this entry:

> *Virtuoso*, an Italian term, lately introduced into English; signifying a Man of Curiosity and Learning; or one who loves and promotes the Arts and

Sciences. In Italy, *Virtuosi* are properly such as apply themselves to the polite Arts of Painting, Sculpture, Turning, Mathematicks, &c. . . . Among us, the Term seems affected to those who apply themselves to some curious and quaint, rather than immediately useful Art or Study: As Antiquaries, Collectors of Rarities of any kind, Microscopical Observers, &c.[33]

Chambers knew that these last three activities were pursuits of the Society, to which he was elected in 1729 by virtue of this *Cyclopaedia*, and its support of Newtonianism.[34] His definition appropriately includes the sciences, but with a negative connotation — "curious and quaint, rather than immediately useful" — that applied to the interests of some members. A more severe insinuation was already present in Isaac Newton's caustic mention of "some great virtuosos": these people, he said, did not "take my meaning, when I spake of the nature of light and colours abstractedly."[35] This remark attests to an incipient tension between mathematical and natural history interests within the Society.[36] Nevertheless, in the late seventeenth century the term "virtuoso" in association with science carried a generally positive connotation. It did not at that time resemble the negative "*macaroni*" tag circulating during Sir Joseph Banks' presidency from 1778.[37] Boyle, Locke, and Hooke were not butterfly collectors. In about 1704, Leibniz used this term in connection with a call for more interaction between "practice and theory" which, he said, could already be seen "among painters, sculptors and musicians, and among certain other kinds of virtuosi."[38]

The English virtuosi pursued Baconian natural history, a category far more extensive than botany and zoology because it embraced almost everything not covered by civil history.[39] For Bacon, natural history was as much a method as a subject. It could be applied to topics ranging from celestial phenomena to technology, thus yielding *histories* of the air, of heat and cold, of sounds, blood, of life and death — as exemplified by his list of 130 "Titles" of "Particular Histories" in his "Parasceve" (or "Preparative to a Natural History"), published in 1620 with the *Novum Organum*, both being parts of his planned *Instauratio Magna*, or Great Instauration.[40] Natural history aimed to describe, collect, and compare instances of species, events, and phenomena, both by observation and experiment. In one sense, this reinforced its lower status as *historia* in comparison with *scientia*, the category that included disciplines (such as geometry) able to achieve systematic knowledge via demonstration from axioms or first principles.[41] This influential Aristotelian classification ensured that natural philosophy, as the study of causes and processes, always trumped natural history. However, in *Descriptio globi intellectualis*, composed in 1612 but unpublished in his lifetime, Bacon insisted that natural history dealt with "the basic stuff and raw material of the true and legitimate induction," thus making it the preliminary and necessary bedrock of natural philosophy.[42] In sketching his *Great Instauration* (1620), he even spoke of watching "over the infancy of natural philosophy in the shape of natural history."[43] In the *Historia naturalis et experimentalis* (1622), he admitted that the method offered in the *Novum Organum* was worth little "without Natural History . . . whereas Natural History without the *Organum* would advance it not a little." Bacon's dedication of this work to Prince Charles (the future Charles I) declared that in "a good and solidly-constructed *Natural History* lie the keys both to knowledge and to works."[44]

Before considering the use of notebooks we need to ask why the English virtuosi would bother at all with this habit. The practice of collecting excerpts from books was a crucial part of traditional methods of humanist and neoscholastic teaching and study.[45] Yet this pedagogy was precisely the target of the antibookish rhetoric found in Bacon and Descartes. When discussing factors that retarded the advance of knowledge, such as prejudice against mechanical arts, Bacon was tempted by the call associated with alchemists for "men to sell their Bookes, and to build Fornaces." Descartes summed up his dislike of "fat tomes" by saying that even his own very short *Discourse on Method* might be "too long to be read all at once." Even a man of books such as Evelyn was not immune to this sentiment: after Boyle's death, he remarked to Wotton on Boyle's "small Library (and so you know had Descartes) as learning more from Men, Real Experiments, & his Laboratory (which was ample and well furnish'd), than from Books."[46]

Some of Aubrey's friends, as treated in *Brief Lives*, exhibited an even more strongly dismissive attitude toward bookish learning. Of the physician, William Petty — "my singular friend" — Aubrey reported that "he hath read but little, that is to say, not since 25 aetat. [years of age], and is of Mr Hobbes his mind, that had he read much, as some men have, he had not known so much as he does, nor should have made such discoveries and improvements."[47] Aubrey confirmed that Thomas Hobbes had "very few bookes. I never sawe . . . above halfe a dozen about him in his chamber."[48] In his manuscript on the "Idea of Education," begun in 1669, Aubrey declared that "A few books, but well chosen, thoroughly digested with constant practice and observation, does the business." He added Hooke and Christopher Wren to his special club of great men who were not "great readers."[49] Petty told his friend Robert Southwell that "you know I am no good book man"; and in his own selection of maxims glossed from Michel de Montaigne, he wrote that "Books seduce us from studying."[50] In drafting the rules of the Dublin Philosophical Society (founded in 1684), he stipulated that its members prefer experiments to "the best Discourses, Letters and Books they can make or read, even concerning experiments."[51]

Sprat came close to casting members of the Royal Society as antibookish. He set up a dichotomy between the modern practitioners of "Experimental Philosophy" and those "Men of Learning" and "Reading," or "studious men," still under the thrall of the Ancients. He put this in terms of opposing aptitudes and sensibilities: "for, it is not only true, that those who have the best faculty of Experimenting, are commonly most averse from reading Books." Later in the work he suggested that many might see the traditional "*Bookish*" scholar and the new experimenters as equally cut off from others and the world: "the one in his *Library*, arguing, objecting, defending, concluding with himself: the other in his *Work-hous*, with such Tools and Materials, whereof many perhaps are not in publick use."[52] Sprat upheld the policy to put doing before talking.[53] Carried away by the rhetorical impulse, he predicted that the spirit of the new philosophy, grounded in practical, manual skills, would survive "the loss of a Library" or the "overthrowing of a Language." Curiously, on the very next page, as if he had recovered his common sense, Sprat mentioned "the *dreadful* firing of the City" of London and the destruction of "as many *Books*, as the cruellest insurrection of the *Goths*, and *Vandals*, had ever done."[54] Sprat's exaggerations played into the hands of critics, such as the royalist cleric Robert South (1634–1716), who attacked the new philosophy in a sermon in Westminster Abbey in 1667 and again at the dedication of the Sheldonian Theatre in Oxford in 1669.[55] Joseph Glanvill told Henry Oldenburg that South was saying that its members "are wholy ignorant of all History, & antiquity, & never have read any bookes."[56] Another scourge of the Society, the physician Henry Stubbe also found ammunition in Sprat for his flurry of publications against the Society in 1670.[57]

This pervasive and influential contrast between those who experimented and those who read books overlooks the concerted book buying, and notetaking, of the virtuosi. Some prominent Fellows of the Society, such as Aubrey, Evelyn, Boyle, Locke, Hooke, Pepys, and Newton, assembled substantial personal libraries.[58] When Pepys took the precaution during the Fire to bury his documents, wine, and "my parmezan cheese" in a garden, he may have guessed that many books were already burned. On Sunday, September 9, 1666, he recorded that the dean's sermon of that day had described the City as "reduced from a large Folio to a Decimo tertio." Pepys found this insensitive. On September 26 he heard of "the great loss of books in St. Pauls churchyard" where they had been stored by the Stationers' Company.[59] This concern about books was representative, but the scientific virtuosi sometimes resorted to attacks on bookish habits as a shorthand way of rejecting older, especially scholastic, ways of thinking. They rarely went so far as Descartes' suggestion that older books, riddled with errors and useless information, could be safely ignored.[60] But they wished to separate themselves from those, perhaps apocryphal, figures who were dependent on books to the neglect of other avenues of information. We see this in Evelyn's assurance to the earl of Clarendon that the Royal Society "does not consist of a Company of Pedants, and superficial persons; but of Gentlemen, and Refined Spirits that are universally Learn'd, that are Read, Travell'd, Experience'd and Stout." He made this remark in a book *about* books—Gabriel Naudé's work on the creation and arrangement of libraries.[61]

Nevertheless, Sprat's formulation of the contrast between books and experiments was so strong that even those seeking to resist it had to start on its terms. Thus Edward Bernard, the Savilian Professor of Astronomy at Oxford from 1673, remarked that "Books and experiments do well together, but separately they betray an imperfection, for the illiterate is anticipated unwittingly by the labors of the ancients, and the man of authors deceived by story instead of science. The happy Royal Society adjusts both together, and I doubt not but, in a short while, will approve itself so great a friend to and near ally to the Universities."[62] As I will suggest in other chapters, books and *notes* also "do well together." Petty's claim to do without many books is contradicted by the citations in his own works, and by his own notes.[63]

Various proposals concerning the good running of the Royal Society stressed that books, as well as observations and experiments, were crucial sources of information. One member suggested that there be "a Curator for Books; whose business should be, to make a diligent search & particular collection, of all that is, or may hereafter be published, of the History of Nature or Art, by any person, at any time, in any Country or Language whatsoever."[64] And even Sprat took care to explain that experiments performed in solitude were later discussed in "Assembly," thereby ensuring interaction with "all sorts of Opinions" derived "from the observations of others, or from *Books*, or from their own *Experience*."[65] This more moderate position was already present in Bacon's writings, despite his frustration with those who relied solely on texts sanctioned by authority. Bacon stressed the value of books as carriers of information in time and space: "the Images of mens wits and knowledges remaine in Bookes, exempted from the wrong of time, and capable of perpetuall renovation in succeeding ages."[66] However, he also maintained that the collection and organization of adequate information from both books *and* nature required assiduous note-taking. This led him, and later the English virtuosi, to think about the kinds of notes and notebooks best suited to empirical inquiry.

Notebooks

According to the *OED*, the word "note-book" entered English in 1568, just over a century after Johannes Gutenberg invented movable type for the printing press.[67] Although loose leaves stitched to form of a booklet could be called a notebook, by the 1600s "paperbook" unequivocally denoted a bound gathering of blank pages dedicated to entries not yet made, thereby recognizing note-taking as a widespread phenomenon requiring material support. However, "notebook" eventually became a generic label for a variety of repositories, often based on different principles. Other similar contemporary terms included "writing-book," "day-book," "ephemerides," "diary," "memorandum," "waste-book," "journal," "ledger" (the last three used by merchants), and "common-place book" (favored by humanist scholars). All these terms preceded "pocket-booke"—in use from 1617—a reminder that not all notebooks were small and portable.[68] Of these, the journal, diary, and ledger have survived in recognizably similar formats into the present; the commonplace book, arranged by subject category rather than by date, has not. We therefore need to be clear about the main purposes of these notebooks.

In early modern Europe, note-taking mattered in ways that we now have to reimagine. Today in modern legal, laboratory, and bureaucratic settings, the making of notes is often controlled by protocols, but in the private sphere it is largely left to individual, and idiosyncratic, choice or habit. Thus in scientific studies requiring participants to make regular diary entries, so-called backfilling is a regular confounding factor: people postpone the task and then rely on memory to make the entries.[69] Of course, such gaps between norms and actual individual practice also occurred in earlier times, but in the seventeenth century, note-taking more explicitly expressed cultural values—religious, administrative, and educational—and, at least in principle, was governed by precepts and methods. The practice was useful and formative in many professional and intellectual pursuits: humanists, merchants, Jesuit pedagogues, travelers, and university dons kept notebooks and urged others to do likewise. When examined, these examples reveal various modes of note-taking—including annotation and excerpting, records of observation and testimony, and registers of spiritual discipline.[70]

Two genres—the humanist commonplace book and merchants' account books—require introduction.

The defining feature of the commonplace book was the allocation of excerpts from texts to "Heads" (or keywords). Until the late 1500s these notebooks were more likely to have been referred to as "books of common places,"[71] echoing the Latin *loci communes*. The Roman authors inherited the notions of *loci* (places) and *topoi* (topics) from Greek philosophy, especially via Aristotle's ten predicaments or categories—such as substance, quantity, quality, and relation—that were supposed to give starting points for all arguments and reasoning. The Romans transferred this concept of "place" from logic to rhetoric, so that in their usage *loci communes* indicated the grouping of quotations or arguments in a "common place," that is, under a certain Head or subject.[72] This gathering of related material was intended to assist memorization and recall. In the illustration (see fig. 1.3) of a "Study" in Jan Comenius' *Orbis sensualium pictus* (1659), the accompanying text allows the option of marginal reminders ("a dash, or a little star") in the book itself, but a scholar is depicted entering selected passages ("the best things") into a small "Manual."[73] In English, from 1578, these notebooks came to be called "common-place books," a term that predated "miscellany" (1638) and "anthology" (1640); by the end of the eighteenth century these three types were often conflated.[74]

Renaissance humanists encouraged the selection of passages on the classical topics of honor, virtue, beauty, friendship, and also those pertaining to Christian concepts such as faith, hope, sin, and grace. The most favored ancient writers were Ovid, Virgil, Horace, Cicero, Juvenal, Lucan, and Seneca (Lucius Annaeus Seneca, the Younger).[75] As Joan Lechner remarked in the first major study of this subject, such topics lent themselves to "the praise of virtue or the dispraise of vice."[76] The method of "commonplacing," which is to say, the grouping of quotations, tropes, proverbs, or arguments under appropriate Heads, became central to humanist scholarship. It was used and recommended by leading authors such as Desiderius Erasmus, Juan Luís Vives, Philipp Melanchthon, and Rudolph Agricola.[77] In his *De ratione studii* (1512) Erasmus advised that every student "should have at the ready some commonplace book of systems and topics, so that wherever something noteworthy occurs he may write it down in the appropriate column."[78] His influential *De copia* (1512) offered a manual of examples.[79] With such advocacy, commonplacing entered the pedagogy of the grammar schools and of undergraduate university training. The core subjects of *studia humanitatis*, such as rhetoric, history, and moral philosophy, provided a

XCVIII.

The Study. Mufæum.

The Study 1.
is a place
where a Student, 2.
a part from men,
fitteth alone,
addicted to his Studies,
whilſt he readth
Books, 3.
which being within
his reach, he layeth
open upon a Desk 4.
and picketh all the
beſt thoſe gø out of
them into his own
Manual, 5.

Mufæum 1.
eft locus,
ubi ftudiofus, 2.
fecretus ab hominibus,
folus fedet,
Studiis deditus,
dum lectitat
Libros, 3.
quos penes fe
fuper Pluteum 4.
exponit, & ex illis
in Manuale 5. fuum
optima quæq;excerpit.

or marketh them in
them with a daſh, 6.
or a little ſtar, 7.
in the Margent.
Being to fit up late,
he ſetteth
a Candle, 8.
on a Candle-ſtick, 9.
which is ſnuffed
with Snuffers; 10.
before the Candle
be putteth
a Screen, 11.
which is green, that it
may not hurt his eye-
ſight; rather perſons
life a Taper,
for a Tallow-Candle
ſtinketh,
and ſmoaketh.

A Letter 12.
is wrapped up,
writ upon, 13.
and ſealed. 14.

Going abroad by
night, he maketh uſe
of a Lanthorn 15.
or a Torch. 16.

aut in illis
Liturâ, 6.
vel ad marginem
Afterifco, 7.
notat.
Lucubraturus,
elevat
Lychnum (candelam) 8.
in Candelabro, 9.
qui emungitur
Emunctorio; 10.
ante Lychnum
collocat
Umbraculum, 11.
quod viride eft,
ne hebetet
oculorum aciem;
opulentiores
utuntur Cereo,
nam Candela ſebacea
fœtet & fumigat.

Epiftola 12.
complicatur,
infcribitur, 13.
& obfignatur. 14.

Noctu prodiens,
utitur Laternâ 15.
vel Face.

Arces

Figure 1.3. A scholar enters excerpts into a notebook, probably a commonplace book. Jan Amos Comenius, *Orbis Sensualium Pictus* (1672), 200–201. Kindly supplied with permission by Rare Books and Special Collections, the University of Sydney Library.

rich storehouse from which anecdotes, adages, metaphors, epigrams, proverbs, and *sententiae* (famous pronouncements by great men of letters and
action) could be selected and placed under a "common" Head. As Ann Moss
has observed, this training "was part of the initial intellectual experience of
every schoolboy," and "every Latin-literate individual started to compose
a common-place book as soon as he could read and write reasonably accu-
rately."[80]

The Renaissance itself has been described as "fundamentally a notebook
culture."[81] This epithet mainly applies to the central role of commonplace
books in fostering a style of writing and thought in which movable units,
such as *exempla* and *sententiae*, were combined and embellished in various
ways and for different purposes.[82] John Marbeck's *A Booke of Notes and Common places* (1581) typifies this conception of material gathered, reduced, and
arranged—in this case alphabetically—and kept for later use. Walter Ong
characterized this behavior as "organized trafficking in what in one way or
another is already known."[83] Yet there was a sophisticated scholarly use of
this method that continued into the seventeenth century. The poet John Milton, who used one commonplace book over three decades, organized material under three main Heads—Ethicus, Oeconomicus, and Politicus—compiling an index for each of these main subjects.[84] This notebook worked as
a partial record of his "industrious and select reading," as a research tool,
and as a resource for composition.[85] When Isaac Newton arrived at Trinity College in June 1661, he set up a notebook for his reading and thinking
about natural philosophy. Before making any entries he preassigned thirty-
seven Heads in one section of a quarto notebook, trying to guess how many
pages each would need, without much success. Consequently, his notes on
these topics (and "questiones") spilled over into various parts of the notebook.[86] Well-made commonplace books like Milton's, and even less efficient
ones such as Newton's, reflected the learning and judgment of their makers. The term "commonplace" had not yet suffered its ultimate debasement
whereby it came to denote a platitude, something unremarkable or not worthy of note.[87]

Somewhat confusingly, another term, "*adversaria*," was often used as
a synonym for commonplace book during the early modern period. The
original Latin term referred to various kinds of collections in which undigested matter was recorded just as it occurred, as in a "waste book" used by
merchants before a systematic record was made in a ledger.[88] In 1728 in his
Cyclopaedia, Chambers reported this usage as well as giving a rough deri-
vation: "Adversaria, among the Antients, was used for a Book of Accounts,

like our Journal or Day-Book. . . . Adversaria is sometimes also used among us for a Common-place-Book."[89] In 1755 Samuel Johnson repeated this definition, giving "adversaria" as "A book, as it should seem, in which *Debtor* and *Creditor* were set in opposition. A common-place; a book to note in." His second illustration picked up the fact that "adversaria" could also refer to the notes themselves, whether or not they were contained in a bound commonplace book: thus he remarked that some "parchments are supposed to have been St. Paul's *adversaria*."[90] Most of the people I discuss treated "adversaria" and "commonplaces" as interchangeable terms.[91] In the late 1600s, Aubrey used the former word in referring to not only notes made from reading but also those made while recording observations of the natural world: "Tis proper for a gentleman to know soils. As they follow their botanics, let 'em make notes of the earth and minerals . . . and let them enter in *adversariis*."[92] This is an example of how the terminology of note-taking extended beyond the world of books; it also indicates a scenario in which adversaria — as raw notes — might be divorced from the method of commonplacing.

The collapse of the Renaissance tradition of commonplaces has been widely reported, and there is no reason to doubt that the peak of its prestige had receded by the mid-1600s.[93] Commonplacing was invariably targeted by those who rejected bookish pedagogy. However, it is difficult to generalize about the demise of the method of commonplaces because its component assumptions — the authority of canonical texts, the value of a *copia* of quotations, the Aristotelian framework of categories, the cultivation of memory and rhetoric — unraveled at unequal rates. And despite a chorus of disapproval, critics of commonplacing were often users of this kind of notebook. Thus in 1642 the preacher Thomas Fuller admitted that "I know some have a Common-place against Common-place-books," but taunted that these people "will privately make use of what publickly they declaim against."[94] Commonplace books could be retained as a "working tool" by those who had shed some of the commitments usually associated with the habit of commonplacing.[95] But this possibility of adaptation and renovation may well have depended on fruitful interaction with the other main kind of notebook: the journal.

The words "journal" and "diary" reflect the purpose of such notebooks to record *daily* events, actions, and thoughts. By the seventeenth century there were several types of journal: the spiritual diary, especially as recommended by the Puritan reformers; the more intermittently used travel journal; scholarly diaries; and the already well-established merchants' account book (or books). The ordering of entries was chronological rather than by subject or theme as in a commonplace book. Thus, in principle, all journals offered the freedom to note information rapidly, either routinely into pre-established categories — such as those used in ships' logbooks and weather registers — or otherwise by immediate entry, without the need to decide on the most appropriate category.[96] Journals were also viewed as portable whereas commonplace books, especially if folio-sized, were regarded as notebooks best left in the study or library. Nevertheless, according to Bacon, there was inadequate use of journals, in spite of the established practice in "Warre, Navigations, and the like to keepe *Dyaries* of that which passeth continually."[97] He remarked that it was strange that "Men should make Diaries" in "Sea voyages, where there is nothing to be seene, but Sky and Sea," but not in "*Land-Travaile*, wherein so much is to be observed." He declared "Let Diaries, therefore, be brought in use," and suggested that the sweep of observations could be extended by not staying too long in the one town or lodging.[98] Significantly, his long list of things to be noticed reads like a set of Heads, such as those provided in the advice about travel (*ars apodemica*) favored by both humanists and Jesuits.[99]

Of course, the mere use of a journal did not guarantee its potential benefits: for example, the tendency to make diary entries post-facto was acknowledged. On November 10, 1665, Pepys recorded that he was now entering "all my Journall since the 28th of October, having every day's passage well in my head, though it troubles me to remember it." At other times, he made a temporary note to be properly entered later: thus on January 24, 1664, he recorded that he went to "my office and there fell on entering out of a by-book part of my second Journall book, which hath lay these two years and more unentered." Hooke noted on January 10, 1678, that he "Wrote notes of Thursday last.[100]

In merchants' account books the principle of immediate entry aimed to avoid likely omissions and errors. This was a feature of the influential method of double-entry book-keeping set out by Luca Pacioli (1446–1517) in his *Summa de arithmetica, geometria, proportioni & proportionalita* (1494).[101] As a Franciscan mathematician, Pacioli regarded the discipline of entering the debit and credit of each transaction as a way of accounting for commercial practices, including usury, before God. There was therefore a strong moral and rhetorical component to his book-keeping that translated easily into notions of personal, and religious, self-monitoring.[102] The so-called Italian Method, with its insistence on detailed daily entries, served both to secure information and to cultivate ethical formation.[103] Pacioli's method required the provision of considerable circumstantial detail, such as the name of the person, the date, hour, and place of transaction, the quantity and prices of goods or loans. Moreover, after an inventory had been compiled, the entries were made in *three* notebooks: the memorandum (*memoriale*), the journal (*giornale*), and the ledger (*quaderno*).[104] Daily transactions were entered in the memorandum (also called a day or waste book) as they occurred. Pacioli said this task could be performed by several members of a family or business, although he added that "some members of your family will understand and some will not." Transactions in the day book were later summarized and entered in the journal. This information could then be properly rendered as both debt and credit in the ledger.[105] Significantly, all this note-taking relied on protocols to be followed by anyone making the entries; by contrast, the commonplace book was usually under the control of one person, its content reflecting that person's interests, tastes, and judgment.

In spite of major contrasts in method and rationale, the practice of mercantile book-keeping served as a resource for thinking about commonplace books. The notion of immediate entry, and the associated option of transferring information across various notebooks, allowed for a more sophisticated use of notes, including those intended for commonplace books. Francis Bacon explored this possibility. During the last week of July 1608, he surveyed his existing paperbooks, which included various kinds of memoranda, journals, legal notes, composition books, and at least two commonplace books. The evidence we have of this audit is contained in one surviving notebook called "Comentarius Solutus," now in the British Library.[106] The nineteenth-century scholar James Spedding rendered this as "a book of loose notes," and Bacon himself described it as being "like a Marchant's wast booke where to enter all maner of remembrance of matter, fourme, business, study, towching my self . . . without any maner of restraint."[107]

In his inventory of these twenty-eight paperbooks, Bacon summarized their purpose and content under the running head of "Transportata," thus indicating that the material had been transferred from other notebooks. Describing one of these, he said "the principall use of this book is to receyve such parts and passages of Authors as I shall note and underline in the bookes themselves to be wrytten foorth by a servant." He asked himself whether it would be better to reduce the current number of divisions, making these "fewer and lesse curious and more sorted to use than to Art." To achieve this he decided that, in addition to the waste book, he needed "another booke like to the marchants leggier booke" in which things worth keeping were entered more carefully. And then finally, selected items could be classified and "thinges of a nature" could be entered "under fitt Titles" in "severall title bookes." We could say that Bacon adopted the spirit of Pacioli: namely, record everything as it occurs and sort it out later. Moreover, he explored ways of combining the journal and the commonplace book, with the former being used in the first stage of recording and the latter offering a mode of arrangement by title or topics. However, Bacon also resolved to reduce the material in his commonplace books, effectively making them less copious and, in turn, more suitable for the "better help of memory and judgment."[108] As we shall see, this aim of refining and re-sorting material was shared by Hartlib's circle and by members of the Royal Society. The potential drawback was the likely inability to return to the original notes.

In all the note-taking discussed so far there was an assumption that material gathered rapidly should be deposited somewhere more secure, for later use. In Pacioli's method, this principle of transferring information was supported by three different kinds of notebook, governed by strict procedures. Such regulation did not occur in the behavior of individual scholars and virtuosi, but the habit of noting material quickly for subsequent re-entering and storage is illustrated by the use of "table books" (or writing tables or tablets). These were the smallest and most portable kind of notebook.[109] They consisted of gatherings of ten or so paper leaves treated with a substance that provided an erasable surface; once wiped clean they could be reused on many occasions.[110] Some of the English virtuosi took notes on how to make their own. Hooke recorded these instructions:

> To make Table booke Leaves. Take a quire of paper, a pound of Cerusse, a penie worth of white starch unboyld. pound the Cerusse very fine, and dissolve the starch in a quart of faire water by stirring it about with a spoon then put the Cerusse into the water and let it stand for two hours then with a large brushe wash over one side of the paper and let it drie, and then after, the other side. when both are drie, rub both sides well with a linnen cloth or scrubbing[?] brush, — when it is well rubd, planish them well over with such a hammer as book binders use to beate theyr books with .[111]

While traveling in France, Locke made an entry under "Table book" in his journal for October 7, 1677, giving this statement of its function: "To blot out what is writ on ordinary table books take pouder of pumice stone in a fine rag & soe rub it out."[112] The next day he continued with details for preparing the substance to be applied to the paper. The recipe and procedure are similar to those Hooke recorded: thus Locke wrote that "when both sides [of the paper] are thoroughly drie take a course cloth & rub it smooth.

For greater perfection beat it as bookebinders beat their books to make it smooth. This one may write on with a silver pen."[113] Boyle evidently considered this piece of equipment valuable: in one of his medical receipt books he mentioned some information "Taken out of my Tortois-shell Table booke."[114] We can glimpse a very different scene in Pepys' story of being shown the contents of a "dead man's pockets" which contained "a table-book, wherein were entered the names of several places where he was to go."[115] Undoubtedly some of the information wiped from these tables (such as appointments and shopping lists) was intended only for short-term use, but other notes thus taken were moved into more permanent notebooks for storage, analysis, and subsequent communication. The possibility of erasure supported the practice of transferring material from table-books to other notebooks for subsequent review and comparison with other data.[116]

Apart from commonplace books and journals, there were less well-defined notes — quotations, observations, or thoughts not necessarily framed either by topic or by the principle of daily entry. In the Italian vernacular of the Renaissance, such loose notes were called *zibaldone*, a word that also referred to other disorderly and miscellaneous assortments, including food. Giovanni Boccaccio (1313–75) seems to have used this term to designate some of his notes. The notion at work here may have been that of a personal notebook in contrast with formalized account books or other registers. Thus Leonardo da Vinci recommended that ideas and sketches should be made "in a little book which you should always carry with you" because "the forms and positions of objects are so infinite that the memory is incapable of retaining them."[117] The appearance of the term "pocket-booke" in the early 1600s thus indicates a new term for a practice already in place — notes made *in situ*.[118] There were also other kinds of notes not committed to notebooks at all, but simply made on loose sheets. The result could be either loss or chaos. Petty confessed to having "many flying thoughts," some of which he scribbled down: Aubrey reported that "I have seen, in his closet, a great many tractatiuncli in MS." Southwell saved many of these in an "Ebony Cabinet wherein I keep, as in an Archive, all the effects of your Pen."[119] However, such loose notes could nevertheless have some structure, such as that implied in lists of various kinds — for example, in wish lists of books, experiments, or inventions.[120]

These lists legitimated short notes, and they clustered information, often from a range of sources, around a topic or problem; they encouraged additions over generations.[121] This possibility was also exemplified in so-called decades (groups of ten items) or, more commonly, "centuries" (100 items) of proverbs, recipes, observations, or experiments.[122] Such compilations often went under the name of "Sylva," suggesting both diverse materials jumbled together and, more optimistically, a collection in which the whole was more than the sum of its parts, as with a forest of trees. Ben Jonson referred to his rough assembly of commonplaces as "Sylva" and "Timber." Bacon's posthumous *Sylva Sylvarum* (1627) comprised ten centuries of numbered observations, experiments, and speculations.[123] The fact that this work was printed together with the unfinished *New Atlantis* allows the insinuation that such an apparently unordered collection would realize its full promise under utopian conditions.[124]

The two main kinds of notebook, the journal (or diary) and the commonplace book, had distinctive features, but there was nevertheless some merging during the early modern period. In his survey of English diaries, Robert Fothergill offered subdivisions of the type, but a more comprehensive account needs to appreciate the mix of genres, especially in scholarly diaries that incorporated commonplace methods.[125] This caution about assuming rigid boundaries applies even to what has often been seen as a peculiar property of the journal: namely, its ability to offer precious insights into the life and personality of its owner. The Dutch historian Jacques Presser (1899–1970) coined the term "egodocument" to describe "those documents in which an ego intentionally or unintentionally discloses, or hides itself."[126] In contrast, some of the notebooks I discuss were not straightforwardly "egodocuments." Those modeled on the commonplace book were certainly different from autobiographies written retrospectively with a narrative endpoint in mind and, unlike diaries, they did not, at least in principle, encompass the activities and thoughts of the individual who owned them. Instead, they gathered excerpts from books written by others. Nevertheless, this does not mean that these notebooks betray no marks of their owners' intellectual and emotional dispositions. Scholars such as Martin Crusius and Isaac Casaubon had already mixed textual annotations, dating, and autobiographical details in their commonplace books.[127] During the seventeenth century this merging continued so that the commonplace book came to include reactions to the textual excerpts assembled. This is how Roger North, a lawyer and music theorist, regarded his manuscript titled "Notes of Me"—as a "life-journal" containing far more than a diary of actions and events.[128] By the early 1700s, Isaac Watts suggested that a commonplace book should reveal the intellectual development of its owner: "At the End of every Week, or Month, or Year you may review your Remarks for these Reasons: First, to judge of your own Improvements, when you shall find that many of your younger Collections are either weak and trifling; or if they are just and proper, yet they are grown now familiar to you."[129] Even without this specific intention, many commonplace books resemble "egodocuments" precisely because they reveal individual styles and preferences—in relation to chosen authors, topics, and quotations; format and detail of entries; and mode of arrangement.

What I have presented so far are the elements of the "notebook culture" in the early modern period. Various combinations of table books, pocket-books, journals, and commonplace books allowed information to be collected and processed in different ways. It is possible to argue that of the last two, the commonplace book was identified with a traditional pedagogy now falling out of favor, whereas the journal was the notebook of choice among travelers, courtiers, and men of the world. We still lack a systematic analysis of a broadly representative sample of seventeenth-century notebooks, especially regarding their continued use beyond the time at school or university. However, F. J. Levy's comment about the English gentry in this period is suggestive: namely, that their education gave them the desire "to acquire and catalogue new information for use as *exempla*: the habit of the commonplace book died hard."[130] One caveat is that individuals who manifested habitual and disciplined note-taking are probably exceptional: many notebooks were started and soon abandoned; and, of course, unknown numbers have not been saved for us to see. I do not claim that English scientific figures were representative of early modern note-takers; rather, I examine their own practice and their reflections on it. These people adjusted the commonplace method to new purposes and blended it in various ways with the journal. Pepys, Boyle, and Hooke kept journals or diaries rather than commonplace books; Aubrey, Evelyn, and Locke used both, sometimes containing a weather diary within the pages of a commonplace book.[131] Before these English virtuosi, Hartlib used diaries that displayed topics as Heads in the margins, thereby feasibly allowing the best of both worlds. And before Hartlib, Bacon considered whether this combination of chronology and categories was best suited to the collection of empirical information.

Bacon sought to make a new kind of note-taking central to the compilation of empirical natural histories. He believed that it was first necessary to jettison the unwanted baggage of commonplacing. In *The Advancement of Learning* (1605), he acknowledged the ancient concepts of "*Places*" and "*Topiques*" and their role in rhetorical training. The "Entrie of Common places," he said, was of great use in study insofar as it "assureth copie of Invention, and contracteth Judgment to a strength"; he suggested, however, that it should also be used "to direct our enquirie."[132] Bacon thus endorsed the use of *"Common-Place Bookes,"* but not the typical practice of his day which, he alleged, severely limited their purpose: "all of them carying meerely the face of a *Schoole*, and not of a *World*; and referring to vulgar matters, and Pedanticall Divisions without all life, or respect to Action."[133] He greatly expanded this discussion in the enlarged Latin edition of the *Advancement*, his *De augmentis* (1623), linking the making of commonplaces to the proper cultivation of memory.[134] However, in the *Novum Organum* and the *Parasceve*, Bacon took a somewhat different stance, one that advocated a kind of note-taking more suited to the collection of empirical information from observations and experiments.

With the aim of modifying the commonplace method, Bacon made two points. First, he shifted attention from *copia verborum* to *copia rerum*. This was not wholly surprising since Renaissance rhetorical theory, as represented in the writings of Agricola, Melanchthon, and Vives, already taught that "things" (*res*) as well as words (*verba*) should be among the examples entered in a commonplace book. As Ann Moss puts it, in Vives' *De disciplinis* (1531) the "things" comprised "lists of famous men, cities, animals, plants, and minerals."[135] Bacon sought a natural history that did not replicate the tedious use of commonplace books; yet this is precisely what he found in the natural history of his day: it was too reliant on textual authorities and often amounted to nothing more than a confused ensemble of parable, folklore, and unverified reports—indeed, akin to some of the material in Jean Bodin's general histories.[136] In wanting to discard "citations and differing opinions of authorities," and "in short, everything philological," Bacon was attempting to shift natural history out of the literary domain in which it was currently pursued.[137] The publisher of *Baconiana* (1679) recognized this:

> as his Lordship noteth, too many of these Histories were at first framed rather for Delight, and Table-talk, than for Philosophy. Stories were feigned for the sake of their Morals; and they were frequently taken upon groundless Trust; and the later Writers borrowed out of the more Ancient, and were not Experimenters, but Transcribers: And such a one was *Pliny* himself.[138]

Bacon's second point was that the emphasis should be on gathering requisite information, not on fine presentation of choice examples. As he explained, "For no one collecting and storing material for ship building or the like bothers (as shops do) about arranging them nicely and displaying them attractively; rather his sole concern is that they are serviceable and good, and take up as little space as possible in the warehouse."[139] In any case, he declared, there was at present inadequate information because no one had "sought out and harvested a forest of particulars and materials" of sufficient quantity, kind, and reliability. The implication was that the commonplace method in natural history should be far more open-ended than usually understood: the information collected was not expected to reinforce larger canonical categories. To some extent, this approach seems at odds with Bacon's complaint about the haphazard ways in which knowledge had so far been pursued, being left to "the turbulent waves of chance, and experience undisciplined and ungrounded"; this was akin to gathering material with "an unbound broom." He conceded that such collections might resemble miscellanies incapable of producing any patterns or discoveries.[140] Nevertheless, whatever the scale of the material, he was adamant that there could be no resort, as there often was in textual studies, to the lazy use of short summaries that often failed to retain the complexities of an argument or subject. He referred to the "Canker of *Epitomes*" that corrodes the body of knowledge, leaving only "base and unprofitable dregges."[141] Instead, he prescribed the careful use of Heads, less as a way of consolidating agreed ideas than as starting points for further inquiry. Admittedly, he continued to use the terminology of humanist commonplacing, but Paolo Rossi goes too far in asserting that "for Bacon this procedure undergoes very little change when applied to the scientific sphere."[142] When Bacon took these concepts into natural history they were significantly transformed: his concept of a "Promptuary" was no longer a treasury of arguments ready to be applied in various rhetorical situations; rather, it was a storehouse of observations and experiments, loosely and tentatively grouped under specific topics of a history, say, that of heavenly bodies, air, or fish.[143] For these reasons, he believed that the note-taking suitable for empirical natural history had to be more rigorous and better managed than the usual commonplacing of his day.[144]

Memory and Notes

Throughout this book I will consider the twofold function of notes as both relieving and prompting memory. Notes can record ideas or information on paper for future use, and they can also stimulate recollection of more than they contain, such as other parts of the text from which they were taken, or the circumstances in which they were made. Today the emphasis is on the first function of notes as records, rather than on their second role as cues to recollect more than the note itself. We use notes to keep information that we do not want to forget or lose, and we do not need to worry too much about finding a specific note because digital technology ensures that the contents of a laptop can be searched and located on the basis of a few sequential letters. The situation was different in earlier times, not just at the technical level but at the attitudinal one.

In the medieval period, as Mary Carruthers has explained, books themselves were seen as containers of knowledge that should be mastered in memory. Thomas Aquinas held that "things are written down in material books to help the memory." The stress was always on using external marks as ways of either fixing images or text in memory, or as prompts to recollection. This is one facet of Carruthers' observation that "medieval culture was fundamentally memorial, to the same profound degree that modern culture in the West is documentary."[145] Standing between these two eras, the early modern period is interesting because memorization continued to be highly valued even though there was no lack of documents. Although some regular textual records were kept in England by the twelfth century, significant agreements were acted out in ceremonies that prompted the memories of reliable witnesses.[146] R. H. and M. A. Rouse have shown that indexing tools emerged from about 1220, and that "after the 1280s, the dissemination and new creation of such aids to study were commonplace."[147] In principle it was now possible for retrieval devices (such as alphabetical indexes) to replace "mental indexing" as keys to the growing mass of documents, especially in particular domains such as diplomacy and canon law. As Ann Blair has persuasively argued, from the sixteenth century general reference works (mainly in Latin) responded to the burgeoning number of books in print. They incorporated tools such as chapter contents and divisions, summaries, and various kinds of indexes that allowed readers to find specific information without having to read extensive passages or chapters.[148] Outside specific liturgical and pedagogic contexts, the need to memorize textual passages diminished, but the cultural importance of memory for individual readers and scholars remained substantial, even though signs of its decline are evident. Even when appropriate search tools were available, the ability to command knowledge from memory was admired. As late as 1710, Humphrey Wanley wrote about "The Library-Keeper's business . . . compiling the Catalogue, & fixing the same in his Memory. And as the Catalogue is what is contained in a Library, So the Library-Keeper should be the Index to the Catalogue."[149]

Well into the seventeenth century, great scholars were celebrated for their ability in conversation to reproduce copious learning with ease from memory. Renaissance polymaths such as Joseph Scaliger (1540–1609), Isaac Casaubon (1559–1614), and John Selden (1584–1654) were renowned for their "table talk," as it was called, and this supported the popular genre of "Anas"—as in *Scaligerana, Casauboniana,* and *Seldeniana.*[150] Scaliger was said to have an "unheard of memory" which, in partnership with his other mental and moral qualities, made him "the perpetual dictator of all sciences."[151] Henry III praised the French historian Jean Bodin for being able to pour forth in conversation "an abundance of most beautiful things from his excellent memory."[152] Bacon traded on this ideal in his conceit about King James I's qualities, among which was "the faithfulnesse of your memorie."[153] There were, however, limits to the praise of scholarly memory: one was not supposed to display parrot-like recital from texts, but deeply absorbed knowledge held in memory and capable of being tapped at will.[154] Furthermore, such learned memory was seen to be founded on study and notes.

The adage that quick memory and strong reason rarely coexist in the same person often informed debate about the appropriate role of the different intellectual faculties. This contrast was embedded in a physiological and medical framework by the Spanish author Juan Huarte in *Examen de ingenios* (1575), a work translated throughout Europe. Drawing on the theory of the four humors—blood, black bile, yellow bile, and phlegm—as derived from the Hippocratic school and elaborated by Galen, he asserted that good understanding required a certain dryness of the brain. In contrast, quick memory needed a moist and hence pliant brain that allowed firmer, and more lasting, impressions.[155] "By this doctrine," Huarte inferred, "the understanding and memorie, are powers opposit and contrary, in short, that the man who hath a great memorie, shall find a defect in his understanding."[156] Aubrey assumed this formulation to be proverbial when, in his character sketch of Robert Hooke, he remarked: "Now when I have sayd his inventive faculty is so great, you cannot imagine his memory to be excellent, for they are like two bucketts, as one goes up, the other goes downe."[157] This dichotomy was glossed through accounts of memory initiated by Descartes' conjecture in *L'Homme* (1630s) that sensations are registered in the porous matter of the brain as impressions. These in turn encouraged the flow of animal spirits along such traces, creating a disposition for later reactivation, or recall to memory.[158] A later variation on this theme is found in *The polite gentleman; or, reflections upon the several kinds of wit* (1700), an anonymous French work translated by Henry Barker. After stating that "a good memory and sound Judgment" are seldom combined, the author claimed that in persons with good memories the "Animal Spirits" were very active, quickly laying down strong traces of sensations and ideas, thereby enabling these to be recalled "in the same Order that the Impressions imprinted in the Brain preserve." However, it was likely that an individual possessing such a capacity would be unable "to banish from his Mind a mighty Number of Ideas it crowds in upon him." The author wanted to affirm judgment over wit, the latter being merely the product of a good memory rather than of sound reasoning. He mocked any unqualified praise of prodigious memory: "If some Men be so *Stupid* as to forget every Thing, there are others no less *Stupid*, for retaining all they read or hear."[159] Such was the predicament of those who "have taken more Care to cultivate their memories than their Minds."[160]

There was, however, a way out of the rigid dichotomy between memory and understanding (or judgment). Aristotle had distinguished between remembering and recollecting. He claimed that memory (*memoria*) was a corporeal faculty also possessed by some animals, whereas recollection (*reminiscentia*) involved a deliberate search for something stored in memory.[161] As Huarte acknowledged, albeit in imprecise terms, Aristotle contended "that memorie is a power different from remembrance," and that the latter was associated with "great understanding."[162] During the seventeenth century, this distinction between memory and recollection was often hidden or obscured (as it is today) by the tendency to use "memory" as a general term; but it was accepted and clarified on occasions by most authors of works on the mental faculties and passions.[163] On the one hand, there was recognition that the physical basis of memory, understood in the Galenic tradition to be located in the third ventricle at the back of the brain, rendered it susceptible to injury and hence sudden, or permanent, loss of content.[164] The young Isaac Newton recorded some ancient and contemporary cases: "Messala Corvinus forgot his owne name. One by a blow with a stone forgot all his learning. Another by a fall from a horse forgot his mothers name & kinsfolkes. A young student of Montpelier by a wound lost his memory so yet he was faine to be taught the letters of the Alphabet againe."[165] More positively, promotion of good health could improve memory: as Fuller said, "moderate diet and good aire preserve Memory."[166]

Recollection, on the other hand, was understood as a process of searching, reviewing, and comparing ideas stored in memory; it was therefore considered as a rational and deliberative activity. For Aristotle, it involved a "hunt," starting perhaps "from something similar, or opposite, or neighbouring" in our thoughts.[167] Hobbes contrasted this deliberate searching with the undirected wandering of the mind, which he regarded as its default state.[168] In his *Treatise of the Passions* (1640), Edward Reynolds insisted that memory, through its powers "which the *Latins* call *Reminiscentia* or *Recordatio*," should be regarded as "a joynt-worker in the operations of Reason." To ensure that memory, via recollection, could fulfill its office of assisting reason, he advised that careful and methodical study of subjects must take precedence over attempts to stock the mind with a "multitude of Notions."[169] The Oxford scholar Obadiah Walker made a similar point: in mentioning that the Spanish philosopher Francisco Suarez (1548–1617) knew Augustine's works "by heart," he added that Suarez himself regarded this as "not *memory*, but *reminiscence*; for it was indeed as much *judgment* as *memory*."[170] From this position, it was possible to construe ways of aiding recollection as part of the proper care of the mind.

In the next chapter, I show how early modern manuals on note-taking maintained that well-taken notes acted as prompts to recollection, and therefore had a function in all intellectual endeavors. However, there was a more ancient argument that all forms of writing encouraged laziness of memory. Somewhat unexpectedly, given his stress on *Reminiscentia*, Reynolds canvassed this criticism without objection: "Whereupon Plato telleth us, that the use of Letters, in gathering *Adversaria* and Collections, is a hinderance to Memorie, because those things wee have deposited to our Desks, wee are the more secure and carelesse to retaine in our Minds."[171] An unspecified alternative here was the ancient tradition that promised to improve the natural ability to recollect by artificial techniques. This art of memory used internalized cues, rather than external aids, such as written notes.

There is no surviving text from classical Greece that describes the art of memory in detail. The earliest extant work, *Ad Herennium*, an anonymous Latin text (c. 88–85 BCE), assumes an earlier tradition. The author explains that the technique involves two concepts: backgrounds or places (*loci*), and images (*imagines*).[172] The practitioner imagined a structure, such as a palace with several rooms, and mentally furnished it with clearly marked places, such as "an intercolumnar space, a recess, an arch, or the like." These places had to be memorized as an ordered series, thus forming a familiar permanent mental background. In the next stage, specially constructed vivid images (*imagines agentes*) were deposited in these places as reminders of the things or speeches or arguments to be recalled. Significantly, the choice of images was an individual affair for, as the text observes, an image "that is well-defined to us appears relatively inconspicuous to others. Everybody, therefore, should in equipping himself with images suit his own convenience."[173] When an individual walked mentally through his or her imagined space in a strict sequence, the images in each of the places surrendered their associated content. The technique, often called "topical," "local," or "place" memory (from *topoi, loci*) thus required the *double* task of remembering both the stable background and a set of personal mental images.[174] The compensation was that the background (with its places occupied by chosen images) was internalized as mental scaffolding, and it then functioned as a visual cue for recollection.[175]

The stories about Simonides, Cato the Elder, Seneca the Elder, and other ancient adepts recount feats of recollection—of names, locations, numbers, and texts—many of which now seem both fantastic and useless.[176] These examples can be misleading; we need to clarify the kind of memory expected. It is almost certain that it did not involve what Ian Hunter has called "lengthy verbatim recall"—namely, "recall with complete word-for-word fidelity of a sequence of 50 words or longer."[177] Ancient memory performances were not checked against a text, although some of them did claim to repeat passages that existed in written form. But as Jocelyn Small and others have explained, the Greeks and Romans were more interested in the "gist" of the story or conversation than in "verbatim accuracy." Neither Plato nor Thucydides claimed to "give the precise words of the speeches" they conveyed.[178] Moreover, both ancient and medieval commentators distinguished between "*memoria rerum*" (memory of things) and "*memoria verborum*" (memory of words), acknowledging that the latter was not so well assisted by the use of associated images. Quintilian observed that the number of words, even in a moderately long passage, exceeded the number of images that could act as an effective mnemonic. He realized that a passage could be bundled under a smaller number of stanzas and other groupings, but was not keen on this method.[179] In fact, he was critical of verbal recitation for the sake of show. With Cicero and other Roman rhetoricians, he favored "*memoria rerum*," which included concepts, themes, and arguments encompassing historical, legal, and natural knowledge. Quintilian recommended that if a speaker sought to "learn a passage by heart" he should read it over "from the same tablets on which he has committed it to writing." He advocated "certain marks" (such as symbols or Heads) to act as prompts, which prevented "wandering from the track." Significantly, he preferred this combination of writing and memory to the Greek "mnemonic system" he had just presented.[180]

During the Renaissance there was growing doubt about the value of "topical" memory trained by these ancient techniques. In 1531 Cornelius Agrippa contended that these mnemonic techniques depended on the foundation supplied by natural memory, which itself might be cluttered by the required stock of places and images.[181] The Italian Jesuit Matteo Ricci (1552–1610) spent twenty-eight years in China and sought to teach the Western art of memory. After listening carefully to Ricci's lessons, the son of the Chinese governor of Macao remarked that "though the precepts are the true rules of memory, one has to have a remarkably fine memory to make any use of them."[182] It is not difficult to find early modern authors reiterating these objections. Thomas Fuller, renowned for his own prodigious memory feats, declared that "Artificiall memory is rather a trick than an art."[183] Hartlib, the London intelligencer who, as we shall see, was constantly on the lookout for mental short cuts, found the traditional mnemonic devices too cumbersome: "Artificial Memorie is not to bee used at all because it is vaine or impious, in the inventing of Images, & also it is burdensome in the threefold apprehension of places, images & and the thing to bee spoken of, & so it dulleth wit and memorie."[184] Petty claimed to be able "at first hearing remember any 50 Nonsensical Incoherent words," but admitted that this was "a thing of noe use but to gett the admiration of ffoolish people."[185]

Although the reputation of the classical art of memory had substantially faded by the seventeenth century, its emphasis on orderly arrangement remained influential.[186] The humanists acknowledged this point when advising on the organization of their commonplace books but, more generally, there was an affirmation that writing was the simplest and best way of supporting memory. It is symptomatic that John Willis, in his popular exposition of mnemonic techniques, conceded that "Writings (I confesse) are simply the most happie keepers of any thing in memorie, and doth for speed and certaintie go beyond any art of Memorie."[187] Here again the terms need to be clarified: if recollection via internal cues is the comparison, then Willis presumably meant that written records acted as external stimuli for trains of recollection—in addition to providing a near-permanent record of specific information. In any case, it was this capacity to prompt that was assumed in the discussion of note-taking. We can see this in Bacon's reflection on the variety of ways to aid memory, one of which was "that a host of circumstantial details or tags help the memory, as writing in discontinuous sections."[188] In other words, notes and other "discontinuous" forms of writing did more than record; they jogged the memory.

It is crucial to appreciate, however, that notes were also valued precisely because they could record ideas that might otherwise be lost—when recollection failed. Bacon's habit of having a secretary on hand during his meditative walks illustrates this assumption. As Aubrey explained: "His Lordship was a very contemplative person, and was wont to contemplate in his delicious walkes at Gorambery, and dictate to . . . his gentlemen, that attended him with inke and paper ready to sett downe presently his thoughts." Hobbes was one of these secretaries and he reported that Bacon "was better pleased with his minutes, or notes sett downe" than those of others "who did not well understand his lordship." Hobbes adopted this regime, and said that when composing the *Leviathan* he "walked much and contemplated, and he had in the head of his staffe a pen and inke-horne, carried always a notebooke in his pocket, and as soon as a thought darted, he presently entred it in his book, or otherwise he might perhaps have lost it." Aubrey also got the extra details: "He [Hobbes] had an inch think board about 16 inches square, whereon paper was pasted. On this board he drew his lines (schemes). When a line came into his head, he would, as he was walking, take a rude memorandum of it, to preserve it in his memory till he came to his chamber."[189] Elsewhere, Aubrey agreed that such "winged fugitives" had to be fixed "or otherwise, perhaps, they may be eternally lost."[190] This use of short notes as an aid to both memory and thought was mentioned by Descartes, Boyle, Locke, and Hooke. It was recognized that trains of thought, either linked deductively or by association, were more secure when noted, and, once on paper, ideas could be marshalled and re-sorted. One of Petty's maxims was that "Writing gives our thoughts a consistence which els would fly away."[191] The stress on preserving fleeting thoughts or chains of argument contrasts with the traditional function of excerpts as take off points for recollection and improvisation on textual themes.[192]

In outlining these ideas about the relation between memory and notes, I have so far relied on the language of early modern authors. How is this relationship understood today? Since the experimental work of Hermann Ebbinghaus and Frederick Bartlett in the late nineteenth and early twentieth centuries, modern psychology has developed a terminology to describe various kinds of memory, such as short-term, long-term, iconic, episodic, semantic, visual, aural, muscle, procedural, and prospective.[193] In the seventeenth century, not all of these memory operations were named as separate processes, although discriminations were certainly made. Thus the Cambridge scholar Meric Casaubon (1599–1671) pithily observed: "Of memories, some are quick, and quickly forgett. Others, the cleane contrary. Some are memories of words; others, of matters and consequenceis."[194] Additionally, the distinction between memory and recollection, so crucial to the practice and precepts of note-taking, connects with recent theorizing about "extended mind" and "distributed cognition." Philosophers such as Merlin Donald, Susan Hurley, Andy Clark, and John Sutton have argued that thinking should not be considered solely as an internal mental process. Instead, a full account of cognitive behavior must escape "the myth of the isolated mind" by considering the total environment in which humans think and act.[195] From this perspective, memory depends not only on the inner chemistry and physiology of the brain, but on the physical objects and spaces with which the mind interacts.[196] More specifically, various artifacts and technologies, including writing in its various forms, serve as cues for individual memory. The term "exogram" has been applied to external prompts such as rhymes, diagrams, knots in string, and, of course, notebooks or electronic notepads.[197] Research guided by the "extended mind" theory has sought to examine the ways in which memory and cognition are distributed across internal mental processes, external objects, and social systems. In a recent application of such a model to historical episodes, Lyn Tribble and Nicholas Keene suggest that features of Puritan religious and educational culture can be understood as reformations of the contexts in which memory performed.[198]

Some members of the Royal Society contemplated a classification of the natural world that might act as a form of "extended," or externalized, memory. John Wilkins' *Essay towards a real character, and a philosophical language* (1668) promised both mnemonic support and a philosophical taxonomy. He proposed to comprehend the world under "40 common Heads or *Genus's*," together with their subdivisions into "*Differences*" and "*Species*," following Aristotle. He set out his "enumeration of things and notions" in "Tables" designed to indicate "both the General and the Particular head under which it is placed."[199] Wilkins declared that his "characters" were formed to reflect "the Natural notion of things" as disclosed by "a just Enumeration and description." In other words, the signs of this artificial language were designed to carry not only a reference to a thing or idea, but also to convey relationships, such as similarity, opposition, and so on. As Wilkins explained:

> But now if these Marks or Notes could be so contrived, as to have such a *dependance* upon, and relation to, one another, as might be suitable to the nature of things and notions which they represented. . . . Besides, the best way of helping the *Memory* by natural Method, the Understanding would be highly improved.[200]

Wilkins' supporters were attracted by the idea that notebooks could be prearranged according to the divisions of his scheme. Andrew Paschall told Aubrey that his revision of the original forty genera could be presented in a pocket commonplace book.[201] In reporting this to John Ray, who had advised Wilkins on the taxonomy of plants, Aubrey mentioned

> some *Tables* that might be made according to those of yours in the *Bishop's Essay*, and fitted to be hung up in Garden-Houses in the Manner of Maps, . . . they might become a fine Ornament in Summer-Houses, and very useful for those who delight in that kind of Knowledge. . . . The same may also be put into a little Pocket-Book, which may be of Use where the larger Tables cannot be had.[202]

It may have been this feature of Wilkins' project that encouraged Aubrey to suggest, in his thoughts on education, that children should not only hear and read proverbs but "also to sett downe in the Reall Character: which will fix it the better in their minds."[203] The Oxford scholar Thomas Pigot, although he disagreed with "Mr Paschall's Schemes," assured Aubrey that these would "be a most admirable worck for a pocket boock[,] common places or other things of that nature," provided that they still "shew the dependence of one thing upon another."[204] However, some contemporaries recognized that this taxonomy would have difficulty dealing with *new* knowledge.

These anticipations of a collective commonplace book are symptomatic of the role of the notebook (in its various guises) as a ubiquitous aid to memory in early modern culture. It remains so today, and it is not surprising that leading proponents of "extended mind" theory should suggest a thought experiment in which an Alzheimer's patient ("Otto") relies on a notebook as an external surrogate for compromised internal memory.[205] This scenario poses a severely damaged biological memory and a notebook that acts as a holder of crucial information—names, dates, and places that Otto cannot remember. Yet beyond such extreme situations, notes do more than retain data in this way; they work in tandem with information and ideas stored in memory. Indeed, in early modern European culture, there was deliberation about the various ways in which this nexus could, and should, operate. In order to capture one of the critical distinctions I will refer throughout this book to *recall* as the ability to remember required information without prompting. The major example of this in the seventeenth century was rote memorization of material, such as Latin vocabulary, grammar, and selected passages; prayer and religious ritual were also expected to be mastered in similar fashion. Although recitation of texts might be instigated by a word or phrase, such performance was rewarded because it relied only initially, and minimally, on the written word. In contrast, *recollection*, as explained above, was conceived as recovery of material not immediately available in memory; recollection was evoked by a note or image that instigated a search through memory for material related, in various ways, to the initial trigger. In modern cognitive psychology, this notion of recollection has largely been replaced by the experimental concept of "cued recall" in which a stimulus (for example, a word, sound, or image) elicits a memory of another item with which it is linked.[206] However, in the framework inherited from Aristotle, recollection in its most advanced form produced far more than an automatic, triggered association. One of the key functions attributed to notes during the Renaissance was their power to prompt recollection of copious material—facts, quotations, arguments—from memory for use in speech or writing. Cultivation of this process was a central principle of humanist learning. I want to consider how, by the seventeenth century, note-taking worked as a means of collecting and analyzing information about the natural world.

In the next chapter I examine early modern views on how notes should be taken, organized, and used to aid memory and manage information and knowledge, both textual and empirical. One of the surprises, especially given the stereotype of reluctant readers that the "moderns" themselves helped to foster, is that some of the most massive and extensive commonplace books were kept by virtuosi closely connected with the Royal Society. Thus although notebooks were central to the bookish culture attacked by some apologists for that new institution, they played a part in the making of early modern science. However, the extension of the commonplace method to the accumulation and analysis of empirical information required significant adjustments. There is a profound difference between using notes to select examples from an agreed corpus of quotations and tropes, and using notes to generalize by working inductively from disconnected, miscellaneous particulars. One adjustment involved questioning the central place occupied by memory, especially rote memory, in contemporary education and learning. This did not mean that memory was necessarily cast into competition with the use of notes, but rather that the relationships between them, sometimes variously perceived in different subjects and contexts, were a serious matter of discussion. Another implication was that both recall and recollection might be challenged by the sheer scale and diversity of the information expected in Baconian inquiries. Instead, in these projects notes might best serve to guarantee *retrieval* of particular, and detailed, material. Some of the English virtuosi faced this choice when their own personal notebooks seemed to outgrow the capacity of memory, even if aided by notes.

[2]

Capacious Memory
and Copious Notebooks

Chi scrive non ha memoria [Who writes, hath no memory].

Giovanni Torriano,

Piazza Universale di Proverbi Italiani (1666)

:::

The Italian proverb above reflects the tension between memory and writing, famously put by Socrates and reported by Plato in about 380 BCE. Socrates announces that reliance on writing weakens memory. He tells the story of the king of Egypt being presented with the invention of writing. The inventor (Theuth) says that "my discovery provides a recipe [or drug] for memory and wisdom." But the king replies: "If men learn this art it will plant forgetfulness in their souls: they will cease to exercise memory because they rely on that which is written, calling things to remembrance no longer from within themselves, but by means of external marks; what you have discovered is a recipe not for memory, but for reminder."[1] From this perspective, note-taking was dangerous, or at least lazy, as Antisthenes (445-360 BCE) remarked in replying to a friend who complained of losing his notes: "You should have inscribed them on your mind instead of on paper."[2] It is a long jump, but an instructive one, from Plato to Sigmund Freud's observation: "If I distrust my memory—neurotics, as we know, do so to a remarkable extent, but normal people have every reason for doing so as well—I am able to supplement and guarantee its working by making a note in writing. In that case the surface upon which the note is preserved, the pocket-book or sheet of paper, is as it were a materialized portion of my mnemic apparatus, which I otherwise carry about with me invisible."[3] For Socrates and Plato, memory is diminished by a reliance on writing; for Freud, notes are external reminders, extensions of memory. How can we situate early modern European ideas about the role of memory (and notes) in this long history?

One useful starting point is Francesco Petrarca's (in English, Petrarch) imaginary conversation with Saint Augustine, as given in his *Secretum*, probably written in 1347. The relevant background is the much-cited Book X of Augustine's *Confessions* (c. 397 CE) in which he presents memory as a capacious Aladdin's cave, a cornucopia of images, ideas, and feelings. He gives a complex and puzzling account of "the great storehouse of memory," at times confident in his ability to command ideas "in the correct order just as I require them," while in other passages describing the memory as "awe-inspiring," its "vast cloisters" unfathomable, mysteriously outreaching the capacities of the human mind in which it seemingly resides.[4] In Petrarch's dialogue, Augustine appears as a premonition of a humanist pedagogue issuing advice on note-taking. Petrarch says he has read the works of Cicero and Seneca but "as soon as the book left my hands, my connection to it likewise disappeared." The saint advises making "notes of important points." When Petrarch asks, "What kind of notes?", Augustine goes on to make it plain that his method of reading is to mark crucial passages—in this case, those offering moral and spiritual instruction—as a signal that these "salutary ideas" should be worked "deep into your memory" and consolidated there "through diligent study." Summing this up, he urges that "when you come upon such things, put marks next to the useful passages. Through these marks, as with hooks, you can hold in your memory useful things that wish to fly away."[5]

If we take Michel de Montaigne and René Descartes—both routinely named as "moderns"—we find various degrees of defection from this recourse to memory. In his *Essays* (1580), Montaigne did not show the high esteem for memory which Petrarch (and Augustine) evinced; instead, he regarded it simply as a natural capacity unevenly endowed among men. He did not question the crucial role of memory in mental life, but admitted that his own was an untrustworthy sieve. He consoled himself with the thought that, lacking a strong memory, he did not talk on and on, or ruin a good story with needless detail. Significantly, Montaigne acknowledged the need to write things down, or to tell someone: "It is beyond my ability to answer propositions in which there are several heads of argument. I could not take on any commission without my jotter." He also remarked that "Lacking a natural memory I forge one from paper."[6] When asked about the weakness of memory, Descartes offered similar advice. In 1648 he told Frans Burman (a Dutch theological student) that "I have nothing to say on the subject of memory. Everyone should test himself to see whether he is good at remembering. If he has any doubts on that score, then he should make use of written notes and so forth to help him."[7] Descartes accepted the limitations of memory and recommended the use of notes as a necessary support, but his blunt response does not tell us whether the notes were envisaged as assisting memory (and recollection), or as dispensing with it.

Petrarch's question — "What kind of notes?" — deserves a more complete answer than he received from Augustine. We can attempt this by avoiding a simplistic contest between memory and writing, focusing instead on debates about the proper use of notes. Historians of the Renaissance have long stressed that the recovery and printing of ancient texts did not lessen the desire to commit choice passages to memory. As R. R. Bolgar put it:

> The whole purpose of the Humanists in transmogrifying Greek and Latin literature into a series of notes was to produce a body of material which could be easily retained and repeated. They made titanic efforts to remember the contents of the note-books they compiled. The Renaissance was the age of memorizing.[8]

However, even profound memory was thought to be powerfully extended by abundant notes. Thus it was said of Joseph Scaliger, doyen of polymathic scholars in the late Renaissance, that "he could have left cartloads of notes on all authors."[9]

We need to examine more closely how notes were thought to aid memory, and to distinguish between the various functions usually grouped under memorizing. Consider the range of memory terms listed by John Wilkins in his *Essay* (1668) on artificial language. In the taxonomy of categories that underpinned this project, he classified "Memory" among the "Internal Senses" and filled this out as follows: "recollect, re-call, commemorate, remember, call or come to mind, put in mind, suggest, record, recount, con over, getting by heart, by rote, without book, at ones fingers end, memorable, memorial, Memorandum, mindful." Wilkins also identified this set of terms by contrast with "Forgetfulness, Oblivion, Unmindfulness, overslip."[10] I think it is legitimate to imagine these memory actions being performed in conjunction with notes of various kinds — notes of sermons and lectures to be learned by rote, of passages or speeches to be conned, of commercial accounts to be entered, of administrative memoranda to be recorded, and of short extracts or epitomes that might help to bring something to mind by prompting recollection of things temporarily forgotten.

There is a nice snapshot of views about the proper combination of memory and notes in Pierre Gassendi's *Life of Nicolas-Claude Fabri de Peiresc* (1657). He recounted that the great French antiquarian and collector

> had a happy memory, and which seldome failed him. For though he complained that his memory was slippery, and weak; yet it cannot be expressed, what a variety of things he remembred, even from his young years, and that not in general only, but also with the particular circumstances of places, actions, words, and persons . . . he could alwaies produce out of his Store-house pertinent matter, which he uttered in choice words.[11]

So far so good: this is the standard Renaissance praise of the polymathic scholar able to draw upon a memory well stocked with learning. However, there is another message a few pages later where Gassendi related that Peiresc was "so unwearied in writing that he presently noted down, what ever he met with." Gassendi explained that this practice was necessary given the range of things Peiresc wished to know, combined with his fear that memory might "let slip many particularities." What follows is an intervention in the classical debate about memory and writing: "Now he [Peiresc] wrote things down in his Memorials, because he then judged they were out of danger of being forgotten, seeing he could not trust his memory as *Socrates* or *Pythagoras* were wont to do; and had found by experience, that the very labour of writing did fix things more deeply in his mind."[12] This is an eloquent statement of the value of notes, even for a mind blessed with a strong memory.

For seventeenth-century English attitudes, John Aubrey's *Brief Lives* is instructive and suggestive. Perhaps as many as half of his pen portraits include comments on the person's memory. If we take Aubrey's opinion as roughly representative, then possession of a strong memory was clearly acknowledged in the late seventeenth century, in much the way it had been in the Renaissance. Aubrey reported of several of his subjects—for example, John Hoskyns, James Long, William Prynne, Edmund Waller—that they had great or "prodigious" memories. He also stressed that memory was a corporeal faculty that could be affected, or ruined, by physical disease, or by other imbalances of the bodily humors, such as melancholy. Thus he reported of

James Harrington that "his memorie and discourse were taken away by a disease"; and of Bishop Seth Ward that "the black malice" of an antagonist (Thomas Pierce, the Dean of Sarum) was "the cause of his disturbd spirit, wherby at length he quite lost his memorie."[13]

Aubrey was also interested in the best ways to aid or support memory. He observed that some of his characters practiced artificial techniques, usually based on the classical art of memory. Thus John Birkenhead (1615–79) "had the art of locall memory; and his topiques were the chambers, &c, in All Soules colledge (about 100), so that for 100 errands, &c, he would easily remember."[14] The allusion here to a decidedly unscholarly application might reveal Aubrey's sceptical assessment.[15] Elsewhere he certainly stressed that other individuals with capacious memories did not need to employ such a specific technique. For example, although Thomas Fuller, was renowned for his ability to "repeate to you forwards and backwards all the signes from Ludgate to Charing-crosse," Aubrey averred that his "natuall memorie was very great, to which he had added the *art of memorie*." But the scholar and lawyer John Selden "never used any artificiall help to strengthen his memorie: 'twas purely naturall." Aubrey proffered his own view that careful ordering of thoughts was just as helpful as deliberate use of "local" or "place" memory. He suggested that the poet John Milton "had a very good memorie; but I beleeve that his excellent method of thinking and disposing did much to helpe his memorie."[16]

Kinds of Memory

In his *Brief Lives*, Aubrey did not directly address the role of notebooks, but one of the figures he included did precisely that. In *The Holy State* (1642), Fuller explained that memory operated in two different ways: "one, the simple retention of things; the other, a regaining them when forgotten." This was a restatement of the distinction, influentially put by Aristotle, between remembering and recollecting. Fuller also stressed that animals ("brute creatures") possess the ability to remember, often exceeding men in "a bare retentive Memory," but "cannot play an aftergame, and recover what they have forgotten, which is done by mediation of discourse." In warning, like many others, that memory was a fragile faculty seated at "the rere of the head," he advised against trusting everything to it: "Adventure not all thy learning in one bottom, but divide it betwixt thy Memory and thy Notebooks."[17] We know that he meant commonplace books, specifically, because he immediately mentioned those who attacked them. Bacon also admitted that there was a "prejudice imputed to the use of *Common-Place Bookes*, as causing a retardation of Reading, and some sloth or relaxation of Memorie."[18] The charge about not reading refers to the lazy practice of relying solely on passages entered in a notebook; but the accusation that common-place books might invite a neglect of memory (a point reiterated in *De augmentis*) appears to question the role of notebooks in aiding memory. What were the possible grounds of such an attack? Here it helps to consider the prevalent cultural convictions about the importance of memorization without reliance on books or notebooks. Then it will be possible to consider the rationale for note-taking itself, and the nexus between commonplacing and the second operation of memory, namely, recollection.

The prevailing suspicion about reliance on writing and note-taking should not be underestimated. In 1697 in his biography of Seth Ward, Walter Pope remarked that "the Bishop had an ill Memory, even when he was in the best of Health, which he empaird, by commiting all things to writing."[19] This view clashed with another ubiquitous dictum about note-taking as an aid to memory, but was itself supported by a widespread respect for the ability to perform "without book." This phrase, which occurs in Wilkins' list of memory actions, referred to the recital of information or knowledge without consultation of a text, or a notebook that might summarize it. Thus in 1612 the schoolmaster John Brinsley declared that students must "have daily some special exercise of the memory, by repeating somewhat without booke; . . . The reason is, because the daily practice hereof, is the only means to make excellent memoryes."[20] In 1559 (English edition, 1563) John Foxe, the Protestant martyrologist, had celebrated the importance of this ability in religious practice, saying that many English people, both women and men, "often read the whole Bible through, and that could have said a great sort of St. Paul's epistles by heart."[21] As printed books became more available, it is possible that the external record they offered made verbatim or rote memorizing even more achievable than before.[22] Moreover, some oral performances required of seventeenth-century undergraduates demanded the memorizing of material such as college sermons, which were heard, not read.[23] This task may often have started with a notebook, but the point of the exercise was to do "without book." John Milton's cohort at Cambridge was expected to memorize the main points in college sermons, and Milton himself read passages to his students as part of Sunday exercises.[24] About two decades later, the influential Cambridge tutor James Duport advised that undergraduates "Be constant at the Church . . . & take notes of the sermon."[25] Of the forty-nine sins the young Newton confided to a small notebook in 1662, number eleven on his list was "carelessly hearing and committing many sermons."[26] Nor were these expectations confined to school or undergraduate training. Meric Casaubon confessed (in about 1668) that he admired Cicero's *De Officiis* [on Duties] so much that "had I such a memorie as some have, I would have learned by hart." He regretted that he was never "so confident, that I durst adventure into the pulpit without my notes," but that if he had his youth again this is what he would aim to do.[27] It was said of the Oxford scholar Robert Sanderson (1587-1663) that he "could repeat all the *Odes of Horace*, all *Tully's Offices*, and much of *Juvenal* and *Persius* without Book."[28]

These attitudes help to explain why some of Bacon's near contemporaries felt the need to defend note-taking per se. Two Jesuit pedagogues, Francesco Sacchini (1570-1625) and Jeremias Drexel (1581-1638), wrote influentially on this topic. They both acknowledged the objections to note-taking, largely attributing such resistance to the continuing effect of the indictment of Socrates, Plato, and Pythagoras. As professor of rhetoric at the Collegio Romano, Sacchini published *De ratione libros cum profectu legendi libellus* (1614), perhaps the first work devoted mainly to the practical, as well as theoretical, features of note-taking. He prescribed careful note-taking as an essential part of study, not only in rhetoric, but across the major disciplines of philosophy, theology, law, and medicine.[29] In reply to the criticisms of Socrates, he cited rival ancient authorities who *did* rely on writing, such as Demosthenes who copied Thucydides several times, and Pliny the Elder who, according to his nephew, always made extracts from books he read, or those read to him by his slaves.[30] Going on the offensive, Sacchini indicated how chapter summaries allowed by writing, and even more so by print, actually helped both memory and understanding of a text. More specifically, he asserted that far from weakening memory, the careful selection of excerpts focused attention and thereby supported retention; the very act of writing something down fixed it in individual memory — a direct rebuttal of the ancient warnings. In Sacchini's view it was not sufficient to mark a passage on the page of a book, especially not by using one's fingernail — an odious habit that did nothing to lighten the memory and also ruined books, so that precisely the most important passages were rendered invisible.[31] Even less-delinquent practices, such as dog-earing pages (favored by Newton) or inserting markers, were ineffective because such behaviors were motivated by a desire to keep the note physically tied to the text, ignoring the usual advice to use a commonplace book of entries gathered from an array of texts, conversations or observations.[32] Reading without noting was likely to yield only superficial knowledge because it was not accompanied by the attention and thought required to make a considered note. The acts of transferring and copying were also crucial. Sacchini advised writing out excerpts twice: first, as they were encountered and then later under Heads in a separate notebook.[33] Of course, he stressed that the aim was to have notebooks not simply full of entries but of material that will be remembered and understood. Those who read for pleasure only, without noting, were forsaking this chance to stock the memory with treasure.

Jeremias Drexel, a teacher of rhetoric in Augsburg, reinforced and amplified Sacchini's defense. Cast in dialogue form, his *Aurifodina* (not published until 1638, the year of his death) has a student, "Faustinus," voicing objections, often based in laziness but sometimes citing classical authorities. The replies of the teacher, "Eulogio," also marshal ancient authors — for example, Pliny and Augustine — who vouch for the fragility of memory, a faculty always subject to bodily infirmities.[34] On this basis, Drexel concluded that memory must be assisted by external supports or techniques.[35] Of course, the classical mnemonic arts claimed to supply precisely such aid, using internalized images arranged in a sequence as prompts to recollection. However, Drexel maintained that there was also an "art" of note-taking, one that was equally useful and more reliable for the long term. This confidence in note-taking rested on two convictions — that copying impressed content in memory, and that notes prompted recovery of copious related material. To be sure, these two operations were sometimes elided. In emphasizing the need to read extracts over so as to fix them securely in memory, Sacchini's advice was tantamount to rote memory training.[36] But generally, the two Jesuits claimed that notes offered the best mode of recollection, via cues provided by short extracts under Heads, as taught by the commonplace method. In this sense, their writings answered the allegation about commonplace books letting memory go on holiday: on the contrary, proper note-taking, under Heads, served to ensure recollection of things stored in memory. One of the English virtuosi made a direct comment on the Jesuit advice. In his manuscript on education, Aubrey endorsed what he called "the admonition of Father Drexilius" about habitual note-taking (*"semper excerpe"*), saying that this created "nest eggs" for the future; but he also implied that the material gathered should not be confined to that offered in books. Rather, notebooks could be "excerpts of observations," thus going beyond the "common way of precepts as the knowledge of a traveller exceeds that which is gotten by a map."[37]

The ideal use of commonplace books involved exercising judgment in the selection and noting of material, and in the choice of appropriate Heads as prompts to recollection. On this point there was a rich ancient legacy, since both classical and medieval authorities distinguished between rote memory of words and the more nuanced memory of things in the world. Learned memory was to be cultivated in tandem with judgment and invention.[38] The metaphor of the bee, as delivered by Seneca, became a commonplace in its own right: the bee gathers material, but it also selects and transforms; so, too, should the good orator, by embellishing and adorning.[39] The use of notebooks, especially by scholars, was not meant to permit slavish regurgitation of excerpts, but rather to encourage a combination of Heads and short quotations as triggers to recover knowledge already stored in memory. In *De Copia* (1512), Erasmus advised how "places," say those for the virtues and vices, could be subdivided into contrasting notions, each of which could be filled with apt quotations, proverbs, tropes, and so on. This binary framework assisted the memory, cueing it to recall material for use in conversation, speeches, and writing, complete with supporting and opposing positions on a theme. As Erasmus boasted, his method supplied examples "ready in our pocket so to speak."[40] But this material had to be truly mastered and produced with *sprezzatura*, thus giving the impression that alternative formulations were within easy reach. It was certainly not the kind of performance that Seneca reported in the story of Calvisius Sabinus, a Roman noble with a poor memory who hired slaves—"one to know Homer by heart and another to know Hesiod"—who fed him with apt quotations so that he could impress guests. Seneca remarked with the utmost disdain that "Sabinus held to the opinion that what any member of his household knew, he himself knew also."[41] Even when one had properly read and absorbed the material, the right balance between memory and notes had to be struck. In discussing the appropriate gestures of orators, Quintilian advised that:

> The hand should not be overloaded with rings, which should under no circumstances encroach upon the middle joint of the finger. The most becoming attitude for the hand is produced by raising the thumb and slightly curving the fingers, unless it is occupied with holding manuscript. But we should not go out of our way to carry the latter, for it suggests an acknowledgement that we do not trust our memory, and is a hindrance to a number of gestures.[42]

Copious commonplace books were not seen as crutches for a poor memory, but rather as foundations that promised a capacious one. The trick was to give due weight to two ideals—*copia* and *brevitas*.[43] The method of commonplacing demanded the ability to select from a large body of material and to make judgments about the most apt Heads for various extracts. Erasmus maintained that a person who can condense material into a pithy form will also be able to elaborate, enriching it with abundant illustrations.[44] Along the way, repeated referral to extracts might result in knowing some passages by heart; but the key point was that a brief note should bring forth a copia of examples and digressions. The aim was not verbatim recall of a specific passage, but learned improvisation on a topic. In this way, the memory promoted by commonplacing was different from that produced by rote learning or, in the language of the day, by heart, "conning," parroting, or "without book." Of course, order and cues assisted both rote memory and recollection. In the case of things learned by heart it had long been observed that the sequence of parts was influential, so that even some of the best performers, if interrupted, would often have to start again.[45] However, in contrast with rote memory, recollection aided by commonplace Heads was not usually expected to produce verbatim recall, partly because embellishment and invention were admired.

Bacon, Sacchini, and Drexel each emphasized that recollection was assisted by proper attention and judgment at the time when both extracts and Heads were selected. The Jesuit pedagogues were prepared to tolerate a two-stage process in which notes (either excerpts or comments) were taken on a loose sheet of paper while reading and later assigned to Heads in a commonplace book. This procedure resembled the advice given to travelers: take notes in pocket-sized paperbooks, or erasable table books, and later transfer the material to larger notebooks.[46] The risk was that these loose notes would never be assigned to a Head. No doubt this is one reason why Jesuit schools allowed only advanced students to postpone the allocation of notes to appropriate Heads in a commonplace book.[47]

The convictions in play can be seen in John Foxe's views on how a commonplace book should be composed and used. Whereas his *Acts and Monuments* celebrated the ability to recite "without book," Foxe believed that commonplacing should develop learned memory prompted by appropriate Heads. His *Locorum communium tituli* (1557) showed how a formal hierarchical structure could underpin the collection and arrangement of notes.[48] This work comprised introductory matter followed by 647 blank pages intended to be filled with copious material under the Heads from the schema.[49] In his introduction, Foxe explained how Heads were to be arranged according to Aristotle's ten "*praedicamenta*," or categories.[50] In principle, each of these could be subdivided so that, for example, "substance" could include Heads descending from God though humanity to the lowest rungs of the great Chain of Being.[51] The revised edition, *Pandectae locorum communium* (1572), was a notebook comprising 1,208 pages (after the introductory material) all of which were blank except the top of every fourth page, where one or more Heads were printed. Foxe's expectation was enunciated in the subtitle: "The studious reader will be able to record here according to his own choice, as in their very own seats and nests, whatever should occur anywhere worthy of remembering in every reading of any author, and to exhibit here afresh, as it were, whatever he may want from the storehouse of memory."[52] An "Index locorum Alphabetarius" at the back gave page numbers for the 768 Heads.[53] These covered topics from "theology, physics, law, medicine, mathematics, etc" that should guide reading and study, and under which a reader might enter salient quotations.[54] However, in contrast with the first edition, the Heads in the *Pandectae* were inserted at the top of the pages in alphabetical order, probably on the initiative of its publisher, the London printer John Daye. As Ann Moss has suggested, this alphabetical arrangement may indicate a concession to changing times in which "the predicament model was thought not to be viable."[55] Foxe regarded this change as a dilution of the underlying framework of interrelated concepts, and his reluctance to scatter subjects and terms that belonged together is apparent: thus "Reminiscentia" remains on the same page as "Memoria" (its traditional partner), although each is indexed as a separate term.[56] Indeed, in these printed Heads he clustered cognate topics, with only the first of each set being determined by the alphabetical ordering: for example, "Adversitas, misereria humanae vitae . . . "(p. 6); "Affabilitas, comitas, popularitas" (p. 9); "Inobedienta, rebellio" (p. 303); 'Vermes, reptilia' (p. 565). Foxe did not want a simple alphabetical index to replace memory of the connections between the various commonplaces.[57]

Even though Bacon showed little enthusiasm for the traditional method that Foxe endorsed, he also stressed that judicious allocation of material allowed one to find it more easily when required, just as it was easier to hunt any wild animal "in an enclosed park" than in "a forest at large."[58] His metaphor was a version of Quintilian's comparison of recollection to hunting or fishing: one must know where to look and how the prey behaves.[59] But whereas Quintilian was mainly concerned with locating "arguments" in their appropriate places, Bacon contemplated the task of searching a wider territory encompassing literary, philosophical, and empirical material. There had to be ways of effecting what he called "*the Cutting off of Infinity. For when anyone tries to remember or call anything to mind, if he lacks a prenotion or perception of what he seeks, surely he seeks, works himself up,

and scuttles about as if in a limitless space. But if he has some certain prenotion, infinity is cut off *immediately*, and memory ranges closer to home."[60]
In his *Directions for the Study of the Law* (1675), William Phillips reinforced
Bacon's message about the need to limit the parameters of a search, warning
students about being lost in "the vast body of Law." The preventative was
to "digest the Cases of the Law" under "Titles" or "Common Places." Such a
procedure laid down a framework into which new information could be integrated, and found again. Thus commonplacing served "the Students ease,
as for preserving and continuing his Memory."[61]

These principles about the use of notes to aid memory and recollection
were addressed to individual note-takers; but they were also supported by
social practice. Although individuals kept personal commonplace books, it
was assumed that the content could be read with profit by others because
there was a broad consensus on the topics under which material might be
gathered, assimilated, and later exploited in public speeches and sermons.
Cicero maintained that speakers might debate "with brilliant originality,"
but they relied on "the commonplaces" which are "for all that really rather
easy and widely current in maxims."[62] As Ann Moss has shown, the phenomenon of *printed* commonplace books, providing ready-made headings,
lists, and even sample quotations, underscores this consensus.[63] Various
manuals for schools bear this out. Thus Brinsley referred to "any ordinary
Theame, Morall or Politicall, such as usually fall into discourse amongst men
and in practice of life." Thomas Farnaby's *Index Rhetoricus* (1634) provided
a two-page alphabetical list of Heads for the classical and Christian virtues
and vices, and the subjects of the liberal arts, a list that he expected students to use in the course of their reading and note-taking.[64] The function of
notes as prompts for recollection was facilitated by this set of common topics. Texts were read, annotated, and selected in terms of a canon of subjects
and themes, as illustrated in Foxe's *Pandectae*. There was a collective dimension to the use of commonplace books because memory was called upon in
social settings—the classroom, the university, Parliament, the law courts,
and polite conversation. We could say, then, that the practice of commonplacing fostered and maintained a collective memory—not in the sense of
shared episodic memories, but rather as a common repertoire of ideas, quotations, and tropes from an agreed body of texts.[65]

At both Oxford and Cambridge, students were required to perform in the
public university disputations held on special occasions throughout the academic year. The topics for such debates were drawn from what was still
largely a scholastic curriculum centered on standard texts in moral and

natural philosophy, such as those of Aristotle. Undergraduates were assigned a side of the question nominated for dispute; they rehearsed both "pro" and "con" in their colleges before the event.[66] Duport's list of "Rules," in use at Cambridge after 1660, advised that "When you dispute be sure you have your arguments by heart."[67] We can surmise that the format of pros and cons worked as a hook for memory, along the lines of Erasmus' arrangement of quotations as comparisons and contrasts. Indeed, a popular manual of the early 1700s recommended making short notes on "Questions in Philosophy" in "a little paper-Book" so that "you will know upon Occasion what Books to consult *pro* and *con*, upon any Question."[68]

Bacon blamed this university pedagogy, with its "scholasticall exercises," for misconceived views on the role of both memory and notes. In *The Advancement of Learning* and the larger *De augmentis*, he claimed that training in logic and rhetoric was prematurely directed at undergraduates whose minds were "empty and unfraught with matter, & which have not as yet gathered what Cicero calls *Silva* and *Suppellex*, that is stuffe and variety of things."[69] A related instance of the gap between pedagogy and real "Life and Action" was the unhealthy divorce between the cultivation of "Invention" and "Memory." The prescribed oral exercises were either "delivered in preconceived words," thus leaving nothing to invention, or given "extempore, where little is left to memory."[70] But in civil life and the practice of professions, Bacon averred that there were "intermixtures of premeditation, & Inventions: Notes & Memorie."[71] He elaborated this point in "A Discourse, touching Helps, for the Intellectual Powers," written between 1596 and 1604 and addressed to Sir Henry Savile. Bacon suggested that the proper balance between using notes and memory only came with the right experience: "For, in most *Actions*, it is permitted, and passable, to use the *Note*; Whereunto, if a Man be not accustomed, it will put him out." However, he insisted that skills should suit the situation: only lawyers needed to cultivate "Narrative Memory," that is, the capacity to recall sequence of events, actions, person and their circumstances.[72]

Holdsworth's Hints

During the seventeenth century, there was agreement that reliance on memory and notebooks should be adjusted to the situation. Richard Holdsworth (1590–1649), Master of Emmanuel College, Cambridge, from 1637, addressed this issue in his "Directions for a student in the Universitie," a study manual copied and used by subsequent generations of tutors and students.[73] The

"Directions" describes the books and the methods of reading and study over the four years of the BA degree. It is clear that the importance of committing various things to memory did not require any justification, but Holdsworth distinguished between rote memorizing and commonplacing. The former was necessary for lodging the grammar and vocabulary of a foreign language in the mind, and writing and note-taking also played a role in this. He cautioned that there was no other way "for the attaining of a language as this getting without book"; it must be rehearsed or it will surely "slip out of your memry."[74] Aware that Latin, already started at school, was often neglected when students began Greek at university, he urged them to write down Greek words "as you doubt you cannot remember," and warned that "Gramers must not be forgotten." However, Holdsworth did not consider rote memory as appropriate for all subjects, warning "that [this] plodding way of conning doth tire and lode the memory rather then beget a readines."[75] For substantive topics and major authors he advised a combination of memory and thoughtful note-taking. He suggested that the student "spend some time in recollecting what he [your tutor] hath explained"; and "read it with short memorial notes." Relying solely on memory was a foolish practice, since "one booke read with Notes . . . brings a better stock of Learning . . . by Noting you make it intirely your own for ever after."[76]

Holdsworth's preference was for notes to prompt memory, not to render it superfluous.[77] He counseled students "to remember something at least . . . in every dispute, Lecture, Sermon, Speech, or Discourse, which you shall heare & when you come to your studie write them downe in one of these paper bookes." The latter were clearly commonplace books in which students should "recorde the best of theyr studies to certain heads of future use and memorie." By "frequent reading them over on evenings . . . they will offer themselves to your memory on any occasion." As for the method of making these notes, Holdsworth stressed the importance of a separate notebook in which entries were made under Heads, rather than notes made on loose sheets of paper or in the margins of books. He exhorted undergraduates to "keep a Note or Catalogue of all that you meet with in your studies, which you understand not; which at any time either your Tutor or some Freind may sattisfie you in. Write them in some paper book rather than in a loos paper, and leave space for the resolution after every one; It will be a content to you after some years to find them there."[78]

Actual practice was another matter. Knowing the worst about lazy students, he conceded that large commonplace books carried a disincentive: the need "to rise evry foot to a great Folio book, & toss it and turn it for evry little pasage that is to be writt downe." He also reported on a way of responding to this problem by dispensing with notebooks:

> I was told of one, who to prevent this toyle, caused a box to be made with as many partitions as he could have had heads in his booke, so that writing his Collection in any bit of paper, he might without more trouble throwe it in to its Topick, & look over each divisio on occasion.[79]

This physical version of a commonplace book probably maintained the assumption of a pre-existing set of Heads (or topics) to which all excerpts could be assigned. However, Holdsworth proposed another option that was, potentially, more radical: namely, to enter extracts in small octavo paperbooks, as the need occurred, while reading. He suggested that if these excerpts were frequently read over they would "offer themselves to your memory upon occasion," but that if one had trouble finding an entry it was possible to compile an index of the topics or authors covered in several paperbooks, referring to these "by different names or the order of the Alphabet." In this way, a "large Index" was able to reduce "such bookes of Collections to a Commonplace book."[80] Yet such a virtual commonplace book, if indexed using mainly authors or titles of books, would not reinforce a focus on topical headings. A similar alternative is found in the Oxford study guide attributed to Thomas Barlow, Bodley's librarian between 1652 and 1660:

> Lastly, let mee adde this advice for the Improvement of your Reason Gett two Common Place Bookes, exactly, & particularly drawne up, with good advice & help; and let one 1. Bee onely for References, write down under every head the Auth: Cap.Page which speakes of the Point, else you will on every occasion bee to seeke, having forgotten when & where you read it. 2 Let the other Book conteine the cheifest matter of your reading, in the words themselves; But the former willbee more usefull than the later, But if you can, doe both.[81]

Here there are signs of retreat from the traditional dictum — approved by Foxe and Sacchini — that excerpts should be allocated immediately to a major Head. In advice to mid-seventeenth-century Cambridge undergraduates, Duport tried to hold back the tide, urging the use of "a little paper Pocket book" so that notes could be made "when you walke abroad, for fear if you should write them in large Volumes, you then lay them aside & never looke on them more." He suggested that the "most remarkable passages" in a book be marked with "a black lead pen & afterward refer them to your common place booke."[82] In *Of Education* (1673), Obadiah Walker put this option bluntly, saying that notes could be taken down as they occurred, "confusedly," in reading, and then "afterwards at your leasure" allocated to a Head.[83] This separation of note-taking from judgments about appropriate Heads violated Jesuit advice about the best way to ensure both memory and understanding; potentially, it shifted the emphasis away from recollection toward retrieval of items of information. Nevertheless, in the case of English manuals aimed at undergraduates, the ideal of a capacious memory was never entirely abandoned. Thus despite allowing some relaxation of earlier strictures, Duport concluded his "Rules" by encouraging students to "Read these precepts over often; at least once a week, that you may better remember them, & put them in practice."[84]

Works such as Daniel Morhof's *Polyhistor* (1688) and Vincent Placcius' *De arte excerpendi* (1689) reviewed the extensive literature on note-taking.[85] This ranged from the reflections of major scholars to undergraduate advice manuals, such as those just discussed. These texts provide valuable insight into the methods and rationale of early modern note-taking; but, of course, they allow generalizations about precepts more so than about actual practice. Nevertheless, the general lines of this advice are confirmed by samples of student and undergraduate behavior: that is, individuals did keep notes, some prolifically, others sparingly; they read with an eye to making extracts that matched various accepted categories, saving these for future study, although they may not have entered them in the prescribed fashion.[86] However, the variety of individual practices makes attribution of precise genre titles somewhat precarious, and it is a moot point as to whether many seventeenth-century notebooks should be called commonplace books. Peter Mack has commented that none of the examples assembled in a set from the Huntington Library is a commonplace book in "the strict sense . . . of a collection of quotations organised under headings."[87]

These deviations from the standard models for commonplace books are not surprising. By the mid-1600s even the declared precepts about note-taking began to allow for considerable individual choice in modes of entering and storing. Drexel gave such licence as early as 1638.[88] In a letter of 1645, Meric Casaubon seemed exasperated by the surfeit of advice on methods of study: "How many have given Direction Concerning Commonplaces? & how few bene fitted by those Directions?"[89] His point was that any recommendation must consider the particular purpose and the capacities of the student or scholar. However, this relaxation in the strictures about how notes should be made did not imply a loss of confidence in the use of notes as stimuli to recollection. Drexel stressed that notes taken in ways that suited the individual were likely to work best. Holdsworth recognized that a mature scholar or reader might be shocked by his own early collections, which typically, he suggested, included "many things uselesse, heterogenus raw, Common, and Childish." Later, commonplace books were more likely to be tailored to a certain profession "whether of Law, Divinity, Phisick, or the like."[90] However, these specifications about various methods of note-taking maintained the conviction that notes assisted recollection.

Nevertheless, notes made in idiosyncratic fashion still had shock value in the late seventeenth century. We can see this in the account by Roger North (1651–1734) of the habits of his brother, John North (1645–83), the polymathic Master of Trinity College, Cambridge. Presenting John's habits as atypical, Roger explained that he "noted as he went along, but not in the common way by commonplace, but every book severally, setting down whatever he found worthily to be observed in that book." In other words, John North went beyond the radical option mentioned by Holdsworth: he simply jotted down points of interest without bothering to place them under a common Head and, *a fortiori*, without any reference to a larger framework of classification. Despite a directive that his papers be burned after his death, one notebook survived. Roger found it to be seemingly without arrangement but of "too great value to be lost"; he copied out his brother's entries which were "out of all order, some with ink but most with red chalk or black lead, clapped down there, *ex improviso*." He then arranged these entries "in a kind of order under heads" for his own satisfaction.[91] Roger did not doubt that John was able to use these notes in alliance with his memory, but his reaction to the apparent chaos manifests a wish for something more orderly and intelligible to others. This was a conundrum of note-taking: what worked to prompt recollection for the maker of the note did not guarantee the utility of the notes to others.

Bacon on Notes, Recollection, and Retrieval

There are two pieces of advice from Bacon that illuminate the issues involved here. He stressed that the benefits of recollection from notes accrue only to the person who made the note, and that, in matters of detail, notes have to be relied on as accurate records, not as triggers of memory. The first of these points occurs in his letter to Fulke Greville (the first Lord Brooke) in response to an inquiry about the delegation of note-taking for a large project.[92] Bacon posed the issue in these terms: "He that shall out of his own reading gather for the use of another, must (as I think) do it by epitome, or abridgment, or under heads of common places." Bacon immediately questioned the value of epitome and abridgment for Greville's purposes, and made it clear that he preferred the third mode of note-taking: "I hold collections under heads and common places of far more profit and use; because they have in them a kind of observation," accompanied by judgment. However, delegation of such note-taking was risky because an assistant was likely to "follow an alphabet and fill the Index with many idle heads," thus leaving "his paper-book . . . full of idle marks."[93] If such helpers had to be employed, Bacon warned, they should use only Heads appropriate to the task and certainly "far fewer" than found in the standard collections. Yet even with these precautions, he questioned the very prospect of entrusting note-taking to another person:

> Therefore, to speak plainly of the gathering of heads or common places
> I think, first, that in general one man's notes will little profit another,
> because one man's conceit doth so much differ from another's; and also
> because the bare note itself is nothing so much worth as the suggestion it
> gives the reader.[94]

Here Bacon went to the heart of the commonplace method: notes serve as a prompt or "suggestion" to recollect more than the "bare note," but this process favors the maker of the original note. Delegation of note-taking greatly reduced the likely benefits. Indeed, although the notes and drafts of scholars were highly prized, there were known instances of severe disappointment.[95] One of these is apparent in the reaction to the papers of the Genevan-born classicist and polymath Isaac Casaubon. Anthony Wood alluded to these as yet unseen papers in his biography of Casaubon's son, speculating that Meric Casaubon enjoyed a head start in philological criticism, "wherein his father's notes might probably have set him up."[96] This may have been so, but when these notes were published in 1710 their sparseness was frustrating.[97] As Mark Pattison conjectured: "Casaubon's notes are bare references, and references not to places in books, but to the thing or word to which he intended to recur. To this vast mass of material his own memory was the only key."[98] This is what confronted Roger North as he tried to make sense of his brother's disorderly jottings.

Bacon's second point occurs in three letters of 1595–96 (printed in 1633) to Roger Manners (1576–1612), the fifth Earl of Rutland.[99] The letters confirm the prevailing assumption about notes as aids to memory, especially via recollection: "To help you to remember, you must use writing, or meditation, or both; by writing I mean making of notes and abridgments of that which you would remember." However, in the context of offering guidance on the best way to remember diverse information gathered while traveling in new countries, Bacon was more precise. He stressed that great care must be taken to ensure that the notes provided a secure record for later use, both for the owner and, potentially, for others. In the second letter it becomes clear that Bacon was urging the use of written notes not primarily as reminders but, given the inevitable failures of memory, as accurate and permanent records that could be retrieved. He urged Rutland to ensure that what he observed "be treasured up, not only in your memory (where time may lessen your stock), but rather in good writings and books of account, which will keep them safe for your use hereafter." This caveat does not deny that the notes might provoke further recollection; however, it implies that Bacon regarded this power as unreliable, especially in such matters of detail mentioned in the third letter that described "the notes I could wish you to gather in your travel." Among the items listed there are "the nature of the climate and the temperature of the air . . . the condition of the soil . . . the diameter or length of the country," and so on. Bacon then anticipates an objection: "If your Lordship tell me that these things will be too many to remember, I answer I had rather you trusted your note-book than your memory."[100] He pinpointed this issue in the first edition of his *Essays* (1597): "Reading maketh a Full Man; Conference a Ready Man; And Writing an Exact Man. And therefore, If a Man Write little, he had need have a Great memory."[101]

A concern with exactness, at least in certain situations, may explain why the ability to do "without book" was losing some of its credit as a mark of memory and learning. Aubrey was impressed by examples of capacious memory, but not if such remembering was unreliable. Of Katherine Philips, he reported that "She was a frequent hearer of sermons; had an excellent memory and could have brought away a sermon in her memory." But of William Prynne, he remarked disapprovingly that "He was a learned man, of immense reading, but is much blamed for his unfaithfull quotations."[102] While living in exile in Holland, John Locke was pleasantly surprised by disputations in the medical faculty at Leiden, as he noted in his journal for October 31, 1684: "And those I saw dispute that they might not mistake had their arguments writt downe. I suppose their studys tend most to practise for in disputeing noe one that I heard urged any argument beyond one or 2 syllogisms."[103] Locke regarded this reliance on the written word as a positive sign; and in the 1690s he used the phrase "without book" in a derogatory way. In

Some Thoughts Concerning Education (1693) he castigated the notion of "exercising and improving the Memory by toilsom Repetitions without Book," and elsewhere, like Aubrey, imputed a lack of care and precision to those who relied on memory in matters of detail.[104] In replying to Edward Stillingfleet, the bishop of Worcester, Locke added forensically that "To show the reader that I do not talk without book in the case, I shall set down your lordship's own words."[105] Of great importance here is the criticism that such learning "without book" could not guarantee sufficient exactness.[106] This had implications for the method of commonplacing which, as we have seen, provided a set of written prompts, such as quotations arranged under Heads, to stimulate recollection of salient material stored in the individual's memory. This entailed the possibility of partial recovery, failure, or the recollection of something other than the item sought. But looked at in another way, as Bacon realized, notes allowed accuracy about details, not just recollection of the main elements. Independent of their capacity to elicit recollection, they could preserve stable records of particular information. These features — exactness and stability — were crucial for copious collections of empirical particulars, but notes with these qualities were only useful if they could be retrieved with ease when desired. As the notebooks compiled by scholars grew into copious, or even elephantine, volumes, retrieval aids became crucial. The next part of this chapter considers how these various preoccupations with memory, recollection, and retrieval were addressed by the English virtuosi.

Some Virtuosi and Their Notebooks

In about 1704, at the age of eighty-four, John Evelyn gave wide-ranging advice to his grandson, including hints on reading and study. He then said that:

> A well digested *Adversaria* as to *common places* should by no meanes be neglected, in which to write down and note what you find most important & usefull in your Readings & not trust altogether to your owne Memory, so in a little time you will find your papers furnish [you] with materialls of all subjects; short notes and References are sufficient for this unlesse wher you meete with some Remarkable passages which may require a larger transcription.[107]

This suggests that the commonplace book was not yet dead, in spite of defections from the ensemble of assumptions that once supported it. For the Renaissance humanists, the principal *raison d'être* of notebooks had been selection of excerpts from major classical authors of Latin prose and poetry. The possibility of reaching beyond books was also accepted: notes could be taken from lectures, sermons, conversation, testimony, and observation. However, Evelyn also indicated how note-taking should be accommodated to the collaborative ethos promoted by Bacon, warning that nothing of value can be done "without Collections, no man being able to build any thing whatever without the help of others which may stand or last no longer than the *Cobwebs* spun out of the bowels of an Insect."[108]

In some respects, the scientific virtuosi continued or adapted the methods of their humanist predecessors. Joseph Levine has nicely captured Evelyn's predicament by situating him "between the Ancients and Moderns," aiming to embrace all the new contributions of the moderns as well as the patrimony of former ages.[109] The English translator of Gassendi's biography of Peiresc dedicated his work to Evelyn, and affirmed that "the compleatly-knowing man must be Janus-like, double-fac'd, to take cognizance of Time past . . . as well as of the late-past, or present times, wherein he lives."[110] This was a formidable ideal, and Evelyn admitted to Pepys in 1682 that the truly "exact" historian "must reade all, good, and bad, and remove a world of rubbish before he can lay the foundation."[111] Evelyn's motto, written at the start of many of his notebooks, was "Omnia explorate, meliora retinete"; he interpreted this as a rationale for the collection of information from both old and new sources leading to a gradual separation of useless from valuable material.[112] In choosing the humanist commonplace book as the instrument by which to advance this goal, Evelyn stretched its traditional purpose. As we have seen, the method of commonplaces was designed for collecting exemplars pertaining to established topics, not for amassing and filtering diverse information.

In the remainder of this chapter, I consider whether the note-taking of some of the English virtuosi can be seen as a deliberate amendment of humanist methods rather than merely as a lazy departure from traditional precepts. In later chapters I focus on figures closely involved in scientific (including experimental) inquiry, such as Boyle, Locke, and Hooke, whereas here I discuss three men who displayed a passion for information gathered from both reading and other sources. The first of these is the courtier and diarist John Evelyn, whom Aubrey described as "one of our first Virtuosi."[113] I will also consider the merchant Abraham Hill, who was a founding member and a Secretary of the Royal Society, and the diplomat Robert Southwell, who became president of the Society in 1690. What kind of notebooks did these individuals keep, and what did they do with them?

Evelyn was a prolific note-taker; but he began as a diarist not as a keeper of commonplaces. From 1631, at the age of eleven, he made rough journals of events "in imitation of what I had seene my Father do," using "a blanke Almanac." Then between 1660 and 1664 he used these notes to write, in part retrospectively, the diary he called "Kalendarium." This opens on October 21, 1620, with his birth at "about 20 minuts past two in the morning."[114] He broke off the process of copying after making the entries for February 1645, resumed the task in about 1680, but did not make new entries until 1684.[115] Even though this project was suspended, Evelyn purported to be a constant note-taker, writing from Paris in 1649 to his friend, Ann Russell, saying that he would not have missed reports about her health because "I keepe a Kalendar of all that passes, and an importance of that nature could not escape without an egregious defect in my Chronicle."[116] It was also about this time that Evelyn began to assemble compilations of manuscripts containing various sorts of information. While in Paris he started to keep notes of sermons in a special volume; he instructed his amanuensis, Richard Hoare, to draw up a *Vade Mecum* of knowledge garnered largely from Johann Alsted's *Encyclopaedia* (1630), and also a digest of medical recipes.[117] This flurry of activity may have caused Evelyn to notice his neglect of commonplace books. The fact that he felt the need to start one says something about his perception of the importance of a record of reading, as distinct from the kind of material committed to his diary. In June 1649 he wrote to his father-in-law, Sir Richard Browne, saying that he planned to direct Hoare on this task: "I my selfe am a little given to bookes & shall have now tyme to reduce my studyes into a method, for which end his assistance in transcribing some things (yet in Embrio) will much ease and please me."[118] The later advice to his grandson about using methodical commonplace books betrays Evelyn's realization that he had been tardy in initiating this habit. Looking back on his life, he regretted "innumerable Insignificant Collections and Atempts, desultory and undigested, cast into no method, some hundred of Authors marked with my blak-lead Crayon, also I intended to have transcribed into Adversaria: But had never leasure."[119]

Evelyn used large folio paperbooks (approximately 22×48 cm), allocating separate sections to major subjects. He divided the first commonplace book, begun in about 1649, into theology (including ethics), the mathematical and physical sciences, liberal arts (including architecture, painting, sculpture, and sententiae), and politics (including jurisprudence), with each of these sections labeled as "tomes" and further partitioned into chapters.[120] In some ways, this arrangement by major subjects was a larger version of

his "Vade Mecum" based on Alsted's division of learning.[121] It also accorded
with advice given in the late 1640s by the royalist, John Cosin (1592–1672),
later bishop of Durham, writing from exile in France. Having recommended
that the Heads for theology be taken from the 170 of these in Daniel Tile-
nus' *Syntagmatis* (1607), he advised that "a Blank leaf or Two be left after
every Generall Head, & a supply may be made hereafter of any thing that
shall occurre, & be fitt to be ranged under it."[122] However, Evelyn's second
main division of the large commonplace book is given as "Hist: philos; Sci-
ent; Mathem: Med:&c," suggesting a reluctance to pursue more refined classification.[123] Although this notebook, titled "Tomus Primus," was the first he
used, Evelyn later altered it to "Tertius" when it was joined by two more folio
volumes, thus making it the third in a series, itself called "Loci communes."
Hoare drew up the earliest entries, presumably taking the selected quotations from marked passages in Evelyn's books, or otherwise at his direction.
The change from Hoare's professional scribal hand (praised by Pepys) to
Evelyn's densely packed script, is noticeable.[124]

Evelyn announced himself as a "modern" in the opening page of one of his
commonplace books:

> Till about 1648, it would have ben accounted a presumptuous thing to have
> attempted any Inovation in the Co-mon course of Learning; they thought
> there could be nothing comparable to the physics, & causes of the philoso-
> phie, & sciences taught in the Universities; & one should have ben cried out
> upon as a vaine projector; to thinke that any thing could be added to their
> Hypotheses.[125]

Yet his identification with the "moderns" did not persuade Evelyn to re-
ject the traditional function of commonplace books as that of storing extracts gleaned from reading. While in Paris between 1649 and 1652, he acquired many books covering a wide range of subjects.[126] We could say that
he adopted Meric Casaubon's notion of "general learning," or "the generall persuite of the whole *Encyclopaedia*."[127] But he then added more of the
new sciences: he told his grandson that natural philosophy, practical mathematics, and anatomy were to be "cherish'd & cultivated."[128] Intensive reading from this wide selection of both ancient and modern authors supplied
the entries for his commonplace books. Still, whatever his respect for the
ancients and more recent humanist scholars, the author Evelyn most often
cited was Bacon. Many of these entries were taken from the English transla-

tion of the *De augmentis* (1623) by Gilbert Watts.[129] At one point, Evelyn has a sequence of thirty-four marginal Heads indicating entries from this book.[130]

Bacon's attitude to the collection of information may have shaped Evelyn's outlook, on at least two points.[131] First, it is feasible that Evelyn regarded a preference for aphorisms over rigid systems as legitimating the way he made his commonplace books. He copied out a passage in which Bacon contended that knowledge "disperst into *Aphorismes*, and *Observations*, may grow and shoot up; but once inclosed and comprehended in Methods, it may perchance be farther polisht . . . but it increaseth no more in bulke and substance."[132] Evelyn's entries usually consist of short excerpts, often linked to multiple keywords, as if suggesting *multum in parvo* (much in little). Second, in some of his notebooks Evelyn was keen to gather information on various topics for the sake of having it in one place, or as a starting point for later sifting. Thus he prefaced the early "Vade Mecum" with this comment: "This Compendium is taken chiefly out of Alsted & de Moulin, &c; and is in many places false written & full of errors."[133] A similar approach is evident in his digests of medical and culinary recipes.[134] His description of one such set of fifty-four entries tagged with approximately 100 Heads, is "Receipts, Compositions & other Medicinal & Artificial Seacrets, Casualy met with in reading & discourse."[135] Like Bacon, Evelyn appears to have tolerated the collection of erroneous material in the interest of accruing a large mass of data on which generalizations could be tried.

When making his own entries, Evelyn abandoned the preallocation of pages to subjects and entered all the extracts from a single book consecutively, in one section of a notebook. The title of one notebook encapsulates this shift: "Adversaria Historical, Physical, Mathematical . . . promiscuously set downe as they Occur in Reading, or Casual Discourse."[136] Evelyn's response to his brother's request for tips on keeping commonplace books reflects the options he had considered. He sent "the frame or Idea of my Adversaria, which after many tryals & reformations, I find to be most advantageous." Presumably, this was the plan for a systematic organization of subjects, but he also canvassed a more pragmatic approach, saying that there were people

> who do not oblige themselves to so much Method & orderly reduction; but transcribe their observations farragiously, & as they come to hand; but for my owne part (who experiment the benefit of it, reading a page or two every day till I have finished, and frequenting this Lecture) I am satisfied with the design. However you may do either with advantage.[137]

He concluded by citing Cicero's distinction between rough memoranda and proper account books—the latter carrying an authority derived from the formal care taken in entering information.[138] But Evelyn thought the order of notes was less important than the placement of them in bound volumes that could be kept and searched. He warned his grandson not to accumulate too many "un-bound Books and loose papers," and even instructed him to burn all his own loose notes: "I would have you burn or other ways dispose of Amongst which are packets & Bundles of Excerpts of all Subjects intended to have been transcrib'd into Books and Adversaria, but growing too numerous."[139]

Evelyn's departure from his earlier subject arrangement may indicate awareness that he needed a convenient way to find specific items in these massive notebooks. When he started to write excerpts from a single book in the one place, he began to assign marginal Latin keywords to these entries, and also to those already made by Hoare. He listed these Heads alphabetically in a large index volume of 113 pages, referring to it as "the Universal Table."[140] Evelyn had a tendency to put a keyword next to almost each proposition or main idea in a quotation, or against his own comments, sometimes repeating these keywords in the one entry. For example, his enthusiastic addition of marginal words in the pages devoted to theology and metaphysics, which Hoare had entered, often resulted in index entries yielding as many as sixty-five page references in one notebook alone.[141] In a series of seven caveats on the first page of the index, he was candid about this: "Note that the same word, or subject, may be in severall places of the same Margent/: v.g.T.II:p:84. Anima is 8 times in the page: therfore for any Subject cast your eye thro' the whole margin of the page."[142] But at least entries were now tagged for retrieval, even though several pages of the commonplace books might need to be scanned to locate a specific entry. Each page of the index has two columns of Heads, with approximately forty-five to fifty-five per page. There are 184 pages set up in this fashion, so the minimum total of Heads is 8,096. But Evelyn was not yet content, and followed this index with one for proper names (many without accompanying page references), another for excerpts containing passages from Scripture, and, finally, one of English legal terms.[143] If some of his scholarly predecessors and contemporaries exhibited what Ann Blair has called "info-lust," Evelyn's obsession with indexing was a related symptom.[144]

The density of Evelyn's index and the sheer size of his commonplace books invite contrast with the more selective, and even portable, notebooks imagined by Erasmus, Sacchini, and Holdsworth. However, it is clear that boundaries of the commonplace book had already been expanded and tested by the early 1600s. Among the most prolific note-takers in England alone, Gabriel Harvey, William Drake, and Samuel Hartlib used standard common-placing for different and demanding tasks.[145] Indeed, Evelyn's efforts can be compared with those of the Jacobean lawyer Sir Julius Caesar (1558–1636), who built his own huge commonplace book on the template of Foxe's *Pandectae*, adding 682 Heads to the 768 it already contained.[146] By the time he had finished in 1629, having possibly used it over almost sixty years, there was hardly an empty space remaining on any of the 1,200 pages in this notebook.[147] As Foxe himself foresaw with regret (discussed above), once the commonplace book turned into a huge compendium of information it was no longer feasible to expect an individual to rely on memory in searching it. Each user might have internalized some of the links between related topics—no doubt Caesar himself was in command of the legal concepts of his day—but when dealing with the full range of topics, only an alphabetical index that picked out all the Heads ensured that retrieval was feasible. In this way, Caesar could use his augmented version of Foxe's "Index locorum alphabetarius" (now with 1,450 Heads) as a "search-engine" capable of locating any one of the entries in his notebook.[148]

It is clear that Evelyn gave thought to the use of his commonplace books. However, it is surprising that the initial plan was so conservative in format, given the allowances already made by Drexel, Holdsworth, Barlow, and others who proffered advice on note-taking. Like Caesar a generation earlier, Evelyn collated information from a standard university set of subjects under traditional Heads, as well as recording material close to his own more concentrated interests.[149] When he began to make entries from a particular book without assigning these to pages devoted to subjects or topics, his notebooks threatened to become receptacles in which material was lost. In spite of being a diary keeper he did not attempt any regular dating of the common-place entries as one plausible aid to organization or retrieval. Not having the advantage of a ready-made notebook of Heads arranged alphabetically—as Caesar enjoyed in Foxe's *Pandectae*—he needed to retro-fit an alphabetical finding aid. The extent to which he used the resulting index to retrieve and consult the greater part of this accumulated bank of information is unclear. What can be said is that, unlike Caesar, Evelyn used several notebooks and gathered material from observation and testimony as well as from books. Sometimes, when making notes from books, he added comments derived from his own experience.[150] He used notebooks to collate information on his projects on agriculture, horticulture, painting, the archiving of manuscripts, his ongoing work toward "Elysium Britannicum," an extensive history of gardening, and his major book, *Sylva* (1664), on the cultivation of trees.[151] But he also worked with loose notes in bundles marked with labels indicating the specific projects, such as "Agr-Agriculture"; "S-Sylva"; "Rx-Medicinal Receipts"; and "Br Concerning the Sagacity of Brute Animals, begun but imperfect." In a note dated after 1697 he confessed that he did not always copy this material as planned: "In this Bundle, or Roll, marked [double circle intersecting] is containd The Excerpts & Collections, casually gather'd out of several Authors &c, and intended to have ben transcrib'd into Adversaria, as applicable to Several Subjects."[152] This might also imply that for material in regular use as part of a project, little cognitive advantage was gained by storing it in a commonplace book (security, of course, was another matter, as he knew). Finally, Evelyn's awareness that his huge commonplace books were ineffective without a correspondingly large index suggests that these notebooks were primarily no longer tools for prompting recollection, still less recitation, of what their owner remembered and understood; instead, they were personal databases from which information, such as citations, could be retrieved and followed up. By 1661 he had read and translated Gabriel Naudé's work on the establishment and arrangement of libraries in which large manuscript commonplace books were shelved with books and catalogued as quasi-reference works: "Neither must you forget all sorts of *Common places*, *Dictionaries*, *Mixtures*, . . . and other like *Repertories*." The French scholar and librarian had already defined such works as containing "Matter ready prepared for those who have the industry to use them."[153] Evelyn believed that his own huge notebooks, a product of his labor, now functioned in this way — as storehouses of information. One of Evelyn's colleagues in the Royal Society took this a step further.

Abraham Hill collected a vast amount of material. Among his papers in the British Library there are nine quarto notebooks, plus an index volume covering the entire set.[154] The preface to the selection of his correspondence issued in 1767 makes the point that he did not publish a major work from "his ingenious labours," yet enjoyed the esteem of those "whose parts and learning were of eminence."[155] For example, Bishop John Tillotson expressed "the pleasure he took in Mr. Hill's conversation, and would frequently term him, his learned friend, and his instructing philosopher."[156] Hill was a regular officeholder in the Royal Society and members knew the range of his reading and learning from his erudite and arcane contributions to council meetings. His comment on January 19, 1681 on the matter of Leibniz's letter to Hooke about "the universal algebra" was typical: the minutes record that Hill remarked "that professor [Johannes] Sturmius had written an *Euclides Universalis* somewhat to this purpose; and that he had the book now by him." On some occasions he seems to have acted as the unofficial research assistant, being asked, for example, to check Samuel Purchas' travel accounts for information about the "art" used by divers in "staying under water."[157] Possibly because of this reputation, several members asked him for information on a range of matters.[158]

Hill's reputation as a walking library must have rested in part on his huge storehouse of notes. He labeled these as commonplace books but they were not compiled in a standard manner. Unlike the notebooks of Caesar and Evelyn, his notes did not begin as material under commonplace Heads; rather, Hill collated loose notes and papers—not always his own—under broad subject divisions and stitched them together in bound volumes. The first two volumes are mainly theological, consisting, in the British Library description, "chiefly of extracts from the Fathers and Divines of later date, concerning the Church, her doctrines, disciplines etc." One volume (MS Sloane 2899) covers natural history and the sciences. This includes papers of Francis Willughby (fols. 6–10) about the generation of insects along with Hill's notes on points it contained. It seems that Hill tipped loose material into subject categories, bound each of these "Collections" (as Holdsworth had said) into notebooks, assigned folio numbers to the pages, and then indexed their content alphabetically in a separate volume. The resulting index volume (MS Sloane 2900) is elaborate, with many subdivisions of main topics, resulting in about 9,000 terms. Each of these gives a reference to the folio number of a specific volume: for example, in the volume containing scientific material, the folio numbers range from 1 to 960. There are no topical Heads in the pages of his notes. This means that Hill departed further than Evelyn from the customary method of commonplacing, thus realizing an extreme version of Holdsworth's radical option of a virtual commonplace book manifested solely through an index of its content.

The one extant commonplace book among Sir Robert Southwell's papers contrasts with those of Evelyn and Hill; its content is far more restricted and it follows standard models more closely.[159] On the first full leaf of this small notebook, he wrote three descriptors: "Terms of Arts," "Diffinetions "; "Doubts." The actual material is somewhat wider than these labels suggest, being mainly drawn from moral and political philosophy. He preassigned a Head to each of the 572 pages, entering these in alphabetical order from "Abstinentia" to "Uxor." The topics range from general subject categories such as Ethica, Philosophia, and Politia, to more specific concepts, such as Aequalitas, Amicitia, Concilium, Officium, and Veritas. Usually there is one Head per page, although occasionally a topic has two designated pages, as does "Conscientia bona" where the entry fills a page and half. But about half of the notebook remains blank, including those pages headed "Brevitas" and "Curiositas," despite the fact that both were given more than one page. These empty pages stand as a measure of the gap between Southwell's anticipation of the topics he would cover and his actual experience; however, some topics needed to be extended over to the next page, where this was possible. In other instances, Southwell continued entries (such as "Politia") at the end of notebook, on unnumbered pages, putting the Head word in the margin of a page marked with the relevant letter of the alphabet. But he was not meticulous: thus an entry on "Office," which could have gone under "Officium" in the main part of the notebook, is also placed in the final pages. The preallocation of topics is an indication that, just as John Foxe hoped, Southwell was able to recall major relevant Heads (in Latin), and no doubt recollect and generate more in the course of listing and writing them. Most of his entries are short excerpts from books, with a citation that would locate the source, but there are also some reports on conversations and of his own reflections. He also used the back pages for topics not forecast in the main section: thus there are entries for "Bookes," "Conquest"; "Decay" and others. On the whole, these addenda seem to be more often topics Southwell had not anticipated than continuations of earlier entries.

Not surprisingly, most commonplace books of this period included an entry under Memory (or *memoria*). Hill's entries on this topic are roughly representative.[160] He collected anecdotes and reports about individuals with powerful memories, noting that

> George Bankes (servant to Sir W. Petty) could repeat out of his memory a sermon verbatim as he had heard it a week before. . . . Longolius had so good a memory that he scarce ever forgot what he read. . . . He did every night recollect what he had learnt that day . . . Suarez could report the very words of the fathers particularly Saint Augustine . . . he could tell the very place where he discursst any point.[161]

Hill cited Willis' *Mnemonica* as a recent synopsis of the classical mnemonic art, together with the claim that this was a "cure for a weak memory." But he seemed equally interested in any method or habit that worked, recording Erasmus' view that "the great helps to memory are repetition communication . . . understanding order care."[162] In another respect, Hill was unusual in not mentioning notes as helps to memory. Some notebooks now held in the Beinecke Library, Yale University, are more typical in including the use of commonplacing under "Memory," and copying out relevant passages from advice manuals such as those of Farnaby, Fuller, Walker, and possibly Holdworth.[163] Southwell conformed to this pattern and, in fact, showed a preoccupation with aids to memory and thought across other entries.[164] Under "Inventio," he recorded a suggestion that when reading a passage that invites a "reason to be given, a speech to be spoken," we should put the book down "& suppose with our selves what is expedient on that occasion"; then we should write down our thoughts and "compare it with what is in the booke."[165] His reference point on these matters was usually Bacon's essay on helps to the intellectual faculties, or his *De augmentis*, as translated by Watts in 1640, also used by Evelyn.[166] Under "Memoria," Southwell paraphrased Bacon on the appropriate role, according to the task and situation, of "Notes and Memorialls" versus "Extemporal Speech." He added his own comment about trying "a good way to improve both memory & invention" with some friends: "a Thame [theme] being given should frame half a dozen topiques & Argumnts on the matter"; these are swapped around and individuals respond to the ones allocated to them.[167] Ironically, the Head for "Meditatio," often seen, like "Inventio," as an aid to recollection, remains blank (see fig. 2.1).

Although we do not have Southwell's thoughts about the use and organization of commonplace books, there in an intriguing memorandum dated November 7, 1664, in which he summarized advice he received from the judge and author Matthew Hale.[168] This recommended the use of commonplace books along fairly standard lines, but also implied that the owner of the notebook should take command of it, making it work for him. According to Southwell's resumé, Hale advised "That I make a Common Place booke Alphabetically as Ab. AC AD . . . And then reduce all things to heads, and if the matter be various to severall heads." Emphasizing a developing relationship between the individual and the notebook, Hale predicted that "although you refer things ill at first yett frequent recourse makes you acquainted with what you have done and you will mend your businesse every day, and in 3 or 4 years grow most dexterous in referring, and quarrell with your first notes, which tho uselesse to an other will serve well your selfe." He also insinuated that fastidious arrangement of the entries was less important than regular reviewing of their content: "You will be particular in things although in your common place booke these are more confused, yett to you not soe."[169] This advice probably came after Southwell started the commonplace book

Figure 2.1. Entries for "Meditatio" (still blank) and "Memoria" in Sir Robert Southwell's commonplace book, c. 1660s. MS Osborn b112, pp. 361–62. Kindly supplied with permission by the James Marshall and Marie-Louise Osborn Collection, Beinecke Rare Book and Manuscript Library, Yale University.

we now have, but several entries resonate with Hale's views on how a person might make the best use of notes.[170]

For some of the English virtuosi, the commonplace book remained the default notebook for reading and study. However, the deviation from standard precepts, already underway among leading humanist scholars, is evident. During the 1600s, these notebooks came to be seen as ways of retaining and organizing a range of information far more diverse, and less anchored to educational curricula, than the moral and rhetorical topics of the classical texts. Entries were made as they were gathered from a source—as they would have been in a journal—and without allocation to pages already dedicated to specific Heads.[171] So what Roger North reported as his brother's iconoclasm—he "noted as he went along, but not in the common way by commonplace"—is also evident in the commonplace books of Evelyn and Hill. This practice may have loosened the knot between commonplacing and memory training, allowing Heads attached to entries to function mainly as index words rather than as standard categories in a traditional "Index Locorum Communium." It is thus telling that Locke's "New Method" for making commonplace books was a novel way of *indexing* them, indeed, one that also radically disaggregated topics. As we shall see in chapter 7, Locke recommended that each person choose Heads that worked best for their purpose, a liberty also granted in Gabriel Naudé's advice on library catalogues, which Evelyn translated.[172] In both cases, there was an acceptance that the owner of the notebook, or library, could order it to facilitate retrieval, rather than to conform to traditional classifications.

On the basis of these trends, it is worth proposing that by the end of the seventeenth century there was a change in the function of notebooks: once repositories of the material that individuals sought to memorize, or recollect, they came to be seen as ways of securing and retrieving information that could never be memorized. No longer expected to act as prompts to recollection, the Heads became tags by which entries could be retrieved from these increasingly enormous notebooks. Nevertheless, as the reflections of Hale and Southwell suggest, the use of notes to assist recollection and thought need not necessarily depend on topical arrangement of material; instead, both regular reviewing of loose notes and rehearsing ideas without notes were powerful practices.[173] However, there were limits to this individual mastery of information. In the next chapter, I consider how the quest for empirical particulars demanded by Bacon presented one considerable challenge.

[3]

Information and Empirical Sensibility

Among the mottos inscribed on the external walls of the British Library there is one from Samuel Johnson, the great eighteenth-century man of letters: "Knowledge is of two kinds. We know a subject ourselves, or we know where we can find information upon it."[1] To ponder this from the pedestrian crossing on Euston Road is life threatening; but in a safer moment it is worth considering what it betrays about Johnson's attitude. There is a confidence in this remark that signals a world in which information on a wide range of subjects was accessible in the vernacular, some of it collated, compiled, or abridged by learned authors such as Johnson himself. Commenting on the practice of transferring "large quotations to a common-place book," he was puzzled why "any part of a book, which can be consulted at pleasure, should be copied."[2]

This attitude seems to support the young Edward Gibbon's contention that the "moderns" were the beneficiaries of ready information. In his *Essai sur l'etude de la literature* (1761), Gibbon observed that his contemporaries enjoyed the luxury of being able to find the information they required. In contrast, the "Ancients," he inferred, lacked an adequate information base, especially for studies of nature, so that "destitute of instruments, and single in their experiments, they were able to collect only a small number of observations, mixed with uncertainty, diminished by the injuries of time, and scattered up and down at random, thro' a number of volumes."[3] The information they managed to collect and hold together was never consolidated, whereas Gibbon's Enlightenment readers had the benefit of a critical mass that allowed them to make solid generalizations and comparisons. Only about fifty years earlier, the option of quick consultation of "reference" works in the vernacular was a relatively new one.[4] The first use of the phrase "to look something up" is credited to the Oxford antiquarian Anthony Wood, reporting a conversation in 1692 in which he and some friends paused to check something: "They decided to look up it [his own biographical register of Oxford alumni], . . . to see what I said of the Presbyterians."[5] Wood's *Athenae Oxonienses* (first published 1691–92) was the kind of book that Johnson took for granted in the 1770s. His expectation about the availability of reliable information rested in part on the presence of dictionaries and encyclopedias that had appeared over the previous century.[6]

We can therefore understand how eighteenth-century authors supposed information was relatively easy to locate. But this was only the case if the relevant material had been gathered and recorded. Gibbon realized this when appreciating the meticulous work done by Sébastien Le Nain de Tillemont (1637–98) on the early Christian Church, which put "within my reach the loose and scattered atoms of historical information."[7] However, for novel topics of inquiry, dictionaries and encyclopedias could provide at best only a starting point. In the seventeenth century, those seeking knowledge of the natural world had to be their own Tillemonts; they had to embark on the laborious and long-term search for material not yet collected or collated, as well as comparing this to that currently available. This was the situation that confronted the English virtuosi. They faced the task of gathering, storing, and processing information from a wide range of sources both for personal use and, potentially, for sharing.[8] In this chapter I explore how their ideas about note-taking intersected with this quest for what we now call *empirical* information. First, however, the early modern understanding of both these terms must be elucidated.

Information

From Chambers' *Cyclopaedia* (1728) we can infer that even by the early eighteenth century the definition of "information" was often still restricted to the technical legal sense: namely, "in Law; see indictment." In his *Dictionary* (1755), Johnson classed "information" as a noun; however, all his expansions involved action: thus "intelligence given . . . instruction . . . charge or accusation exhibited . . . the act of informing or accusing."[9] This usage confirms a link between the noun "information" and the transitive verb "to inform" (Latin, *informare*, to form, shape, and instruct). In Johnson's illustrations, which he often took from seventeenth-century authors, there is no strong sense of "information" as something waiting to be checked, read, or stored; instead, the stress is on deliberate acquisition and expeditious communication. Geoffrey Nunberg has suggested that this need to find and collect information, as "intelligence," continued into the mid-nineteenth century, so that before then one "could not really speak of information in an abstract way."[10] Today we do this without pausing: we talk about information in terms of quantity, agreeing that there is more of it now than in previous times.[11] In the strictly technical definition used in mathematical theories of communication, "data" and "information" do not necessarily contain "semantic information"—understood as "instructional" (how to do something) or "factual" statements about the world.[12] But in everyday speech we conflate unprocessed data with information that has been found and selected for a specific purpose or argument. Thus we are happy to refer to data gathered about finance or weather as information, in the same sense as we refer to information discovered and selected by historical scholarship about the French Revolution or the rise of Cubist painting. Furthermore, we take it for granted that much of the information we need is already assembled and stored, waiting for immediate or future access and analysis.

When the virtuosi spoke about "information," they were referring to particular facts, reports, passages in books, or news of certain events obtained from various sources. In this period, information was often linked to a geographical location and only reliably gathered and evaluated at that place. Depending on the type of information, this might be the royal court, the vicinity of Parliament, the law courts, the university, coffeehouses, ports, or markets.[13] Aubrey recorded one such location as "Joseph Barnes' shop, the bookeseller (opposite to the west end of St. Mary's), where the newes was brought from London."[14] One consequence of treating information as "news," either as resulting from a specific search or by chance, was that each piece needed to be assessed and warranted, according to its specific source. Thus Boyle could say "I have received Information of it from more than one Eyewitness."[15] With his contemporaries, he also used the plural form: "informations."[16] When discussing the pursuit of civil knowledge, Bacon said that it depended on the gathering of "good informacions of particulars touching persons," their nature, actions, customs etc.[17] John Beale told Boyle that "if I had a residence in Gressam, I should rejoyce to be a collector, & compiler of others fuller informations."[18] This use of the plural underlines the active sense of conveying something from a range of sources, each of which had to be assessed, credited, accepted, or rejected.

For people living in early modern Europe, one likely response when faced with a question of information was to check their personal notebooks to see if they had already recorded something pertinent. These notebooks typically contained excerpts, epitomes, and summaries from books, journals, and pamphlets; matters heard, seen, or reported by others; and perhaps notes copied from the notebooks of friends. Individuals made such entries partly (but not only) because they could not rely on being able to return to the source. Thus from the early 1680s, Evelyn copied items from public notices and newspapers into his diary.[19] A high proportion of the entries in Locke's commonplace books are extracts taken from books he did not own, made before he had assembled a large personal library.[20] Information so gathered was dependent on specific actions: borrowing a certain book, going to a particular place, asking some person, or hearing a conversation while traveling. Such note-taking involved hard work and mental discipline, and there was a particularity and contingency in some of the material thus acquired. An intriguing glimpse of this process can be seen in a remark to Locke in a letter of 1704: "I will not forfeit your opinion, and am therefore newly returnd from getting the best information I could at the Bank, in Order to satisfy your demands, for I dare not trust my memory. . . . I am there told, that since 25 March 1702 the following dividents have been made."[21] A list of these dividends followed. This letter betrays the limited availability of accessible information (either in manuscript or print) in many fields of commercial and everyday activity, and the consequent need to memorize or note whatever bits one managed to winkle from various sources. It would be most surprising to find anyone in the 1600s speaking of a "torrent of information," as an early nineteenth-century writer did with respect to publications on the use and effects of smallpox vaccination throughout Europe.[22] A plausible corollary is that information was not yet conceived as a readily accessible body of data, collected by agreed methods and at regular intervals, and able to be found and consulted at any time.[23]

Those engaged in the supply of news and information—diplomats, travelers, spies, and scholars—were known as "intelligencers" or "informants." Bacon made the point that there were higher kinds of knowledge that kings "with their treasure cannot buy," and of which "their spials [spies] and intelligencers can give no news."[24] However, members of the early Royal Society were aware that they could not afford to ignore these circuits of information.

Beale told Oldenburg that "Communications must run through all the veines of your Maine worke."[25] Writing to Boyle, he stressed that those engaged in the new science had to distill "sound information" from the white noise generated by "the printing-presses." Beale urged that the "royall Society" must send out the "most seasonable information" and in turn receive it from the "Innes of Courts," the "Exchanges, Westminster Hall, & all places of considerable resorte. The like from Pauls to the Universityes, & all over England."[26] Two weeks later he confided with Boyle about what the diplomat Sir Henry Wotton (1538–1639) had told him about tracking the channels through which information passed—by leaking misinformation. He would

> send out all his traine of Attendants, To trye, How far wee could rayse such rumors as Hee directed; & what returne wee could make of current Newes, provoking our Emulation to prove our Wits & Capacityes by our successe, allotting our particular Walkes, Ordering what fame wee should correct, & giving us the forme & style (in some Variety) of our reports & of our Enquyryes. Wee could soone find out the chiefe Newesmongers, & Receivers in Westminster-hall, the exchanges, famed ordinaryes, Taverns, Carriers Innes, Stationers, & concerned relations in Courte, Innes of Courts, shops of Trade, (The barbers & baladry not neglected).[27]

Returning to this point, Beale said there were also lessons to be learned from the communication networks of the Jesuits, so that "we may receive usefull Intelligence, & Experimental Informations from all parts of the world."[28]

Paradoxically, it seems, this awareness of the difficulty in acquiring some information coexisted with loud protests about the "multitude" of books.[29] After the advent of the printing press in Europe (c. 1450), humanist scholars expressed anxiety about the burgeoning number of books, arguably with more cause to do so than, say, the Roman Stoic author Seneca the Younger, who believed that "the mass of books burdens the student without instructing him." Seneca did not regret that "forty thousand books were burned in the library at Alexandria."[30] From the 1550s, as Ann Blair has demonstrated, compilers of large reference works claimed to abbreviate and consolidate information by copious note-taking from the ever-rising flood of books.[31] Leibniz's protest in 1680 is therefore a late announcement of a crisis already felt: he complained about that "horrible mass of books which keeps on growing," so that eventually "the disorder will become nearly insurmountable."[32] Yet there was uneven availability of some books. In preparing his *Dictionnaire historique et critique* (1697), Pierre Bayle announced that "the prodigious scarcity of books, very necessary to my design, stopped my pen a hundred times a day."[33]

The English virtuosi were at least as worried about *lack* of information as about the burgeoning number of books. Their cry was that despite the plethora of books there was too little information of the right kind. Joseph Glanvill contended that there were too many books of the wrong kind. He advocated radical culling "to remove the Rubbish . . . to throw aside what is useless, and yields no advantage for Knowledge or for Life."[34] However, as in other ways, Bacon's legacy was powerful. Rather than worry about the over-abundance of books, he urged an assessment of "what parts of Learning are rich and well improved; what poore and destitute." In his view, "the multitude of Bookes makes a shew rather of superfluity than penury." The remedy was "not suppressing or extinguishing books heretofore written, but . . . publishing good new books, which may be of such a right kind."[35] Although Bacon called for the investigation of nature rather than the study of texts sanctioned by reputation and tradition, he wanted *more* books that offered empirical and experimental information. Then it might be possible to avoid the situation that Locke found in medicine, where "books multiplied without the increase of knowledg."[36]

As founder and first editor of the *Philosophical Transactions* (from 1665), Oldenburg received letters from people excited by the fact that this was the start of a cumulative collection of papers and reports, available, in principle, throughout the European republic of letters.[37] In order to appreciate this positive response it is necessary to recognize that the regular and large-scale collection of scientific information was a relative novelty. Certainly, there were earlier impressive examples of massive information collection in some of the absolutist States: Philip II of Spain (ruling between 1556 and 1598), who sometimes signed 400 documents in a single day, was nicknamed *el rey papelero* (the king of paper).[38] The diplomatic *relazioni* pioneered by the Venetians provided intelligence on trade and politics, and the regular reports from Jesuit missionaries in several countries to their superiors in Europe began to produce vast archives by the early 1600s.[39] As with parish registers of birth, deaths, and marriages, these were cases of routine collection of information, usually governed by agreed protocols. This was a feature of the London Bills of Mortality, issued weekly from 1603 to warn of plague epidemics.[40] Similar records of the weather were rare, although one exception, the daily register kept in Florence between December 1654 and March 1670 by the Accademia del Cimento, may have been known to Oldenburg.[41] He hoped that his *Transactions* might produce a similar supply of regular information across the many areas of natural knowledge embraced by the Royal Society.

Empirical Information

The history of "empiricism" has been troubled by anachronisms introduced mainly by nineteenth-century accounts. In 1857 Kuno Fischer, the German historian of philosophy, unsurprisingly saw Immanuel Kant as resolving the contest between rationalism/idealism and empiricism. He cast Bacon into that history by making him the father of "the English empirical philosophers."[42] Since Bacon was the acknowledged patron saint of the Royal Society, the history of philosophy gained a new chapter: the physical sciences became identified with empiricist epistemology. One of the reasons Fischer's book was translated into English was that it resonated with the debate between J. S. Mill and William Whewell. They construed empiricism and idealism as rival twins, clashing in domains from science to morality, from aesthetics to politics. John Passmore once remarked that if Whewell had not lived, Mill would have had to invent him. Mill admitted that he needed to see a strong statement of an idealist philosophy of science before expressing his own empiricist account.[43] This is what Whewell's *History* and *Philosophy* of the inductive sciences provided.[44] But this nineteenth-century conflict between empiricism and idealism (or rationalism) does not capture the preoccupations of seventeenth-century natural philosophy and natural history. Whereas nineteenth-century thinkers argued over the best philosophical account of the *success* of science, early modern virtuosi faced the challenge of how to defend the very *raison d'être* of empirical and experimental inquiry.

The words "empiricism" and "empiricist" were not used by the English virtuosi. The variants in circulation during the seventeenth-century were "empirick" or "empericall," often with reference to the Greek medical schools, as influentially cast by Galen.[45] Bacon said that "Those who have dealt with the sciences have either been empirics or dogmatists"; his original Latin has "Empirici," a clear reference to one of the ancient medical schools in contest with both the "dogmatici" and the "methodici."[46] Such usage was invariably pejorative, thereby reflecting Bacon's own opinion: "it is accounted an errour, to commit a naturall bodie to Emperique Phisitions."[47] We can see this colloquial, derogatory sense in Locke's letter of March 1678 to someone who sought his medical and spiritual advice: "Shall I not passe with you for a great empirick if I offer but one remedy to the three maladies you complain of?"[48] The physician Thomas Browne said that such people could make extraordinary promises because their remedies were unconstrained by theory and learning.[49] More specifically, when declaring a strong interest in "experimentall philosophie," Meric Casaubon warned that this subject could be a honey pot for those claiming access to the secrets of nature and so acting "more lyke an Empirick, or Mountebanck, than a serious man, or philosopher."[50]

In early modern England, the adjective "empirical" did not describe a well-defined epistemological position delineated against a clear alternative. Rather, it indicated a preference for "particulars" over "systems." In his "Proemial Essay" (1661), Boyle expressed suspicion of "systems" and "super-structures" not founded on observation or experiment.[51] In his *Experimental Philosophy* (1663–64), Henry Power ridiculed excessive attachment to systems as entirely at odds with the spirit of the new science:

> Me-thinks, I see how all the old Rubbish must be thrown away, and the rotten Buildings be overthrown, and carried away with so powerful an Inundation. These are the days that must lay a new Foundation of a more magnificent Philosophy, never to be overthrown: that will Empirically and Sensibly canvas the *Phaenomena* of Nature.[52]

How, then, might a word with negative connotations in contemporary medical discussions be used positively, as in Power's claim that the new science would "empirically" study natural phenomena? Part of the answer lies in the complex story whereby the term "empirical" (and later "empiricism") became associated with information acquired by experience, under-stood as input from the senses—just at the time when these three words acquired increasingly positive meanings, especially in English.[53]

In his *Keywords*, Raymond Williams identified the words "empirical" and "empiricism" as "among the most difficult words in the language." More recently, Anna Wierzbicka has suggested that one reason for this is the overlap of these terms with a new use of the term "experience" that crystallized during the seventeenth century—one distinct from the original, and continuing, meaning of experience as an accumulation of skill and knowledge over a lifetime. By the 1600s the word "experience" was also used as a "countable" word because it referred to conscious awareness of *specific* events or feelings occurring in discrete situations.[54] This point had special significance in scientific circles due to the displacement of the traditional Aristotelian notion of common experience, conceived as the way things usually happened, and understood in terms of bodies following their essential natures: stones fall, air rises, oak seeds produce oak trees. As Peter Dear has argued, the "matters of fact" sought by the new experimental communities were not generalized statements of how some aspect of the world generally behaves, but reports of how, in one instance, the world *had* behaved.[55] Although this point has been stressed in discussions of the emergence of the new *experimental philosophy*, it is important to recognize that observation was also considered as a crucial source of experience.[56] Boyle alluded to "my Own experience" when reporting both observations and experiments.[57] Locke declared that the opinions contained in his *Essay* were as sound and comprehensive as "my own Experience and Observation will assist me."[58] But such observations had to be made first-hand, or accepted from reliable witnesses. William Harvey stressed first-hand experience when he asked his "Reader" not to rely on "other mens commentaries, without making tryal of the things themselves," and to trust the "faithful testimony of they own eyes."[59] Hooke made it clear that this kind of experience involved "ocular Inspection and a manual handling, and other sensible examinations of the very things themselves."[60]

This emphasis on first-hand experience was often linked with note-taking. When Boyle, Hooke, or John Ray mentioned a specific experience, either observational or experimental, they invariably had a note of it. Certainly, they also relied on memory, but they underlined the importance of notes as a more secure record for future consultation and transmission to others. Ray vouched that he and Francis Willughby "did carefully describe each Bird from the view and inspection of it lying before us." He stressed that it was such accurate description that allowed Willughby to discover "certain Characteristic notes of each kind" of bird, whereas previous authors had been content with "only general notes."[61] A case in point, as he told Aubrey, was (the younger) John Tradescant's discovery of "sundry rare Exotics out of Virginia": their significance was obscured "by reason of the brevity & obscurity of some of his descriptions containing only generall notes, & for that he tells us not whether they are of plants already described or not."[62] Christopher Merrett gave fairly precise descriptions of his experiments on grafting the barks of different trees: "In the midst of *March An.* 1664, I made a Section of the Rinds of *Ash*, and of the Tree, falsly called *Sycamore*." [63] Of course, taking such notes required procedural discipline and prospective memory: as Hooke confided to his diary for July 12, 1675, "I have forgot most particulars of this week."[64] In one of his proposals about the governance of the Royal Society (probably written in the 1670s), Hooke actually turned this point about forgetting details into an advantage, saying that the oral reports at its meeting were relatively easy to keep secret, "memory alone not being able to carry away exact particulars."[65] But his more considered advice was that experimental details be recorded on the spot, "as soon as the Observations or Circumstances occur," not only to compensate for "Frailty of the Memory" but also because of the need to review "some of the meanest and smallest Circumstances."[66] This new stress on first-hand experience, connected with a specific time and place, was thought to put pressure on individual memory and the ability of the mind to process what contemporaries called "informations" of the senses.

"Informations" from the Senses and the World

We need to start with early modern assumptions about how sensations and experiences from the world were assembled by the mind. The relevant model derived from Greek and Roman medical theory and its philosophical glossing via Aristotle, Avicenna (980–1037 CE), and Aquinas. Still influential during the seventeenth century, this was a physiological theory in which *sensibilia*—the impressions or sensations from the five senses—are turned into images by Imagination, preserved by Memory, and compared by Reason (or Judgment).[67] In this schema of the mental powers, or "inward wits," the sense impressions are delivered to the front ventricle of the brain where the "sensus communis," a power (*virtus*) capable of comparing these various sensory inputs, transforms them into perceptions. Bacon accepted this model, which he took to be Aristotelian, when he said that "the images of individuals are taken up by the sense and fixed in the memory. They pass into the memory as it were whole, in the same form in which they crop up. The mind recalls and reflects on them, and, exercising its true function, puts together and divides their portions."[68] Memory was regarded as a corporeal inner sense located at the back of the brain; it was charged with storing "informations" from the other senses (and the "sensus communis") for the work of imagination and judgment (see fig. 3.1).[69] Bacon agreed that memory and reason each performed its "proper office," with the function of memory being near the start of "the intellectual process."[70] External impressions made on the senses were fixed as images in memory "just as they present themselves," until "the mind recalls and reviews them."[71] Boyle and Hooke also thought within this framework; moreover, they considered its implications for the kind and quantity of information demanded by the new empirical sciences.

Figure 3.1. Memory viewed as an inner sense, located at the back of the head. This position is indicated as no. 8, barely visible, and explained on p. 87. Jan Amos Comenius, *Orbis Sensualium Pictus* (1672), 86–87. Kindly supplied with permission by Rare Books and Special Collections, the University of Sydney Library.

(86)

XLII.

Senſus externi
& interni.

The outward and
inward Senſes.

There are five
outward Senſes;
The Eye 1.
ſeeth colours,
what is white or black,
green or blew,
red or yellow.
The Ear 2.
heareth Sounds,
both natural,
Voices and Words;
and artificial,
muſical Tunes.

Senſus externi
ſunt quinque;
Oculus 1.
videt *Colores,*
quid album vel atrum,
viride vel cæruleum,
rubrum aut luteum, ſic.
Auris 2.
audit *Sonos,*
tum naturales,
Voces & Verbi;
tum artificiales,
Tonos Muſicos.

The

(87)

The Noſe 3.
ſenteth ſmels
and ſtinks.

The Tongue 4.
with the roof of the mouth
taſteth ſavours, what is
ſweet or bitter, keen or ſharp,
ſowr, or harſh. (ting

The Hand 5.
by touching diſcerneth
the quantity
and quality of things,
the hot and cold,
the moiſt and dry,
the hard and ſoft,
the ſmouth and rough,
the heavy and light.
The inward Senſes
are three.

The Common-ſenſe 7.
under the forepart of the
(head,
apprehendeth
things taken from the
outward Senſes.

The Phantaſie 6.
under the crown of the head
judgeth of thoſe things,
thinketh and dreameth.

The Memory 8.
under the hinder part of the
(head
layeth up every thing
and fetcheth them out:
it loſeth ſome,
and this is forgetfulneſs.

Sleep,
is the Reſt of the Senſes.

Naſus 3.
olfacit *Odores,*
& Fœtores

Lingua 4. cum Palato
guſtat *Sapores,*
quid dulce aut amarum,
acre aut acidum,
acerbum aut aſperum.

Manus 5.
dignoſcit tangendo
rerum Quantitatem
& Qualitatem,
calidum & frigidum,
humidum & ſiccum,
durum & molle,
læve & aſperum,
grave & leve.

Senſus interni
ſunt tres.

Senſus communis 7.
ſub ſincipite,
apprehendit
à Senſibus externis
perceptas res.

Phantaſia 6.
ſub vertice,
dijudicat res iſtas,
cogitat, ſomniat.

Memoria 8.
ſub occipitio,
ſingula recondit
& reponit;
quædam deperdit,
& hoc eſt oblivio.

Somnus,
eſt Senſuum requies.

G 4   *Anima*

Boyle understood "Informations of the Senses" as the inputs (to modernize) that can be registered and stored in memory for the subsequent use of reason.[72] He spoke of the "ways of information" as including sensations from all the senses—vision, hearing, smell, taste, and touch: "For the Knowledge we have of the Bodies without Us, being for the Most part fetched from the Informations the Mind receives by the Senses."[73] Each of these bodily senses provided distinct and separate "informations" that had to be checked against each other. He warned that "Weather-glasses & our Sensories may give very differing Informations about the Temperature of the Air turn'd into Wind, by being blown out of the same pair of Bellows."[74] He referred to this checking, comparing, and noting as "reflections on the Information of the Senses." This was one marker of the difference between men and those "Creatures destitute of Reason," and so, if "Impressions be onely receiv'd and not improv'd," then "we faultily loose both one of the noblest Imployments, and one of the highest Satisfactions of our rational Faculty."[75]

Whatever the urgency about making notes of *individual* experiences, Boyle explicitly acknowledged that all accounts of natural, or experimentally produced, phenomena had to be interpreted against each other. In the *Christian Virtuoso* (1690–91; written in the 1680s), he contended that dutiful inquiry should not solely rely on individual experience but seek information from diverse sources. In this way, the "indefatigable industry of the modern *Virtuosi*" had produced "a much fuller information of the number and *Phenomena* of the fixed stars and planets."[76] These investigators did so by opening themselves to "all those ways of Information, whereby we attain any Knowledge that we do not owe to abstracted *Reason*."[77] Unlike other pretenders, the "modern virtuosi" carefully collected and examined information from diverse sources:

> That, when, in this Discourse, I speak of an Experimental Philosopher, or Virtuoso; I do not mean, either, on this hand, a Libertine, tho' Ingenious; or a Sensualist, though Curious; or, on that hand, a mere Empirick, or some vulgar Chymist, that looks upon nothing as Experimental, wherein Chymistry, Mechanicks, &c are not employ'd; and who too often makes Experiments, without making Reflection on them, as having it more in his aim to Produce Effects, than to Discover Truths. But the Person I here mean, is such a one, as by attentively looking about him, gathers Experience, not from his own Tryals alone, but from divers other *matters of fact*, which he heedfully observes, though he had no share in the effecting them.[78]

Boyle asserted that this kind of person—he included himself and other adherents of the corpuscular theory—avoided the errors of philosophical systematizers by canvassing the widest possible range of information.

This more extensive array of information was "Historical Experience." Boyle said that this was wider than *"Personal Experience"* which an individual "acquires immediately by himself, and accrews to him by his own Sensations." We can see that Boyle was aware of the new stress on experience as personal and *"Immediate,"* because he said that he wanted to speak with more "latitude," as when we say "that Experience teaches *us*, who perhaps were never out of *England*, that the Torrid Zone is Habitable, and Inhabited." He noted that neither common nor learned usage confined "the word *Experience* to that which is *Personal*."[79] Boyle realized that the method of making preliminary histories of particulars in various fields would not immediately form a coherent picture. He explained that the quest for empirical information required both time and a special habit of mind: "the Temper of Mind, that makes a Man most proper to be a *Virtuoso*" was that of "Docility." Boyle imagined a persona for the virtuoso, largely in contrast with that of the system builder. Thus, for him, this ideal inquirer did not have "a great Opinion of his own Knowledge"; realizing the difficulty of studying nature "he will easily discern, that he needs further Information, and therefore ought to seek for it."[80]

Boyle stressed that the "ways of information" included books, conversation, testimony, observation, and the reports of experiments done by others. He treated all these as appropriate avenues of empirical "matters of fact," provided that adequate precautions were taken. He did not necessarily insist on first-hand (or "Personal") experience; some information could reasonably be sought in books that recorded the observations of others. In the introduction to a proposed work, he wrote that "for the Principall things I am to acquaint you with being matters of Fact, and consequently such as would be the most properly prob'd out of the Bookes of observations of our Naturalists & Physitians, and out of other Historicall writeings."[81] But those who restricted their range of sources were at fault—as Boyle alleged with regard to recent information about the poor quality of air in deep mines: "those that philosophize only in their studies, . . . have not received information from any that visited the deeper parts of the Earth."[82] He was happy to extract whatever useful knowledge books contained, while also being hungry for the particulars that usually had to be newly observed or collected from the testimony of others. He defended this position against Hobbes who, maintaining a traditional distinction, argued that all current experimental inquiry conducted under the auspices of the Royal Society was merely *Historia*, not *Scientia*.[83] Emboldened by Bacon's authority, Boyle affirmed the importance of making observations and experiments, "for they enrich Natural History, without which Natural Science is in vain sought for."[84]

Hooke believed that the new philosophy must consider the process by which information was received by the senses, memory, and reason. He made this point in both the *Micrographia* (1665) and the "General Scheme," probably composed in 1668. In this latter work he argued that each of our senses gave us distinctive information about the world. The problem was that the impressions carrying it did not easily combine, even in the "common Sense," to provide a coherent notion of phenomena, such as a "Sonorous Body" like a vibrating string that affects the eyes and ears in quite different ways. Hooke's solution was to ensure "a comparative Act of the Understanding from all the various Informations" received directly from the senses, and also "more mediately by various other Observations or Experiments."[85] Similarly, in a manuscript entitled "Philosophicall Scribbles" (probably written in the early 1680s), he stressed the role of "an *active faculty*" by which the mind works on the "information receiv'd from the senses" in the form of "impressions." Hooke explained that it did this by "comparing and compounding one with an other, noting wherein they agree or differ."[86] However, in order to make comparative inferences and generalizations that allowed the transition from particulars to universals, reason had to draw on past experiences and ideas held in memory. Hooke was well aware of the ancient anecdotes, common wisdom, and contemporary medical advice about the limits and physical vulnerability of this inner corporeal faculty.[87] Although selective retention might help, he observed that memory tended to preserve useless "things":

> The like frailties are to be found in the *Memory*; we often let many things *slip away* from us, which deserve to be retain'd; and of those which we treasure up, a great part is either *frivolous* or *false*; and if good, and substantial, either in tract of time *obliterated*, or at best so *overwhelmed* and buried under more frothy notions, that when there is need of them, they are in vain sought for.[88]

If memory was fragile and unreliable, then the mass of information demanded by Baconian natural histories posed a major problem.[89] There were good reasons for this concern—as we can see in Bacon's comments.

Baconian Information

Shortly after the publication of the *Novum Organum* (1620), Bacon receieved a letter (now lost) from Redemptus Baranzano, a young professor of philosophy at Annecy in France. Baranzano asked about the multitude of details required for the projects Bacon outlined. Indeed, he may have been alerted by passages such as the one acknowledging that the human faculties would be daunted by the abundance of material: "the army of particulars is so vast and so scattered and dispersed that it dissipates and confounds the intellect."[90] Bacon replied that it did not matter "if the description of Instances should fill six times as many volumes as Pliny's history."[91] In his *Naturalis historia* (77 CE), Pliny the Elder had boasted that his work gathered "20,000 Things" drawn from 2,000 volumes by 100 authors.[92] Bacon promised that something could be done about this, as I discuss below, but first we need to understand what he meant by this constant plea for more "particulars."

In surveying the deficiencies of contemporary intellectual inquiry, Bacon asserted that the search for underlying causes of natural phenomena had been thwarted by "too untimely a departure, and too remote a recess from particulars."[93] Clearly, one referent is the contrast with universals in traditional logic, as taught from Aristotle's works.[94] Arguing in the *Novum Organum* against the premature leap to "axioms of the highest generality," Bacon advocated a cautious movement from particulars; he amplified this by indicating that he meant details drawn from the senses and experience.[95] There is an apparent similarity between "particulars" and what Bacon called "singular" (or "irregular") phenomena; however, these two concepts should not be conflated.[96] The "irregular" phenomena comprised Pliny's *mirabilia*—rare, unusual, and extraordinary objects and events such as monstrous births, dwarfs, giants, tremendous storms, and individuals with prodigious memories.[97] These were examples of what Bacon called singular (or "*Monadic*") instances. In contrast, "particulars" were specific details about regular occurrences in nature that must be collected if sound generalizations were to be induced. The chronic problem, as he diagnosed it, was that the eye-grabbing capacity of the rare and wondrous diverted proper study of the seemingly "commonplace and trivial"; yet it was just these more usual, everyday phenomena that needed fuller investigation. As he put it, "we need information less often about things which are not known than attention to things which are."[98] The corollary was that if more particulars were gathered about the rare or monstrous "waifs and strays of nature," it may turn out that "every irregularity or singularity will be found to depend on some common form."[99] The key point is that Bacon called for a "multitude of particulars," sought out in a systematic fashion such that temporary generalizations would "readily point out and specify new particulars, and so render the sciences active."[100] His "catalog of particular histories" gave some illustration of how such details might be derived from minute descriptions in response to queries about, say, "normal, stormy and abnormal rains."[101] Moreover, Bacon's demand for observation and collection of particulars was not confined to the curious, the precious, and the unusual; rather, it included what he called everyday phenomena and seemingly trivial things ("*ad res vulgarissimas*").[102]

How was this vast collection of "particulars" to be managed? Judging from Bacon's classification of knowledge, as given in the *Advancement of Learning* (1605), it would seem that memory had the task of retaining this material.[103] The faculty of "Memorie" governed the domain of History, including natural history, and one of its functions was the "Custodie or retayning of Knowledge," albeit supported by "Writinge."[104] However, in the greatly enlarged *De augmentis* (1623), Bacon accepted that his program would overtax memory. Accordingly, he canvassed various remedies: a reconsideration of the arts of memory, the use of lists, inventories, and tables in various kinds of "paper books," and a new philosophical language, possibly founded on a classification of simple, or radical, notions.[105] Most fundamentally, he stressed the importance of writing, in its various forms, as a way of recording and processing information:

> The great help to the memory is *writing*; and it must be taken as a rule that memory without this aid is unequal to matters of much length and accuracy; and that its unwritten evidence ought by no means to be allowed. This is particularly the case in inductive philosophy and the interpretation of nature; for a man might as well attempt to go through the calculations of an Ephemeris in his head without the aid of writing, as to master the interpretation of nature by the natural and naked force of thought and memory, without the help of tables duly arranged.[106]

This function of writing was also linked to Bacon's notion of "*experientia literata*" (literate experience), by which he meant the assembling of preparatory information, sorted in various ways to assist memory and thinking.[107] Note-taking was part of this.

Although Bacon admitted that the number of empirical details would become overwhelming, he maintained that this risk had to be taken. Trying to assuage, he suggested that no one should "dread the multitude of particulars" because these were "just a handful compared with the fabrications of wit" generated by textual commentary and "fanciful meditations." Such fabrications were not limited by the "evidence of things" and produced "endless perplexity." In natural histories, however, the appearance of miscellany in the early stages could be tolerated because sound results would only derive from a copious collection of particulars.[108] This was an invitation to take notes freely, without too much concern about sources or reliability; indeed, Bacon was happy to record things that are "downright unreliable" but are "bandied about" either through carelessness or "from figurative use." If recorded, these errors could be "publicly proscribed lest they do any more damage."[109]

Nevertheless, while calling for such expansive collection, Bacon cautioned against the circulation of bits of information without at least temporary allocation to a category.[110] He advocated "collections under heads and commonplaces" in preference either to sheer miscellanies or biased and misjudged epitomes — for example, of a book or a biographical or historical topic. In the letter to Fulke Greville (discussed in the previous chapter), Bacon stressed that such Heads "have in them a kind of observation, without the which neither long life breeds experience, nor great reading great knowledge."[111] He also contended that the gathering of material for natural histories was far less daunting than the content of libraries of civil and canon law. Jumping somewhat prematurely ahead of his own emphasis on the amassing of particulars, he presented himself as "a faithful scribe, [who] takes down and copies out the very laws of nature." In this case, "brevity is natural, for it is practically forced upon me by the things themselves; whereas the numberless host of opinions, tenets, and speculation goes on for ever."[112]

Bacon's quest for empirical particulars required a mode of note-taking that departed from the usual guidelines for commonplace books. As indicated in chapter 1, he announced this in the *Advancement of Learning* and, later, in the *Novum Organum* he argued that the focus on the wondrous or remarkable was not sufficient for his program. In the *Parasceve*, he spelled this out more fully, calling for the end of "everything to do with oratorical embellishment, similitudes, the treasure-house of words and suchlike emptinesses."[113] Bacon named four classes of material that must be gathered and recorded, issuing, as he put it, "concise instructions" because otherwise people would assume that "it would be pointless to write them down." These were: (1) quite ordinary and everyday matters; (2) things "vile, illiberal, and repellent"; (3) things "frivolous and childish"; (4) things that "seem to be far too subtle" and of "no use."[114] He urged that apparently obvious and well-known phenomena must be recorded, not just the marvelous and not merely those immediately relevant to the "art in question," but also those that "crop up in the process," such as lobsters turning red while cooking as do bricks. This fact was not relevant to the art of cooking but it did pertain to the "nature of redness."[115]

Bacon believed that "experience undisciplined" could be mitigated by careful note-taking. On this score, he declared that the moderns must improve upon "the ancients." In contrast with Gibbon's view about the ancients suffering a lack of information, Bacon asserted that they collected "a vast body and abundance of examples and particulars, and arranged it by subjects and titles into their notebooks." However, the problem was that "they thought it would do no good and serve no purpose to make public their drafts, notebooks and records of particulars, . . . they got their tools and ladders out of sight once the building was up." Bacon regarded this disinterest in the raw data as typical of the ancients' rush to "the most general conclusions or principles of the sciences," and their subsequent adherence to these doctrines despite "new particulars" or anomalous examples.[116] His message was that notebooks were indeed crucial to his project and that they should be kept, and updated, as correctives to the very generalizations that they served to support. In this way, he said, they would play a part in "transferring, putting together and applying" known facts "by means of what I have called literate experience."[117]

Leading members of the Royal Society knew that this was a task easy to announce but difficult to accomplish. A review of Sprat's *History* in the *Philosophical Transactions* spelled out what it meant to accumulate Baconian particulars.[118] The reviewer, most likely Oldenburg, explained that Sprat had effectively silenced the "importunate demand" of those who doubted the productiveness of the Society. He explained that although much of the Society's work had "not been exposed to open view, yet their *Registers* are stored with a good number of *Particulars* they have taken pains about." He then listed nine ways in which such information had been gathered—such as by queries, testimony, experiments, and observations—and even counted the data assembled under these.

> The *Particulars* upon which *Heads* are more numerous, and of greater moment and variety, than perhaps Detractors and Cavillers imagine or expect: they exceed indeed the number of 700; of which the *Experiments* and *Observations* both together amount to above 350; the *Relations*, to about

150; the *Queries, Directions, Recommendations,* and *Proposals,* to above 80; the *Instruments,* to about 60; the *Histories* of *Nature* and *Art,* to above 50; and the *Theories* and *Discourses* to as many.[119]

Oldenburg also underlined Sprat's assertion that those who had taken on this laborious work "should be Judges by what steps and what pace they ought to proceed."[120] This might be taken as an endorsement of Sprat's declaration that the Society had not prematurely sought to fill "the world with *perfect Sciences,*" but this review's careful discrimination of the classes of particulars does not so easily match his depiction of the preliminary work as consisting merely of "*immethodical Collections* and *indigested Experiments.*"[121]

Empirical Sensibility and Its Challenges

The gathering of information from a wide range of sources, and the comparison and corroboration of this with extant knowledge, is a crucial aspect of what I call the empirical sensibility of the English virtuosi. As Peter Anstey has recently argued, these figures certainly endorsed "experimental" over "speculative" philosophy, but we should not overlook their affirmation of other kinds of empirical information—sensory experience, testimony, books, and observation, as well as experiment.[122] This broader set of sources and data reflects the sense in which Boyle defined "Experience" to include diverse avenues of "Information," as discussed above.[123] The common denominator here was a stress on empirical information of "particulars," as understood in Bacon's project for the collection of natural histories. This outlook went hand in hand with a suspicion of premature systems, although it recognized the indispensability of some temporary organization of material, as provided by Heads or Titles. Although this empirical sensibility involved assumptions pertaining to what later became epistemological disputes between empiricism and rationalism, I think it was sustained by different preoccupations: namely, the function of methodical note-taking in the collection and collation of empirical information, and the challenge of making personal notebooks serve collaborative goals. We can see these preoccupations in play as leading members of the Royal Society sought to manage the content and format of the information they accumulated. Their efforts were motivated in part by the need to have information that allowed comparisons, and by the commitment to the long-term collection and storage of data.

Bacon's call for large-scale gathering of observations required the storing of information for comparative analysis by people not responsible for its initial collection. The rationale of the Royal Society invited this situation by casting a wide net for information from all sources. In the third year of the *Transactions*, Oldenburg reported on the gathering of observations, experiments, and inventions "scattered up and down in the World" but now available in London, "this Famous *Metropolis of England*."[124] Indeed, this program may have been one of Shadwell's targets in *The Virtuoso* (1676), especially when he lampooned Gimcrack's collection of curious facts and phenomena that relied on reports of travelers in exotic places: "I keep a constant correspondence with all the virtuosos in the north and northeast parts. There are rare phenomena in those countries. I am beholden to Finland, Lapland, and Russia for a great part of my philosophy." He also mentioned the "queries" Gimcrack sends with travelers, or to correspondents, requesting answers to various questions — a direct reference to the practice of Boyle and other Fellows.[125] For this ammunition, Shadwell only had to read other notices in the *Transactions*, such as "Inquiries for Turky" or similar lists for Greenland and the East Indies.[126] The unfair insinuation was that intelligence, news, and gossip thus assembled were treated without due care.

The early Royal Society did, however, experience difficulty in striking a delicate balance between encouraging communications from correspondents and shaping the ways in which observations were made and noted. Oldenburg tried to coach his correspondents about the kind of information required. He made it clear that the Society wanted matters of fact, not unsupported hypotheses, and certainly not metaphysical speculation involving "the forms, qualities and useless elements of the schools."[127] He often sought out quite specific information by explaining its role in a larger program. Thus when asking Sir William Curtius to "obtain for us a specimen of the vitriol" from Lower Saxony, he wrote: "You cannot be ignorant of our object, to compile an inductive history, both natural and artificial, and at last to build a solid and fruitful philosophy upon that basis."[128] Although the emphasis was on empirical information, it is important to realize that Oldenburg did not exclude the use of books. He stressed that that "we have [not] discouraged or refused the Essays of some famous Philosophers, learned Philologers and *Antiquaries*; whose Disquisitions, Readings, and Reasonings, have extended farther than their Experiences." But he underlined the "liberty" of the Society to select and extract the most useful and solid information from such sources.[129] This can be taken as confirmation that the Society urged careful reading, noting, and excerpting from both manuscripts and print for comparison with observations and experiments. All this was in keeping

with Bacon's notion of "experientia literata" and its importance in the preparatory stage of inductive inquiry.

Boyle offered some directions for this preliminary collection. He distinguished between experiments that required both skill and special instruments—such as those performed by Hooke as curator—and observations
that might be made by many individuals. However, he made the proviso that
this be done with some care: "That those that are not qualify'd or dispos'd
or at leisure to undertake any methodical part of the History may be invited
to assist those that do, and may be both invited & incourag'd to write Promiscuous Observations, and directed how to write them in the most useful
manner."[130] However, for any effective comparison of information, notes had
to be taken in a manner that allowed for use by others. In April 1666, Boyle
urged "some Inquisitive men" to "make *Baroscopical* Observations in several
parts of England (if not in forrain Countries also;) . . . that by comparing
Notes, the Extent of the Atmospherical Changes, in point of Weight, might
be the better estimated." However, it is evident that Boyle could not assume
the operation of standard conventions for empirical note-taking. Thus he advised "that it will be very convenient, that the Observers take notice not only
of the *day*, but, as near as they can, of the *Houre* wherein the height of the
Mercurial Cylinder is observ'd: For I have often found, that within less than
the compass of one day, or perhaps half a day, the Altitude of it has so consid-
erably vary'd." He also requested that observers give notice of the "*Situation
of the place*," and that "those *Virtuosi*" seeking Oldenburg's instructions be
told "to set down in their Diarys not only the day of the month, and the hour
of the day, when the *Mercuries* height is taken, but (in a distinct *Columne*) the
weather, especially the Winds."[131] The bald directness of the hints indicates a
lack of consensus about these points, even though such agreement was crucial for any useful comparison of information.

In the mid-twentieth century, the sociologist Edgar Zilsel identified a distinctive notion of intellectual progress associated with Renaissance arts and
crafts, and then with the sciences: namely, the idea of piecemeal additions to
a larger intellectual structure, one not fully grasped by the various contributors.[132] This entailed a long-term commitment. However, in the 1660s it was
easy for critics of the Royal Society to construe slow information gathering
as indefinite postponement. These jibes about sluggish progress found their
target because the payoff from Baconian natural histories was cast in an indefinite future. A program that did not bring quick conceptual results had to
be defined and defended. Glanvill, for example, tried to do so by asserting

that there was more promise in "the Repository of this Society" than in "all the Volumes of Disputers." He dismissed those who "do not comprehend the vastness of the Work of this Assembly," and stressed that natural history "must proceed slowly, by degrees almost insensible"; it was a labor of generations.[133] This was a repetition of the official line in Sprat's *History* (1667): namely, that tasks which are unmanageable by a few hands are more easily done by "the joynt force of a multitude: many that fail in one *Age*, may succeed by the renew'd indeavours of *another*." There cannot, he affirmed, be a proper judgment about the impossibility of things "unless they have been often attempted in vain, by many *Eyes*, many *Hands*, many *Instruments*, and many *Ages*."[134]

In 1679 Evelyn felt the need to show that progress was being made, and he called as "witness" the "Journals, Registers, Correspondence, and Transactions."[135] He had earlier put his understanding of this position to Beale: "The Members of the R. <u>Society</u> bring in Occasional <u>Specimens</u>, not Compleat <u>Systemes</u>, but as Materials & particulars which may in time amount to a rich & considerable Magazine, capable of furnishing a most august & noble structure."[136] Thus from an institutional perspective, one answer to criticism was that the body of "Materials & particulars" would accumulate, ready for analysis in the future. At the individual level, however, there was an appreciation that a life of empirical collecting and note-taking had its costs. In 1668 Evelyn confessed to Beale that "I have treated myne Eyes very ill near these 20 yeares, during all which tyme I have rarely put them together, or compos'd them to sleepe before One at night."[137] He also reflected on the psychological load incurred by an individual seeking to amass information, even on a single topic — in this case the completion of his encylopaedic work on gardening:

> When againe I consider into what an Ocean I am plung'd, how much I have written, & collected for about these <u>20 yeares</u>, upon this fruitful & inexhaustible subject (I mean of Horticulture) not fully yet digested to my mind, and what inseperable paines it will require to insert the (dayly increasing) particulars into what I have already in some measure prepar'd, and which must of necessitie be don by my owne hand; I am almost out of hope, that I shall ever have the strength & leasure to bring it to maturity.[138]

In responding to this challenge the virtuosi were able to draw sustenance from ancient models.

The Ancient Note-taking Legacy

Hippocrates' famous first aphorism supplied both a rationale and a possible inspiration for the note-taking of the English virtuosi. In its Latin form it reads "*vita brevis, ars longa*"—that is, "life is short, art is long." In the full original Greek version it announces that "life is short, the art is long, occasion sudden, experience dangerous, judgement difficult."[139] This was usually reduced to the proposition that any art or skill requires more time than the short life allotted to each person. It became an immensely protean trope, with glosses by authors such as Pliny, Seneca, Plutarch, and Petrarch. Seneca called it a "complaint" from "the greatest of doctors."[140] We can see that it posed the question of how any art (in the sense of *techne*), such as medicine, or any of the empirical sciences, could be mastered in one short life. We can also imagine two plausible reactions: one could try to abbreviate art (or sciences) or somehow extend one's life.[141] Regarding the *first* option, Galen reported with annoyance that one of the Greek medical sects, the "methodists," scolded "the man who has said that life is short and the art long." They inverted the aphorism, saying that "art is short and life is long," arguing that "if one does away with all those things which have been wrongly taken to further the art, . . . then medicine is neither long, nor difficult, but rather is easy and clear and can be learned as a whole in a matter of six months."[142]

The second option was viable if one's own life of experience could be extended by access to the accumulated experience of generations. Galen gave some support to the "empirics" in their preference for experience acquired through observation over dogmatic theoretical systems. However, he argued that it was not adequate to rely solely on the capacity of memory to retain the most frequently repeated patterns of cause and effect in diseases and their remedies. Instead, he stressed that appropriate experience included "those things which one has seen for oneself," or "the report of those things which have been seen," or "are as if they had been seen" by others.[143] Galen therefore endorsed the Hippocratic practice of taking histories of individual patients.[144] The "use of history," as he put it, was necessary "because of the vastness of the art, since one man's life will not suffice to find out everything. For we accumulate these things and collect them from all sources, turning to the books of our predecessors."[145] Using history meant availing oneself of collective past experience, provided that this experience had been duly observed and correctly noted. Galen employed the Greek term *hypomnêmata* when referring both to commentaries and notes on the Hippocratic texts;

but is possible that early modern authors understood this as analogous to the copious notes they often made on both books and medical cases.[146]

Bacon mentioned "life's shortness" as one of several obstacles to the progress of knowledge.[147] Indeed, he accepted the need "to remedie the complaint of *vita brevis, ars longa*," and his response echoed Galen's. He called for a return to the "serious diligence of *Hippocrates*, which used to set downe a Narrative of the speciall cases of his patientes and how they proceeded"; and he urged that such "*Medicinall History*" be carefully pursued.[148] Bacon's position was in keeping with the recognized importance of medical "observationes" assembled by excerpting and commonplacing from various texts. Thus the German physician Johannes Schenck (1503–98) explained that his aim was "to collect in one volume those new and wondrous things that the most celebrated physicians observed not so much by means of doctrine as by means of experience (*experimentum*)."[149] Bacon knew that physicians in his own day kept case histories, but judged these to be neither "so infinite as to extend to every *common Case*, nor so reserved, as to admit none but *Woonders*."[150] As we have seen, this was a point in his critique of all current natural histories: although these appeared "large in bulk" they mainly contained "fables, antiquities, quotations, idle controversies, philology and ornaments." Once these were taken away such histories would "shrink into a small compass," whereas the ones Bacon imagined would be vast and ongoing.[151] He reassured his readers that extensive data gathering was not "beyond the capacity of mere mortals" because "its completion is not confided entirely to a single age but to a succession of them."[152] The implication was that the immediate beneficiaries might not be the note-takers of Bacon's own time. His answer to Hippocrates' complaint was a call for collaboration over long periods of time so that in the future individuals would benefit from a legacy of experience captured in notes. It is also reasonable to infer from Bacon's other comments about the mind being naturally "impatient of investigation," that acceptance and tolerance of this postponement was a quality now required for the pursuit of natural knowledge (see fig. 3.2).[153]

The English virtuosi took this cue, but had to confront its challenges.[154] In *Some Considerations Touching the Usefulnesse of Experimental Naturall Philosophy* (1663), Boyle recognized that "Hippocrates begins his Aphorismes with a complaint, that Life is Short, but the Art long." He also cited Paracelsus' acknowledgment that the "Art of Medicine" required knowledge of other subjects, such as chemistry and philosophy.[155] Later, however, in his *Christian Virtuoso* (1690–91), Boyle pointed out that current medical knowledge was

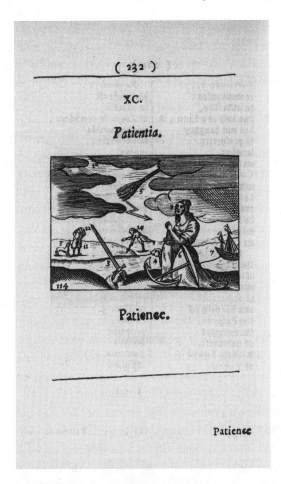

(232)

XC.

Patientia.

Patience.

Patience

Figure 3.2. For Bacon and Boyle, "Patience" was a necessary virtue for empirical note-takers; but "impatience" (indicated as nos. 11 and 12) was the natural tendency of the mind. Jan Amos Comenius, *Orbis Sensualium Pictus* (1672), 232. Kindly supplied with permission by Rare Books and Special Collections, the University of Sydney Library.

in part a "kind of Historical Experience, consisting of the personal Observations of *Hippocrates*, *Galen* and other Physicians, transmitted to us."[156] In his *Micrographia*, Hooke contemplated how the medical histories kept by a physician, either in memory or in notes, might be amplified into a collective database:

If a Physician be therefore accounted the more able in his Faculty, because he has had long experience and practice, the remembrance of which, though perhaps very imperfect, does regulate all his after actions: What ought to be thought of that man, that has not only a perfect *register* of his own experience, but is grown *old* with the experience of many hundreds of years, and many thousands of men.[157]

The problem, in Hooke's view, was that individual memory could not be trusted to retain knowledge so that it could be passed on to others.[158] A cooperative effort was thus required, not just to produce information, but to ensure that all contributions were adequately noted and recorded. Even then, there was a risk that the legacy of one generation to the next might merely be a greater volume of unprocessed information, such as disconnected particulars. Indeed, Evelyn's dismay was that even after twenty years of collecting in one subject, he had to confront information collected *by others*, and then, as he put it, to "insert the (dayly increasing) particulars" into whatever patterns or generalizations he had discerned.

This may be one reason why the first option of devising short cuts held its own attraction. Petrus Ramus (1515–72) and his followers claimed that their use of branching diagrams crystallized the key concepts of a subject, allowing it to be learned in the shortest time.[159] Versions of this strategy were appealing to some seventeenth-century figures, such as Jan Amos Komenský (more usually, Comenius) and his pansophist followers, who sought to reduce knowledge into condensed form as the basis of a quick key (*clavis*) to new knowledge. The assumptions underlying these hopes were quite various, but we can discern a later, sophisticated, version in the thought of Leibniz. In a paper of 1680 he asked Louis XIV to arrange for the "the quintessence of the best books" to be extracted, and to add to them the observations, not yet recorded, of the best experts of each profession.[160] However, Leibniz also had another motive for this collection and reduction of material. Three years earlier, in "Towards a Universal Characteristic" (1677), he asserted that a dictionary of the most common ideas, analyzed and reduced in this manner, would "not require more work than is now already expended on lectures or encyclopaedias." He predicted that a "few selected persons might be able to do the whole thing in five years."[161] He reasoned that since all knowledge can be expressed by the letters of the alphabet it should be possible to

> calculate the number of truths which men are able to express, and that we can determine the size of a work which would contain all possible knowledge, in which there would be everything which could ever be known, written, or discovered; and even more than that, for it would contain not only the true but also the false propositions which we can assert, and even expressions which signify nothing.[162]

In the next chapter we shall see that the prospect of quickly distilling the essentials of knowledge was entertained in Samuel Hartlib's circle. While acknowledging the importance of particulars, Hartlib's correspondents were encouraged by Bacon's assurance that such a foundation would eventually uncover "*the simple formes or differences of things*, which are few in number."[163] Hartlib and his friends wished to hasten this process — not by justifying less intensive note-taking, but rather by gathering and collating diverse information with a view to crystallizing fundamental ideas and propositions in a relatively brief time. As one of Hartlib's correspondents, the young Boyle had some acquaintance with this "short art" option; but he soon decided that slow compilation, storage, and transmission of information was the only proper response to Hippocrates' aphorism.

[4]

Taking Notes in Samuel Hartlib's Circle

In the anonymous utopian tract *Macaria* (1641), a "Schollar" converses with a "Traveller." In the opening scene they meet in "the Exchange," which puzzles the Traveller who says "I conceive you trade in knowledge, and here is no place to traffick for it." He suggests that they "go into the fields" where he can relate "strange newes, and much knowledge, and I have brought it over the sea without paying any Custome." Keen to oblige, the Schollar replies, "We Scholars loves to hear Newes, and to learne knowledge." This work was written by Gabriel Plattes but published by Samuel Hartlib, the leading London intelligencer who corresponded with Boyle, Petty, Aubrey, Beale, and others before the establishment of the Royal Society.[1] In one sense, the scenario in *Macaria* invites Bacon's protest that even the learned were content with "certain rumours, tidings or vapours of experience" and, consequently, that "philosophy is run exactly as if it were some kingdom or state" that relied on "town chatter and gutter gossip." Bacon warned that "loose and vague observation yields unreliable and untrustworthy information."[2]

Hartlib was a keen reader of Bacon, but he was not overly fussy about the distinction between trustworthy reports and gossip. His mantra was that information and knowledge derived as much from observation and testimony as from books and, consequently, from diverse sources in various places. He believed that all information, once collected, should be collated and indexed; this method would show up contradictions and duplications. In order to coordinate material, Hartlib planned an "Office of Publick Addresse" to gather "Informations in matters of Religion, of Learning, and of all Ingenuities."[3] In a statement of 1648, he said that the aim was to communicate "the Informations of things profitable to be taken notice of in a private or publick way."[4] One of his models was the *Bureau d'Adresse* convened, with support from Cardinal Richelieu, by Théophraste Renaudot in Paris from 1633.[5] In fact, Hartlib was trying to launch two "Offices": one for practical information ("Accommodations") in London, to be directed by Henry Robinson, and one for "Communications" about knowledge, in Oxford, which he would run himself. In the one dealing with "Accommodations," Hartlib anticipated the need to update news, services, and requests for labor or employment. Thus he declared that "the Register is to be discharged of it [old information] within foure and twenty houres . . . lest fruitlesse Addresses be made to any concerning a Matter already dispatched."[6] However, this rapid removal of material was not an option when it came to the ideas and observations collected under "Communications." These had to be carefully stored and arranged for later use, and the "Office" was intended to achieve this by centralizing information: "For therein, as in one Magazin or Market-place, all things Necessary, Profitable, Rare, and Commendable, which are extant in several places, and scattered here and there, are brought together."[7] As it turned out, Hartlib himself became this central point.

Hartlib's livelihood and reputation depended on a voracious appetite for news from the Republic of Letters, and an ability to handle it.[8] In introducing *A Discours of Husbandrie* (1652), he declared that, although lacking the relevant experience, "I finde my self obliged to becom a conduit-pipe thereof towards the Publick," communicating "the Experiences and Observations of others."[9] This metaphor gives the impression of passive reception and transmission, and hence does not adequately capture the full ambition and potential of his activities. Hartlib's petition to Parliament in 1660 summarized his life-long efforts on behalf of learning, thus justifying his government pension of 100 pounds a year from 1649 as an agent for "the Advancement of Arts and Learning."[10] But in this petition, Hartlib merely specified the pursuit of "rare collections of manuscripts," "the best experiments of industrie practised in husbandry and manufactures"; and the maintenance of "a constant intelligence in matters of pietie, vertue, and learning, both at home and abroad."[11] He did not indicate sufficiently his vigorous selection and management of information. Yet his ability to provide useful communication rested upon the personal archive of material that he deliberately ordered under Heads with the aim of reducing it to essentials. In so doing he showed how information gathered in the Baconian spirit was a demanding task, perhaps more so than Bacon himself realized or stated.

In modern scholarship Hartlib and his circle of friends and correspondents have received a mixed press. For Hugh Trevor-Roper, Hartlib, John Dury, and Jan Comenius were "the real philosophers, the only philosophers, of the English Revolution." Yet they were also exponents of a "vulgar Baconianism" that "lacked the range and power of the true Baconian message."[12] Margery Purver asserted that Hartlib and Comenius conflated Bacon's plans for the renewal of learned institutions with their own dream of pansophical colleges and, accordingly, she denied them any precursor status with respect to the Royal Society.[13] Marie Boas Hall acknowledged that Hartlib was "a not unimportant source of communication" before the formation of this and other scientific societies, but concluded that "most of the information he collected and the discoveries he published were very minor."[14] Against such lukewarm appraisals, Charles Webster redeemed Hartlib's circle as an intellectual network, driven by a potent combination of Protestant millenarianism, social utopianism, and Baconian vision. Their reforming program involved empirical science, its institutional arrangements, and its social applications—all constituting a "Great Instauration" of knowledge.[15] John Beale's dedication of *Hertfordshire Orchards* (1657) can be read as a participant's appreciation of how Hartlib embodied these prospects. In Beale's words, he was "the zealous sollicitor of Christian peace Amongst all Nations . . . [and] the sedulous advancer of Ingenuous Arts and Profitable Sciences."[16]

After Webster's work, there can be no doubt about the religious commitments that motivated the plans and activities of Hartlib and his close associates.[17] In English Protestantism of their day, the notion of a particular "calling," in addition to the general calling of all men to follow Christ, was well accepted. The former, as the bishop of Lincoln, Robert Sanderson, expressed it, was one "wherewith God enable us, and directeth us, and putteth us on to some *special* course and *condition of life*, wherein to employ our selves, and to exercise the *gifts* he hath bestowed on us."[18] Individual talents and vocations contributed to a divinely supervised plan for the common good, one that enabled the application of science to the relief of man's estate in this probationary world. A distinctive emphasis of Hartlib's circle was the insistence that this outcome depended on sharing skills and insights imparted by God to individuals; in effect, there was a "duetie of Communication."[19] In *The Reformed Librarie Keeper* (1650), Dury declared that current knowledge of the "Sciences" had to be quickly directed toward the "publick Use," not individual interests; otherwise "the increase of knowledg will increase nothing but strife, pride and confusion."[20] Hartlib feared that it might be too late. In a diary entry of 1640 he noted with regret: "If wee had but gathered all that which is knowen and done already O what a world of profitable matters should wee enjoy. But now all things are confounded within themselves, the first and last degrees are neglected and the middle untowardly followed."[21] Together with Comenius and his pansophist followers, Hartlib and Dury believed the second coming of Christ would occur in their lifetime—as announced in Joseph Mede's *Clavis Apocalyptica* (1643). They cited the prophecy in the Book of Daniel (12:4) about the quickening of knowledge as one of the final signs, but this went hand in hand with urgency about organizing current knowledge as the basis for a swift set of discoveries.[22]

Seventeenth-century Protestant beliefs about the consequences of the Fall of Adam and Eve are relevant here. Peter Harrison has argued that English Protestants envisaged a Baconian reformation of the sciences that compensated for the diminished power of post-lapsarian senses and intellectual capacities. Avoidance of a priori speculation and adoption of careful methods of observation and experiment would allow some recovery of the knowledge that Adam once commanded. However, as he indicates, this scenario was complicated by the notion of apocalypse espoused by Hartlib, Comenius, and some of their circle. The fast-approaching millennium did not allow sufficient time for the slow progress Bacon envisaged.[23] Yet these Protestant reformers needed to make significant scientific and technical advances before the last days; such an increase in knowledge was necessary as both a sign and precondition of the imminent millennium.

Such advances required feats of memory and reason—faculties supposedly damaged by the Fall. Accordingly, Hartlib did not dwell on the weaknesses of the mind, but rather affirmed that the ability to classify and reduce information made it possible to identify the essential properties of nature available to Adam, and to combine them in new ways:

> For although there was nothing imperfect in Nature before the Curse, yet all the imaginable perfection, which the seminal properties of the Earth contained, were not actually existent in the first instant; the kinds were each distinct themselves, without any defect, but what Marriages and Combinations there might be made between them, and what the effects thereof would be, when the proper Agents and Patients should meet, I suppose was left to his industry to try.[24]

The point made here is that although "we now come farre short of that knowledge" it was possible to make considerable progress by analysis, and recombination via technical intervention, of the fundamental natures of things. Indeed, the implication is that natural phenomena—not produced by Adam in Eden because he did not need to—might now be revealed. Hartlib and his friends acknowledged the limits of human faculties but they asserted that various shortcuts—keys, epitomes, compendia, indexes, shorthand, artificial languages—would accelerate discoveries. Examples of this conviction abound: Beale's confidence that he could teach a new form of shorthand based on universal characters "in one weeke";[25] that an artificial language of the kind proposed by John Wilkins could be quickly learned; and that "the reall Character may be easily taught in few dayes."[26] There is also John Pell's belief that all mathematical knowledge could be condensed and codified "by one man," and Comenius' assurance that a single book capturing the essentials of all knowledge could be composed by his generation.[27]

It is not easy to reconcile such optimism with the prolific note-taking within Hartlib's circle. There is a juxtaposition of two seemingly incompatible assumptions: a call for collections of copious information *and* the conviction that memory could master its essential elements. In one sense, this combination was not unusual, since it underpinned standard Renaissance rhetorical techniques. The *copia* of tropes and quotations had to be counterbalanced by *brevitas*—hence the use of pithy aphorisms and parables that put an idea in a nutshell. However, the Baconian imperative to collect "particulars" threatened to result in *copia* of data without the advantages of *brevitas*. Yet Hartlib did not despair that such a mass of particulars would overwhelm memory. The benefit of ordered information was that it functioned as a prompt for the recollection of related material. Hartlib stressed that judgment and understanding were best served if salient material could be easily located. In an early entry he endorsed the view of the French jurist Jacques Cujas (1520–90) that the ability to find (or recollect) a text or idea when required was more valuable than being able to remember a stock of things stored in memory. Hartlib decided that "a Rational Reminiscentia to remember things apposite when wee would have them is far better than a bare Memoria of confused things at random which the Localists are able to doe. But that is the best Memory that helpes most the judgement."[28] Hartlib's note-taking was part of an attempt to gather, condense, and order information in ways that helped both memory and reason. In this way, one might dare to cheat Hippocrates' dictum: the proper use of notes could shorten the route to many arts and sciences.

Hartlib's papers provide an embarrassingly rich and daunting resource for the historian.[29] After his death in March 1662, Hartlib's surviving letters, diaries, and other manuscripts were purchased by his friend, William Brereton. In 1667 John Worthington, who had been corresponding with Hartlib since at least 1655, found them at Brereton's house in Cheshire.[30] He knew from Hartlib that some papers had already been lost in an "accident of fire" early in 1662.[31] Worthington recounted his discovery: "At my late being in Cheshire I met with two trunks full of Mr. Hartlib's papers, which my Lord Brereton purchased. I thought they had been put in order, but finding it otherwise, I took them out, bestrewed a great chamber with them, [and] put them into order in several bundles."[32] Apart from some letters and other papers that Worthington extracted, the bundles disappeared from Brereton's house and were not recovered until 1933 when the historian of education G. H. Turnbull found them in a London solicitor's office. As he reported, there were sixty-eight bundles, possibly as sorted by Worthington. Turnbull left these to Sheffield University and they have been in its archives since May 1963.[33] In more recent times, other papers have been added and the second edition of the Hartlib Papers on CD-Rom includes seventy-two bundles of material.[34]

By the mid-nineteenth century, some appraisals of Hartlib drew on portions of his papers, which were then in the British Museum, having found their way into the hands of the omnivorous collector Hans Sloane. In 1865 Henry Dircks incorrectly listed all the publications associated with Hartlib as if he was their sole author; he estimated this output as "two duodecimos, two octavos, and about twenty-eight quarto treatises of various bulk and character, but mostly they are short pamphlets or tracts."[35] Nevertheless, Dircks put his finger on the salient point: "He [Hartlib] was in himself a kind of imaginary institution, of which he represented the proprietors, council, and all the officers; the funds too, being wholly his own."[36]

Ephemerides

Hartlib's Ephemerides were among the papers not seen until the early twentieth century. These are notes in manuscript diaries amounting to about 300,000 words. Although Hartlib may have begun the first diary as early as 1631, the extant diaries cover the years 1634–35 and 1639–43, followed by another gap until 1649 before continuing without interruption until 1660.[37] In most cases the year of the diary is marked on the cover or first page. The first notebook among those now remaining is titled "Ephemerides Anni 1634/1635" and covers those two years.[38] The notebooks are made from sheets of paper (roughly A4 size) folded to form quires of between six and eight pages. Each quire has a signature (thus A1–A6 for a quire of six pages), and the quires are gathered in unsewn notebooks.[39] The notebooks consist of sheets (approximately 29.5×20 cm); they are folded into booklet form so that the pages are approximately 20×15 cm. Each page is ruled so as to leave a clear margin of about 3 centimeters on both the left side of the verso and on the right side of the recto. An empty border measuring about 2 centimeters is left at both top and bottom of each page, often slightly larger at the bottom. There are entries on all pages, except the verso of the front page and sometimes the last page. There is no regular internal dating of entries, although some mention a date: for example, "The 14 of January a great thundering and lightning at London."[40]

Hartlib's entries run on immediately from each other, separated only by a horizontal line that does not always extend across the entire page. The margins are reserved for Head words indicating the content of the adjacent entry. These Heads appear next to the first lines of the entries; very occasionally, and only in the case of the right margin, a word from the entry encroaches on the margin. For the early years, quite general Heads such as "Eruditio," "Libri Selecti," "Critica librorum," and "Desiderata" predominate; later, these are joined by more specific Heads such as the names of informants, or subjects and concepts, such as "Physica," "Chymica," or "Experientia." The margins of the early diaries are crammed with Heads, especially when both topic and informant are given, as is usually the case with medical and scientific information.[41] Hartlib's own reflections are sometimes put under "Pensa," although this does not seem to happen until the 1650s.[42]

The Ephemerides has been aptly described as a "diary of information."[43] There are two levels at which that label can be explicated. First, the Ephemerides is a massive collection of information from a range of sources and correspondents on subjects from biblical studies and Protestant unity to chemistry and horticulture. Second, the entries offer a running audit and commentary on the process of gathering, collating, indexing, and communicating information. They contain notes on the making of paper, pens, ink, shorthand systems, and various methods of excerpting, entering, and arranging material. In short, many of the entries are notes on note-taking. They are accompanied by musings on a theory of knowledge that supplies a rationale for this note-taking.

In Hartlib's time the use of diaries as a means of spiritual accounting was well accepted.[44] When adopted within pansophist circles, this practice was meant to encourage intellectual discernment as well as moral awareness. Comenius recommended keeping a "vitae Ephemeris" in which to make notes of good books and authors, important things heard in conversation, and one's own reflections.[45] Hartlib also made notes to himself about what he was trying to collect, often in terms of kinds of knowledge rather than specific topics. Thus in 1634 under the heading "Ephemerides," he thought that such notes needed to contain both experimental observations and various ideas. Some of his reflections are certainly compatible with Comenius' rationale. In 1648 he said that "the great meanes to come insensibly to a universal knowledge and experience is to keepe diaries exactly of all whatever wee heare or see by way of converse out of Books."[46] But Hartlib's diary was also a tool of his employment as an intelligencer; consequently, his collection of diverse information resembles a database as much as a private record of reflection.[47] Furthermore, from an early stage, the entries were not limited to reports or extracts from sources; they also included "desiderata"—lists of things to be found or done: "Desideratum I wish Ainsworth had done all summaries upon the whole Bibel as hee hase done upon the 5. bookes of Moses."[48] Unlike Comenius, Hartlib added scientific information to such lists.

Hartlib exhibited a thirst for tips that might hasten the gathering and collating of notes. Most contemporaries were conscious of the materials required to make and use notebooks, but Hartlib is probably distinctive in collecting so much information about this. His Ephemerides is sprinkled with entries about types of paper, ink, and glue—all with a view to making note-taking easier, faster, and more effective.[49] There are many entries about "table-bookes," with special interest in better ways of treating pages so that they could be wiped clean for reuse: "The neatest best Table-Bookes are those of slaite which hee [Boyle] found at Amsterdam. For they will never weare out. 2. no need of Paper between them. 3. shew very white what's written upon them. 4. not need such tedious wiping out by spunges or cloutes." Under the heading of "New Table booke," he described one that allowed "paper to bee wipt off a hundred times with a sponge and warme water and to write upon it again which wil make a far better new Table-book."[50] It is likely that Hartlib used such table books for rough notes, possibly of conversations, which he later transcribed into his Ephemerides, although he certainly also made entries directly from books and letters. More so than Locke or Hooke, who both recorded instructions for making erasable surfaces, Hartlib also sought ways of copying information for distribution within his circle:

There is a kind of black Paper under which might bee laid white then a black again et [and] then a white etc. and writing hard upon the first the impression will bee seene legible upon all the rest. This may bee made more practicable for ordinary use to compendiate the multiplication of Copies as also in Printing.[51]

Of course, faster ways of writing were desirable, so Hartlib canvassed all the latest methods of shorthand: "Willis Stenography is more fundamental et formal et easier applicable to other Languages but Sheltons is more speedy et expedit." Hartlib added (apparently without irony) that another person had improved Shelton's system "but hase no time to set it downe." He later observed that Shelton's was the "least burthening of memory" and that "Mr. Boyle intends to make himself perfect in it."[52] Petty's double writing machine supplied another option: "an Instrument of small Bulke and price, easily made, and very durable, whereby any Man . . . may write two resembling Copies of the same thing at once."[53]

The Ephemerides must be considered in the light of what Harold Love called "scribal publication."[54] In the seventeenth century, the printing of books and pamphlets coexisted with the use of manuscript as a medium for the communication of news and ideas. Love mainly considered the circulation of scribal poems, musical scores, and political opinions. Hartlib's activities are a prime instance of the ways in which medical, technical, and scientific information could be kept in manuscript form, either to be withheld from, or exchanged within, a small community. The papers he collected can be viewed, in Love's words, as "scribally withheld knowledge," insofar as the swapping of these manuscripts helped to consolidate a self-defining community, distinct from the kind formed by the readership of printed publications.[55] Hartlib's note-taking shows not only how he gathered and communicated information, but how he used the various media available to him. He collected manuscripts from a range of authors, made notes of *oral* information that might otherwise be lost, and extracted facts and ideas from *printed* sources, turning all of this into *scribal* form in his notebooks. From there the cycle continued, as he passed on such knowledge and information to others, sometimes after adding cognate material. This process enabled the juxtaposition and collation of diverse sources.

Hartlib's notes also constitute a rich storehouse of information in manuscript that fits the parameters of what Don Swanson called "undiscovered public knowledge."[56] This term describes the scenario in which separate pieces of data, or ideas, existing in books or journals might, if brought together, constitute a new empirical finding or a conceptual advance. As long as these necessary components remain isolated and unknown to "any one person," we can speak of "undiscovered," yet "public," knowledge. Hartlib occupied the center of a correspondence network that allowed him to combine scribal and printed information that might, in principle, yield as-yet-unknown relationships. He certainly increased the possibility of making such connections by stockpiling information not yet captured in either scribal or printed documents. For example, he recognized that craft skills passed on by "tradition" could be lost:

> Trades are not recorded to any purpose by any body, much lesse those mysteries and slights [sleights] of particular tradesmen which have beene hitherto preserved only by Tradition. If such a disastre should befal them . . . the world should bee bereaved at once of such great treasures irrecoverably. Thus all Inventions should bee recorded though they grow out of use. Hence wee lament justly the negligence of former Ages in not setting downe all the particular meanes which they used in Navigation before they knew the compasse.[57]

Hartlib was keen to acquire the manuscripts, both notes and correspondence, of great scholars. From 1634 this aim featured in his statements of policy and intention. In 1639 he made a catalogue of Bacon's extant manuscripts, listing twenty-five items, with the intention of obtaining some of them.[58] The aim was to give safe custody to papers that might be lost, but it was also based on the belief that these often held novel information or ideas not contained in books. In 1634 Hartlib jotted down a remark from the nonconformist minister Thomas Goodwin that there were "Few good English bookes. the best things are kept in Mens studies, in MS," and, moreover, there were "very few good bookes in the world," perhaps up to 400. One practical inference was that "Travellers if they bee given to writing out of things should spend this labour in transcribing of Library MS. of which there might bee a more permanent use then of their Loci Communes and other reading-collections."[59]

The *notes* of scholars were especially prized because these came with value added—namely, the labor of epitomizing and commonplacing done in the course of the person's own reading. Hartlib recognized, however, that notes of this kind were inseparable from their owners who relied on them as intellectual tools: "Mr Goodin uses to write all his Sermons yet hee does not speake as hee writes them. Wil not part with his Notes."[60] However, a dead scholar was a source of notes, and Hartlib was always on the lookout: he was not bothered by Bacon's caveat that the main value of notes accrued only to their makers. In 1634 he was happy that a clergyman had left "a Catalogue of Preaching heads," and that another person had "left his Commonplaces."[61] In the same year he was excited to hear about "a perfect Copie of Richersons [Alexander Richardson] Notes."[62] A similar interest is evident among his younger friends: thus Hartlib welcomed Boyle's request "to looke especially after the Adversaria of Learned Men as of greatest price and by which you may know freely what they are. The Adversaria of Borell Iustinus van Ashen and Moriaen hee counted above gold."[63] With characteristic enthusiasm, Aubrey reported that some of the papers of the physician Jonathan Goddard (1617–75), who had died intestate, were "in the hands" of Sir John Bankes; they included "a kind of Pharmacopoeia" and it was "possible his rare universall medicines aforesayd might be retrived amongst his papers." In contrast, Aubrey remembered William Harvey's devastation at the loss of his "anatomicall observations" for "a booke *De insectis*" when his lodgings at Whitehall were "plundered."[64] Once the papers of an individual were obtained, they could be collated with others, not only to make such papers easier to interpret but also to identify overlaps with information from various people and sources.

Collating and Finding

What did Hartlib take note of? As a sufferer of both gout and bladder stones, he was naturally eager to seize upon medical cures and recipes for these ailments, but his notes betray a gulf between this continual trawling for remedies and any efficacious results. There is a sadness in Hartlib's hopeful note of 1657 that "Mr Boyle promised to impart a most stupendious way of curing the stone," as well as in many other entries about possible cures, such as eating twenty raisins twice a day.[65] We can almost feel his pain in a letter a few months before his death: "the stone is like a bull enraged, that will not fall with one blow."[66] It must be said, however, that Hartlib's interest in collecting such material preceded his own health crisis (from about 1656).[67] As early as 1635 he noted that "Cuffler knowes an excellent Medecin of the Gout to draw out the paine of it presently by making of blisters with bagges of salt." Several pages later he scribbled with increasing expectation: "Alberti hase already a most excellent Receipt for the Gout which hee got from the same Gentleman which is to communicate the other unto him." Between January and June 1648, Hartlib heard about "Helmont's stone wherby hee cured the stone in bladder or kidney."[68]

For these topics, and as a general principle, Hartlib was eager not only to collect information, but also to collate it. Hence his excitement in 1650 about the possibility of a medical compendium: "Mr Haack desideratum of which hee hath written more largely to Mr Morian, is that there should bee made once or compiled a Booke of Medicinal Cookery (Ein Medicinisch koch-buch) for common use."[69] In 1661 Worthington hoped that Hartlib was well enough "to peruse those many bundles of papers your study is furnished with, so out of them you might extract such select passages as would make a Silva Silvarum, or a Collection of Memorable Things." Hartlib did not need Worthington to tell him anything about the desirability of such collections, or about arranging the material "under several general heads."[70] Some of his earliest notes concern the importance of reducing knowledge and infor-mation to manageable proportions, either to ease memory or to assist re-trieval. Admittedly, such remarks often indicate what he aspired to rather than what he actually did—as Worthington's request possibly implies—but they do show that Hartlib recognized the tension between amassing infor-mation that might otherwise be lost or not fully exploited, and working on it to refine some core facts and notions.

As seen in chapter 2, the Jesuit pedagogues had allowed advanced stu-dents to choose their own method of entering and indexing, provided that they kept taking notes. Hartlib's Ephemerides is an example of how one as-siduous note-taker did just that, combining the chronological order of the diary form (albeit without regular dating of entries) and topical Heads that might be indexed. Hartlib was also able to canvass a range of more recent adaptations from Protestant theologians in Germany and the Netherlands.[71] Through Comenius, Hartlib may have heard about the concerted group note-taking efforts that Bartholomaeus Keckerman (1573–1609) encouraged in his students; these were continued by Johann Heinrich Alsted, Comenius' teacher.[72] Obviously, Hartlib did not possess similar pedagogic control over his many correspondents, but he did advocate the utility of commonplacing in making collations of information and knowledge that might be of public use. His aim was to extract useful knowledge from all extant books, both an-cient and modern, by condensing and gathering material under convenient Heads. One benefit of such collation was that points of agreement, contra-diction, and duplication might emerge. As we shall see, this was a step in the process of managing information, indexing it for ease of retrieval, and com-municating it to others.

Finding a starting point was crucial: which books and other sources should be checked? In Hartlib's early diaries there are numerous notices about "Libri Selecti," or often more specifically, say, "Libri selecti Theolog-ici."[73] He also wanted catalogs from which these books could be chosen or, better still, selections already made by others. He knew that the Protestant preacher Peter Streithagen (1591–1653) "Hase made Loci Communes or Refer-ences of English Authors above 40. Quires of Paper close written and goes on daily."[74] Some years later he made a memorandum for himself:

> Acquaint Mr. Streithagen with this purpose of mine and let him know that I
> expect from him a great furtherance of this worke in one or 2 things which
> will bee of singular good use. The first is that I may have a Catalogue of the
> English Bookes which hee doth know and hath made use of in compiling his
> High-dutch common-places.[75]

He wished others would survey and describe information, including mate-rial collections, at their disposal. Thus in 1634 he noted with interest the "Cat-alogue of Herbes" in John Tradescant's garden, and he later agreed with Wil-liam Rand that "Tradusken should bee induced to make an exact catalogue of all his Museum."[76] He was quick to identify others who were engaged in collation or abbreviation of specific material in any subject of interest to his circle. For Hartlib, proverbs were already distillations of larger bodies of opinion "in reference to common life," so collections of these were useful.[77] More generally, he heard that John Milton was "not only writing a Universal History of England but also an Epitome of all Purcha's [sic] volumes."[78]

Once material was collated it had to be stored for retrieval on later oc-casions. The commonplace method led directly to indexing. Our modern "index" is an abbreviation of the earlier "index locorum communium" (index of commonplaces) that displayed the main Heads treated in a book, usually giving chapter numbers as a way of locating particular passages. Hartlib's use of marginal heads in the earliest of his Ephemerides allowed for index-ing, and there is some evidence that he was able to return to entries to ex-tract material.[79] It is likely that he did so with the help of a separate general index incorporating these marginal keywords. Indeed, in 1635 he recorded approvingly Thomas Goodwin's account of how this might work: "The things wee Meditate are to bee written by way of Miscellanies the Head in the Mar-gent and after an Index for those heads."[80] But Goodwin's advice on indexing was coupled with an injunction to focus on a select set of books and the core notions these contained. In recounting this view, Hartlib said that "Selected bookes are to bee often read over"; with reference to theological subjects he ventured that "the choisest things should bee mad[e] or compiled out of [the

Church] Fathers."[81] Such a reading strategy made it more feasible to recall the limited set of chosen Heads and the material they contained, but it also predetermined responses to new material, absorbing it within a previously agreed framework.

Hartlib canvassed other opinions on how "all manner of Learned Men" collated and indexed their notes. However, it seems that much of this advice endorsed the reduction of material to a small set of stable Heads. Thus he noted that the humanist scholar Justus Lipsius composed his notebooks by "making himself exceeding perfect in some Authors and then referring all whatsoever hee read unto them by adding or omitting."[82] Similarly, the young German scholar Joachim Hübner suggested that Comenius' "Metaphysica" might "serve for fuller Common-places and more comprehensive of Heads to which all our Readings and Meditations may bee far better reduced then to any other systeme whatsoever."[83] There is an indication that Hartlib himself sought to establish a standard set of Heads in certain subjects, starting with those used by exemplary readers. In the case of "Practical Divinity," he recorded Dury's wish that "Mr Streithagen" be asked to "impart unto us the list of the Heads of his Common-places either Alphabetically or in any other Order, which hee thinks fit, with a quotation of the Places of the Authors who handle those matters," and further, that he "get any body to write out . . . the List of the Heads with the quotation of Authors and Places."[84]

The "Conferences" sponsored by Theophraste Renaudot's *Bureau d'Adresse* might be interpreted as offering a different option: namely, the idea of a pragmatic set of temporary Heads. Following Bacon's *Sylva Sylvarum*, these conferences comprised a century (100) of topics, starting with "method," "principles," and the basic elements of fire, air, water, and earth, before becoming more miscellaneous.[85] In 1640 Hartlib made a list of their advantages, including "for Ars Colligendi or extracting of Notions" and "for *Loci Communes*, common-places."[86] But some of his closest colleagues did not regard these looser topics as a license for open-ended inquiry. In 1641 Hartlib entered Hübner's comment that:

> The French Conferences give the Breviat and substance of all common
> Learning as much as is had. By which may appeare the smal quantity of
> that knowledge which wee seeme to have. It may bee it will serve for fuller
> Common-places and more comprehensive of Heads to which all our Read-
> ings and Meditations may bee far better reduced then to any other systeme
> whatsoever.[87]

The implication here is that a relatively small number of "comprehensive" Heads might manage the content of this collaborative exercise. In a rather rambling paper of 1649 about the Office of Address, Dury was even more prescriptive, insisting on a fixed number of Heads, perhaps drawn from Bacon's *Advancement* (1605) or from Alsted's *Encyclopaedia* (1630):

> To these Heads then wee suppose that all Thinges; that can bee Knowen or conceived may bee reduced, whether they bee Reall or Notionall Objects of Knowledge. For our Aime is only to set downe the Generall Heads of Objects, as they are most comprehensive of Particulars in all Respects: For all the Reall Objects and whatsoever may bee Knowen of any of them, may come under some or all of these Frames of Notions, so that whatsoever any man shall have discovered in the way wherin hee doth excell; and shall bee willing to impart unto others, may bee brought home to one of these Heads.[88]

Dury's edict was meant to apply all knowledge, but would this use of pre-established commonplace Heads work in the case of empirical and experimental knowledge of the natural world? And did Hartlib hear from anyone who said that it would not?

In the case of scientific information, chemistry was among the first of the sciences to come within Hartlib's orbit.[89] As early as January 1634 he listed "Experimentales Observationes" among the kind of information he was seeking but, as with other subjects, he saw no problem in starting with books or conversation. In 1639 he recorded what he had heard about Gabriel Plattes who was in the process of reading books that might contain chemical knowledge; as Hartlib phrased it, Plattes' dictum was "Labor first to collect the true Experiments out of all Bookes. Then let your best wits philosophat upon them."[90] Hartlib's interest was reinforced by Cheney Culpeper, a lawyer and gentleman, whom he met, according to his diary, on April 13, 1641.[91] Culpeper wrote to Hartlib in appreciation of Plattes' work, requesting that it be properly arranged, and hoping that "none of his thoughts (not yet published) should be loste, & therefore if you withdraw him from his other imploimentes to a reCollection & orderinge of them, you will (I conceive) muche advance the publicke."[92] Hartlib gave Plattes a posthumous chance to explain what he had being doing by including Plattes' essay, "A Caveat for Alchymists," in a collection of chemical and medical essays of 1655. In this, Plattes declared that at "any mans Judgement ought to be grounded by a Concordance of the best books, before he fall to practice," and that no one should attempt this "Art" of chemistry "by his own speculation and practice, without the help of books."[93]

After this acquaintance with Plattes' approach to chemistry, Hartlib heard about the more concerted effort to commonplace a large range of books on natural philosophy. This was the quest of the club established in 1648 by John Wilkins, the Warden of Wadham College, Oxford. In 1652 his colleague, Seth Ward, an astronomer and theologian, explained that "our Clubb which consists of about 30 persons" had

> gone all over all or most of the heads of naturall philosophy & mixt mathematics collecting onely an history of the phenomena out of such authors as we have in our library . . . our first business is to gather together such things as are already discovered and to make a booke with a general index of them.[94]

Hartlib followed the work of this group, and a note he made in 1651 suggests that he was in communication with some of the members: "The Club-men have cantonized or are cantonizing their whole Academia to taske mee to several Imploiments and amongst other's to make Medulla's of all Authors in reference to Experimental Learning."[95] This approach was consonant with that of some of his other correspondents, such as William Rand, who said that "a man should reduce [Joan Baptista van] Helmont into commonplaces, collecting all homogeneal passages upon one & the same subject under one title through the whole booke."[96]

Hartlib therefore encountered two kinds of advice about collecting and collating information. Hübner and Dury pushed for a tightly limited number of Heads under which all new material could be entered. Plattes and the Oxford club surveyed books, also collating information under Heads that made sense to them, but possibly with the option of adding new categories as they proceeded. Hartlib's sympathies seemed to lie with the second approach, although in either case he realized from his own diary-keeping that sub-Heads were necessary to aid both the collection and retrieval of information. Even with respect to a single major author, Hartlib recognized the need for a good search tool:

> It's great pitty that no accurat Index is made yet upon Seneca Opera. They containe a world of Historical particulars concerning Nero, Caligula, etc. who are mentioned by him in several places and always with some new

circumstance or other. But except wee have a true Index Rerum of all how shall wee finde them?[97]

Once information on a wider range of subjects was being collected, the need for a larger (and less memorizable) number of Heads became almost unavoidable.

Harrisonian Indexes

Given these challenges and requirements we can appreciate Hartlib's excitement when a new invention for indexing on a grand scale fell into his lap. During the spring of 1640 his diary is laden with entries singing the praises of "Harrisons booke-Invention." He described this as "nothing else but an excellent and the compleatest Art that ever yet hase beene devised of a commodious and perfect art or slight [sleight] of excerpendi."[98] It promised, he believed, "to give a perfect Index upon all Authors or a most Real and judicious Catalogue Materiarum out of all Authors to represent totum Apparatum Eruditionis which is extant in what Bookes soever." This first entry concerning Harrison's new device runs to some thirty-two lines, the longest for that year (see fig. 4.1). Over the next few years he often returned to this invention, considering its application to the problem of condensing and indexing knowledge and information. Although he viewed it as useful "for every mans study and his library," its great potential application was in collaborative endeavors: "One perfection of it is that it can never bee perfect. For it is Opus Generis Humani rather than one mans which must bee perfected by every Nation."[99]

Who was the inventor of this marvelous thing, and what did it do? Some years ago, Noel Malcolm rescued Thomas Harrison (1595–1649), an Oxford graduate and schoolteacher, from near oblivion (and misidentification) to show that he was the author of a Latin manuscript now deposited in the British Library.[100] Harrison proposed a specially designed cabinet for an "Arca studiorum" (The ark of studies) that housed notes and extracts on slips or "slices" of paper (see fig. 4.2). These loose pieces of paper were to be placed on hooks fixed to the interior walls of the cabinet. Each hook could be reserved for a letter of the alphabet and the slips could be placed according to their topical headings.[101] This innovation solved the problem of guessing how much space to leave for entries in notebooks with pages allocated to previously selected Heads. Harrison's notes were separate entities that could be moved from one category (or hook) to another. Notes could be lent out to

Figure 4.1. Samuel Hartlib's *Ephemerides* of 1640 with entries concerning Thomas Harrison's indexes. HP30/4/47A. Kindly supplied with permission by the University of Sheffield Library.

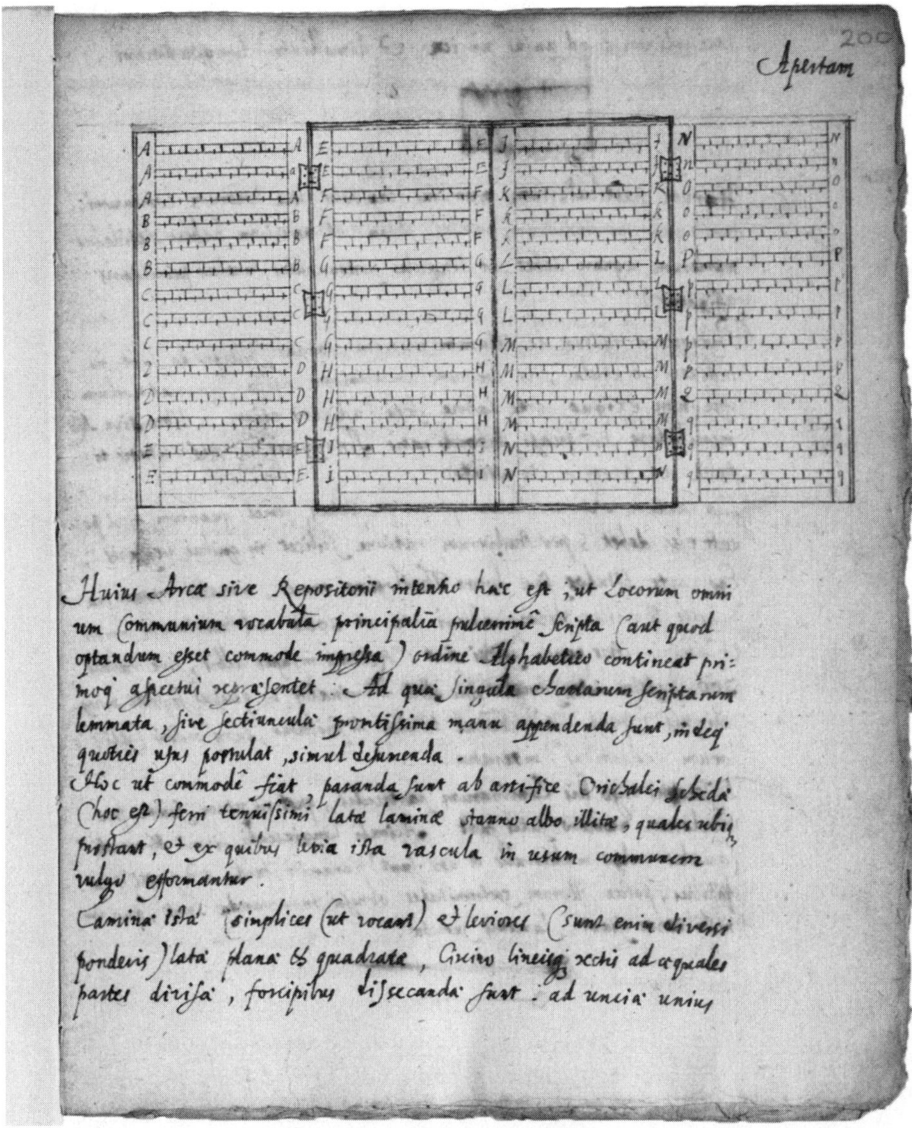
200
Apertam

Huius Arcae sive Repositorii intentio haec est; ut Locorum omni
um Communium vocabula principalia pulcerrimè scripta (aut quod
optandum esset commode impressa) ordine Alphabetico contineat pri:
moq aspectui repraesentet. Ad quae singula Chartarum scriptarum
lemmata, sive sectiunculae promptissima manu appendenda sunt, indeq
quoties usus postulat, simul desumenda.
Hoc ut commodè fiat, parenda sunt ab artifice Orichalci schedae
(hoc est) ferri tenuissimi latae laminae stanno albo illitae, quales ubiq
prostant, et ex quibus levia ista vascula in usum communem
vulgo efformantur.
Laminae istae (simplices ut vocant) et leviores (sunt enim diversi
ponderis) latae planae et quadratae, circino lineisq rectis ad aequales
partes divisae, forcipibus dissecandae sunt: ad unciae unius

Figure 4.2. Thomas Harrison's "Arca Studiorum sive Repositorium," in its open position. BL, Add. MS 41846, fol. 200r. By permission of the British Library.

friends and additions to the collection could be solicited. Hartlib heard of Harrison via Hübner and, together with Theodore Haak and John Pell, met the inventor in London.[102] Hartlib circulated the news to his inner circle, but the wider scholarly world did not know of this invention until the Hamburg jurist and professor of rhetoric Vincent Placcius printed the still-anonymous manuscript in his *De arte excerpendi* (1689).[103] Placcius added his own suggestions for what he called a "*scrinium literatum*," or literary chest, and announced it as a technique suitable for an advanced method of note-taking that would orchestrate the collective notes of scholars in the Republic of Letters (see fig. 4.3).[104] He was wary of the loose, unindexed, slips of paper used by his teacher, Joachim Jungius, and he suggested that entire pamphlets could be hung on the hooks (see fig. 4.4).[105]

Hartlib was enthusiastic about the potential uses of Harrison's method, but his response was calibrated by recourse to his own storehouse of information on the ways of making and keeping notes. There is no doubt that he saw the use of loose slips, freed from notebooks, as the masterstroke: "The ground of it is a passe-port with as much paper upon it as you please. Upon it there bee slices of paper put on which can bee removed and transposed as one pleases which caries a world of conveniences in it." In listing the "perfections" of this invention he cited "Mobility to transpose your notions" in a manner that allowed finding them again "without distraction in Meditation"—possibly an admission that the labor of recollection might not bring reliable results. Even so, he wondered what other methods Harrison might have trialed before settling on the version he now presented: "hee should set downe all the ways which hee hase tried" in case they offered further ideas.[106]

Hartlib was aware that others had speculated about ways of detaching notes from notebooks and rearranging them for compositions. For example, "[Matthias] Bernegger lays and spreads so many Authors all along or upon a long kind of board et repositorium and no doubt Thuanus had a special slight [sleight] in the writing of so vast a History, whereby hee could expedit himself and save abundance of time."[107] Some had tried to skip the note-taking stage by cutting extracts directly from books and reassembling them as desired: "[Theodor] Zwinger made his excerpta by using of old bookes and tearing whole leaves out of them, otherwise it had beene impossible to have written so much if every thing should have beene written or copied out."[108] Hartlib mentioned Pell's observation that Conrad Gessner's *Pandectae* "describes the Manner of making Indexes somewhat like to that of Harrison."[109] He later recorded that someone "hath invented a Mechanical Way for

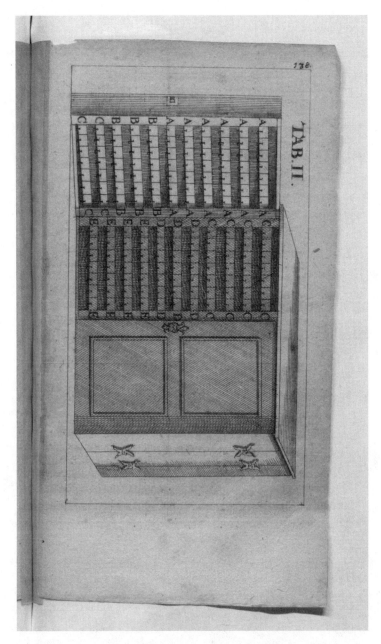

Figure 4.3. Vincent Placcius' design of a *scrinium literatum* based on Harrison's "Arca Studiorum." Vincent Placcius, *De arte excerpendi* (1689), 138. Shelfmark: 1089h.15. By permission of the British Library.

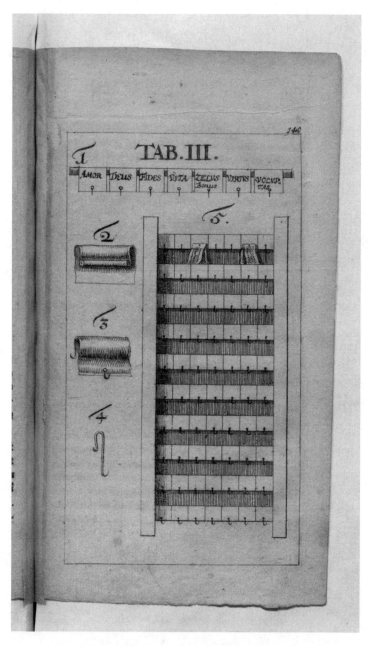

Figure 4.4. Vincent Placcius' depiction of subject labels and hooks based on Harrison's "Arca Studiorum." Vincent Placcius, *De arte excerpendi* (1689), 140. Shelfmark: 1089h.15. By permission of the British Library.

common-places with removeable Notes as Dr Gawden. The Manner of which hee is to describe in writing." But he had already concluded that Harrison had gone "far beyond that of Gawden."[110]

Whatever the precursors or contemporary rivals for the innovation of movable notes, it was the *indexing* capabilities of Harrison's method that appeared as a breakthrough.[111] This is why Hartlib referred to "Harrisons booke-Invention" and to the "Harrisonian Indexes."[112] The physical design of the "Arca" promised easy retrieval of notes or extracts from the hooks on thin brass plates labeled with Heads.[113] Hartlib's use of the term "Mechanical Common-places" suggests that he viewed Harrison's endeavors as a technical supplement to the well-established practices of excerpting and commonplacing. Hence he often discussed it under the Heads of "Ars Eclogandi," "Ars Excerpendi," and "Eruditio."[114] The examples of Heads that Harrison gave in his proposal were quite standard ones, such as "Amor dei, Deus, Fides, Vita aeterna," although when speaking more generally he also mentioned natural and civil subjects. Nevertheless, both Harrison and his admirers among Hartlib's circle were thinking on a scale that went beyond the parameters of humanist commonplacing for basic pedagogy and rhetorical training. The technology of loose slips, seen at its most advanced stage in Harrison's invention, facilitated a different way of thinking about the units of information. For a start, the scale envisaged is indicated by Harrison's remark that even for his own purposes he had already allowed for a minimum of 3,000 Heads, catering for future additional ones with a quota of 300 blank plates.[115]

Having arrived in London in the autumn of 1641 (at Hartlib's invitation), Comenius was told about "a new and wonderful invention" and wanted to "meet Harisson and learn fuller details from him about his project."[116] He reported to friends in Leszno that Harrison had begun the groundwork:

> I hear (but this is only hearsay evidence) that he already has a list of some 60,000 authors whose works he proposes to index. My friends here think that a considerable number of students from the two Universities will be deputed to assist Harisson. Under his supervision they might index the authors assigned to them.[117]

There is no doubt that Comenius' eagerness reflected the potential synergy between Harrison's proposal and his own plans for "compiling a pansophic work"; indeed, he said that this point was being considered by the Parliamentary commissioners presently looking into the funding of this invention. Comenius was enthralled by the prospect of distributing "into one Index all the authors of merit in any given language" because it would assist his own encyclopedic project, already sketched in manuscripts and elementary textbooks, such as *Janua linguarum reserta* (1631).[118] He aimed to produce a complete compendium of knowledge, including knowledge of nature, which would take the form of a "Pansophicall Booke." Thus he announced that "a Booke should be compiled, for the containing all things which are necessary to be knowne and done, believed and hoped for by man, in respect of this and the life to come."[119]

Harrison's invention appealed as a likely aid in such a massive task, but, in fact, what it offered was in principle more radical than Comenius implied. It promised not only a way of composing a total book but a machine for dis-aggregating all books. As Hartlib summarized it: "Hee hase a way to cut of all superfluitys and impertinencys from the Propositions. The maine drift is first to make a particular Index of words and Propositions upon every Author, then out of all these particular Indexes to make a general one of unrepeated things."[120] As Noel Malcolm has observed, Harrison's conceptual rationale for his "Arca" is less transparent than his description of its physical components.[121] The basic units of knowledge to be collected and distributed in the cabinet were variously named as propositions, axioms, and notions — the salient point being that they did not duplicate each other. As Hartlib put it, "Hee knows an exact way that no Axiom, etc, shall bee twice repeated."[122] On this point, it is telling that Hartlib linked several notes about Harrison with others on mathematics and logic, sometimes on the same page of his diary, thus giving him a place in the history of logic:

> The use of Propositions hase never beene taught truly in Logick. Only
> to make a syllogism of it which is a poore thing. The maine is that our
> concepts must bee resolved into them. Aristotle, Zabarella written well of
> them but did not observe it themselves as Harrison very well notes.[123]

Harrison offered a reduction of knowledge to simple elements; the less explicit, but implied, corollary was that as well as producing a compendium of existing knowledge, new knowledge could be discovered by recombining these units in new ways.

The provision of multiple indexes was the enabling device. In glossing Harrison's intentions, Hartlib identified three kinds of index: for words, for *sententiae* or adages, and for things (*realia* of various kinds).[124] He added that it was useful to distinguish between different kinds of knowledge encountered in books, mentioning philological, historical, and scientific.[125] Hartlib understood that "when the Indexes bee compleat the Propositions or Notions may bee collected out of them," adding that Harrison "hase written a special Logicke for the whole Art of Collecting. Hee is a most Reall man." Somewhat later he concluded that "Harrison intends for every booke an Index of words . . . and an Index of Propositions so contracted as hee hase invented. Then of all those Indexes one universal Index."[126] Potentially, this index could divorce ideas from both authors and book titles, although Harrison did make provision for the citation of sources on the slips.[127]

Harrison's main concern was not what an author said, but the indexing of core, or simple, notions. Comenius was pleased that Harrison made it feasible "readily to ascertain the opinions of divers authors on any specific point of interest," but a more powerful indexing could separate authors from content.[128] The extent to which this clashed with traditional humanist or Jesuit commonplacing is evident in the way Hartlib admitted that it removed the need to read a book. As he put it, Harrison's "Way seemes in it's perfection . . . to give the Notions out of all Authors so that every body may owne them by right tho hee goe not to the Authors." Hartlib conceded that some readers might wish to "receave satisfaction from the Authors themselves." Several entries later he put this point more strongly: "Although by the bare indexes a great use will bee made by every body yet it will bee done far more excellently and substantially by those that have read the Authors themselves. The other will oftentimes bee very raw and incipid."[129] We can ponder what might have happened if Harrison had been let loose in a major library. Indeed, he was nominated in connection with a plan for a public university library in London: "to make Harrison the Librarie keeper to make Index upon it and to correspond about it with all other Librarie Keepers."[130] It is likely that such an index, having disemboweled all the books, would not need to indicate their position on the shelves.

How were these Harrisonian indexes to be made? Attempts to answer this are hampered by the fact that the basic units were not clearly defined. In Hartlib's view, Harrison regarded the selection of simple propositions from books as menial: "The making of Propositions hee counts but a drudgery worke." This is surprising because the task of crystallizing such fundamental units required educated judgment. Comenius accepted a division of labor when he mentioned the help of students, but he also said that this would be under Harrison's "supervision." Certainly, Harrison did identify a higher-level task, understood by Hartlib as "Contemplation" aimed at "conforming of many places together which is the highest act of judgment to make discourses etc."[131] This process is not specified in Harrison's original proposal, but some of his remarks in papers associated with his petition to Parliament are relevant. For example, what are we to make of his claim in 1648 that he had now made "neere 100000 observations" using "500 Sheetes"? [132] The number is presumably too large for a supply of ready-made Heads (previously given as 3,300). It is not clear whether he was referring to identifications of basic propositions or to the initial quotations from books.

It is pertinent to ask whether Harrison's units were systematically arranged. Hartlib and his associates believed that orderly reduction and summary of knowledge would show up duplication and repetition, and also reveal gaps to be filled.[133] This conviction is clear not only in Comenius' desire for a comprehensive book or encyclopedia, but also in his insistence that it should have "an exact Index of such things we are ignorant of, whether they be those whose knowledge is altogether unattainable, or those that are left for further search."[134] It is not clear whether Harrison envisaged a classification that ordered his fundamental units in some systematic fashion. There are some clues in his memorandum of 1647 to Parliament that advertised the ability of his invention to accept and arrange new information(s) "as they come into their proper sorts and order, without the lease confusion."[135] This suggests some pre-established categories. The Parliamentary Committee mentioned "those Tables invented by Mr Harrison."[136] The word "tables" might also imply a schema of some sort, as in Beale's reference to "a Mnemonicall Table to appoint a distinct place proper and peculiar to every distinct Character."[137] But Harrison's "tables" might just as easily mean a "register" or the indexing apparatus.[138] We do know, however, that Harrison was annoyed that the Committee emphasized the general utility of his invention, its ability to gather and find ("*inveniendi*"), rather than its capacity to arrange material. He therefore reiterated that his "artifice" promised "not only inveniendi but also aptissime disponendi," thus invoking the rhetorical term "*disposito*," or orderly arrangement.[139] Hartlib also mentioned "Modes of Excerpendi disponendi etc" and thought that Harrison "mainly hase studied order and this hee calls also Contemplation to order exactly every thing etc." Although it is difficult to discern what this "order" referred to, Hartlib had some reason to think that one was intended.[140]

Outside Hartlib's circle, the most likely attraction of Harrison's invention was its ability to retrieve documents rather than any capacity to distill basic notions. Harrison himself stressed the former when pleading for Parliamentary support, saying that he understood "that it is the pleasure of the House" that its "Passages and Transactions . . . should bee embodied into one and fitted with a convenient Repertory for a ready Representation of whatsoever in such a masse of occurrences may be particularly sought for."[141] Culpeper told Hartlib that his Office of Address would not function well "without such an index" as offered by "Mr Harrisons invention."[142] However, Hartlib was preoccupied with the philosophical assumptions of Harrison's work, in part because these resonated with his own concerns about how information should be gathered and refined as the basis of sound knowledge.

Historia, Scientia, *and* Memoria

Hartlib was consciously eclectic, aiming to combine the best insights from a range of authors, mainly English and German. In 1639 he reflected that "it is not good to enslave ones-selfe to any kind of Method or Meditations but to observe a certaine generosity and liberty in all our Studys. This will bee found to bee far more profitable."[143] While Bacon was clearly one of his favorites, this did not prevent Hartlib from expressing preference at various times for, say, Giacomo Aconzio (or Acontius), Joachim Jungius, or Joachim Hübner.[144] Over many years his diary reveals an assessment exercise framed by a consideration of the proper balance across the concepts of *historia*, *scientia*, and *memoria*. Hartlib's benchmark was the conviction that secure and stable knowledge (*scientia*) had to be grounded in the material gathered in natural histories, stored either in memory or, more reliably, in writing for analysis and judgment by various people over time.

We can infer Hartlib's intellectual disposition from his notes and comments on the major thinkers of his day. His inclinations were apparent quite early. In a letter of September 13, 1630, he told Dury that he was sending something from "my by-collections," so named because he happily rejected "ordered Systemes." Hartlib favored Bacon's "Aphorismes as the onliest way for deliverie of Knowledge . . . For discourse of Illustration must bee cut of." He suspected "the shew of a Totall . . . Whereas the Aphorismes representing a knowledge broken doe invite men to inquire further."[145] Five years later he declared that "the only way to write for the encreasing of Learning is to write Truths by way of Aphorismes. Systematical Method is like a bag or sack which come bound up."[146] Like Bacon, Hartlib viewed such aphorisms not as pithy compressions of authoritative notions, but as spurs to further investigation and collection of particulars.[147]

In speaking of "particulars," Hartlib and his correspondents pressed the need for what we would call empirical information, gained from observations and experiments, but also, as I have stressed, from books, testimony, and conversations. He used the term "particulars" when rejecting excessively deductive methods, professing instead that "the more new particulars wee meet withal the more knowledge will bee enlarged."[148] In 1640 he noted that

> Those abstract Axioms which are made in Philosophy breed but a slavish assent in men. For they must beleeve only that they are true not knowing out of what particulars they come to bee so. . . . It is a very hard matter to come once to know throughly that a thing is certain and true. Therfore wee must labour principally in the Historical part of all things.[149]

This assertion was pitched against both the scholastic pedagogy of the universities and, less aggressively, against modern philosophical systems, including that of René Descartes.[150] Even Bacon, on Hartlib's appraisal, had underestimated the magnitude of effort and time dictated by his program.

The question was how to gather, order, and store information, especially when much of it was at first necessarily disconnected. Although Hartlib sponsored Comenius' visit to England, he had already noticed that the Czech reformer "play[ed] so much upon . . . [the point that] all things are reducible to certaine maine heads and principles from whence all other particulars can bee deduced."[151] In contrast, Hartlib responded that the collection of "a great Copia Rerum" must take precedence over the quest for an "Ars Universalis."[152] Referring to the ideal of "Pansophia," he cautioned that first "wee must labor to get more particularia. Else wee shall repeate the selfesame Notions by several expressions and advance not a whit further. Wheras when wee must write of particulars wee must always bring some new matter or other." Later that year he regretted "a great fault in Comenius and others to strive too much for compendiousnes and brevity wheras some things must of necessity bee handled at large."[153] The endemic problem was that

> Every body will flie presently to abstractions and generals but they are loath to meddle with the gathering of the Experimental History because it is more troublesome then the other, [and] because a man cannot here compendiat as in the other but must bee as large as the things require themselves.[154]

Hartlib stressed the importance of "the Historical Experimental part of Learning" as the secure basis for inferences and theories in "all Arts and sciences." The other advantage of this empirical foundation, provided it was properly secured, was that even if "deductions of [from] them should perish yet the true ground always remaining all things could easily bee regained."[155]

Harrison promised much, but failed Hartlib's test. One appeal of his invention was that it promised a way of storing and sorting information; yet soon after welcoming the prospect of making an index of "unrepreated things," Hartlib expressed doubts about its application to all subjects. Although he was sure that Harrison's was "absolutely the best Method in philosophical matters," he doubted that it could "succeed in Historical Collections where all circumstances must bee as well noted as the bare substance is much doubted." He noted Hübner's comment that, unlike Bacon or Edward, first Baron Herbert of Cherbury, "Harrison cares for neither [external or inner things] but for Bookes." This concentration on books as the prime source of information to be indexed would not pick up the technical inventions, dependent on oral transmission, to which Hartlib referred in the midst of the earliest entries on Harrison.[156] About a year after making his first breathless note about Harrison's invention, he admitted that:

> The nature of making Propositions may bee divers. Harrisons way is too abstract. Item collections must bee made of Testimonies and for eloquence which must bee taken out verbatim by whole Paragraphes to which Harrison is not so subservient. Item it must bee shewen that the benefit of the Author themselves remaines entire.[157]

The great expectations surrounding Harrison's invention were thus complicated by unresolved disputes about the best way of collecting and condensing information. The key decision was whether to reduce material radically for ease of memory and manipulation, or to store and index multifarious particulars for later inspection and analysis. In two significant contributions, both originally addressed to Hartlib, the mathematican John Pell and the physician and political economist William Petty underscored some of the issues at stake.

Pell's Idea

John Pell's *An Idea of Mathematicks* was contained in a letter to Hartlib of July 1638, and circulated among some of his close friends.[158] It was printed as a folio broadsheet in 1638.[159] In October 1639 Haak sent it to several people, including Marin Mersenne and Descartes.[160] In February 1682 Hooke published the Latin version in the *Philosophical Collections*, together with comments from Mersenne and Descartes.[161] The English version appeared as an appendix in Dury's *The Reformed Librarie-keeper* (1650) and again in the second edition of 1651.[162]

In arguing for a number of aids to the study of mathematics, Pell assumed the desirability of pithy reductions to essentials. He thought the prospects for such a task were good, at least in principle because, as Bacon had observed, "there is a great difference in Deliverie of the *Mathematiques*, which are the most abstracted of knowledges, and *Policie*, which is the most immersed."[163] Nevertheless, Pell complained that mathematics suffered similar problems to other subjects: it too was drowning in "that multitude of bookes, with which the world is now pestered." His response was to offer various "*meanes*" or aids for both the student and the adept, such as a catalog of mathematical works; advice on "the *best* bookes" on various topics and the order of reading; and "a *Public Library*, containing all those bookes, and one instrument of every sort that hath been invented." Writing about six years before Harrison's proposal was circulated, Pell gave hints on how such a large indexing project might be achieved: for example, by including only the oldest books if later ones merely repeated them; and by the suppression of worthless books, thus making a universal catalog quicker to compile. But Pell wanted to publish some small books of his own. He proposed three new works: *Pandectae mathematicae* (a compendium); *Comes mathematicus* (perhaps a *vademecum* or pocket-book), and *Mathematicus* αὐτάρκηξ (roughly, a self-help manual). Having outlined the rudiments of his *Idea*, constituting an encyclopedia of mathematics, the underemployed Pell made a job pitch: "I see not why it might not be performed by one man, without any assistants, provided that he were neither *distracted* . . . nor *diverted*."[164]

Pell offered an illustration of how knowledge might be cast into a nutshell, or a "pocket-booke."[165] His aim of fitting "the usefullest Tables" and "Precepts" into one small notebook, as "briefly as may be," was central to a desire to "be no longer tyed to bookes." As he explained, the point of his various "meanes," such as epitomes, tables, and rules was that a mathematician "utterly destitute of bookes or instruments" might solve any problem "exactly as if he had a complete *Library* by him." Even the suggested *Pandectae*, themselves summarizing and improving other texts, might be further refined by individuals because "men would easily see how to contract *these Pandects* into a *pocket-booke* for their ordinary use." The next step is significant: Pell sought to cast mathematical knowledge into a form that would allow the user to lay all necessary axioms and principles "up in their heads, as to need *no booke at all*." He admitted that such a goal "will perhaps seem utterly impossible to most," but asked that the whole repertoire of external aids be seen as acting in consort to "fortifie the imagination, to prompt the memory, to regulate our reason."[166]

In order to understand Pell's intentions it is important to realize that what he said about mathematics was less ambitious than his related scheme for applying algebraic method to all knowledge. In February 1638 Pell sketched a method of "synthesis" that began by setting down "all the prime truths which are no conclusions (& therefore are naturall to us) in a due order."[167] Continuing this conjecture about six months later, he proposed a way of reducing all complex ideas to units that could be treated algebraically: "If we had all our simple notions set downe, we had as perfectly all the thoughts of men as if we had all our simple sounds, we have perfectly all the words and speeches of men potentially."[168] Compared with this prospect, his suggestions for the abbreviation and ordering of mathematics seem moderate — except that his remark about self-sufficiency as reliance on memory rather than books requires clarification. It is unlikely that Pell meant that individuals should memorize mathematical tables; rather, he contended that the careful categorization of topics, axioms, and theorems would facilitate solutions to any mathematical problems. His undelivered *Pandectae* may have included tables such as those offered in his *Tabula* (1672) of square numbers, comprising thirty pages of densely packed columns; his annotation of this table suggests how he would use the patterns displayed to determine whether any given number is square.[169]

Pell's approach therefore accords with the interest, manifested in Hartlib's notes, about ways of using tables and other displays to aid memory and recollection of rules, systems, and procedures. As early as 1635, commenting on the reading methods of John Brook, Hartlib said that "Mr Brooks Method is admirable in this respect because it shews how one may remember many hunderts of Rules without charging the Memory. The best et easiest helpe of memory that ever I knew, and that both for the Practic and Theoretic disciplines."[170] And in 1655 he noted that "Mr Mercator [the Danish mathematician, Nicolaus Mercator] given Mnemonica to Mr Brereton how to remember Astronomical Tables without Booke etc."[171] He also recorded various plans by Petty to do something similar for music.[172] The point in all these cases was that order assisted memory and that careful arrangement of data made it possible to perceive patterns.[173]

The reactions of Mersenne and Descartes to Pell's *Idea* are instructive. Both French thinkers endorsed his aim of reducing knowledge into shorter, more convenient, forms. Indeed, Mersenne thought Pell had not gone far enough, telling Haak that rather than a summary of all available mathematical books, there should be a more rigorous selection of "the best and most worthy." By this strategy, "all pure and mixed mathematics could be included in these twelve volumes."[174] Similarly, Descartes welcomed the goal of collecting "in one book" all the mathematical knowledge "now scattered in many volumes."[175] However, while endorsing Pell's notion of a "self-sufficient" mathematician, understood as one who possessed a universal method of problem solving, he did not envisage this as nonreliance on books. When speaking of the various printed sources suggested by Pell, Descartes imagined the user consulting these to locate "previous discoveries, should they be useful to him at any stage." But he went on to express a position that broke substantially with the deference to memory that Pell implied. As Descartes told Cornelis van Hogelande:

> There are indeed many matters which are much better kept in books than memorized, such as astronomical observations, tables, rules, theorems, and in short whatever does not stick spontaneously in the memory at the first encounter. For the fewer items we fill our memory with, the sharper we will keep our native intelligence for increasing our knowledge.[176]

For Descartes, then, external aids such as notebooks relieved the load on memory, enabling reason to work more freely.

Hartlib took Descartes' reaction (conveyed to him by Haak and Hübner) as confirmation of his own doubts about Harrison's approach.[177] Pell's *Idea* resembled Harrison's indexing in so far as it reduced knowledge (in this case, mathematics) to smaller units. However, it was more specific in suggesting that these units were patterns and operations that could be committed to memory, so that a person may be self-sufficient, requiring neither books nor indexes of things extracted from them. Descartes questioned the value of this dependence on memory and thus reinforced Hartlib's view that neither Pell nor Harrison was sensitive to the requirements of observational and experimental inquiries. In 1640 Hartlib wrote that "the Notion in Cartes letter de Idea Pelleana is very excellent" in that it distinguished between a "Historical" and a "scientifical Part" of knowledge. Hartlib judged that it was the former, the historical "particulars," that must be foundational: "It is a very hard matter to come once to know throughly that a thing is certain and true. Therfore wee must labour principally in the Historical part of all things."[178] This path would facilitate the progress from *historia* to *scientia*.

Petty's Advice

In William Petty's *Advice* (1647) we find a program that formalized some of the key preoccupations of Hartlib's Ephemerides.[179] Petty reiterated the point that duplication and repetition must be avoided, regretting that "some are now labouring to doe what is already done, and pushing themselves to reinvent what is already invented." He nominated eight desiderata, including the following: "all the Reall or Experimentall Learning" should be "sifted and collected out of the said Books"; "able Readers" should be appointed and suitably instructed "with certaine and well limited Directions." In this way, it would be possible (as Comenius hoped) to compile a single book: "Out of all these Bookes one Booke, or great Worke, may be made, though consisting of many Volumes." The eighth and final point may have reflected what Petty knew about Harrison (albeit without mention of his name), namely that "the most Artificiall Indices Tables or other Helps for the ready finding, remembring, and well understanding" be contrived. He recommended the compilation of "various indices besides the Alphabeticall ones," suggesting indexes of information on the qualities of natural elements, the properties of artificial components, and known technical operations. His expectation was that these paper tools would encourage the combination of material from various sources, possibly in novel ways: "There ought to be much Artifice used, that all the aforementioned Indices may handsomely referre One to anothers, that all things contained in the whole Book may be most easily found, and most readily attend the seekers of New Inventions."[180] Petty may have thought that these cross-referencing devices, hinted at in Bacon's concept of *"experientia literata,"* were superior to Harrison's single, large alphabetical index.[181]

After his initial remarks pertaining to general education in schools, Petty focused on a plan for a medical research hospital, which he called a "Nosocomium Academicum."[182] A large portion of his *Advice* was devoted to a detailed prescription for a methodical gathering of information toward a natural history embracing botany, chemistry, surgery, medicine, and the practical trades that produced materials and artifacts necessary for the work of the hospital. Petty outlined the roles of physician, vice-physician, chirurgeon, apothecary, nurses, and their various students and assistants. He imagined almost everyone taking notes. These were to be kept as "journals" or "histories"—of symptoms, cures, anatomical procedures, and experiments. As Petty explained, "The Vice-Physicians proper charge is to see the History of Patients most exactly and constantly kept." Students were to make notes of what they observed, so that when attending the apothecary they should "keep an exact History of all rare and Unusuall Accidents, happing in his Operations." There was to be a regular "Journall of all notable Changes of Weather, and fertility of Seasons." Nor were books neglected: very much in the style in which Plattes approached chemistry, books were to be read, abbreviated, collated, and compared with "the things themselves," as studied through observation and experiment. Overseeing all this note-taking was the physician, who "must be a Philosopher, skill'd at large in the Phaenomena of Nature." His duty was to be "acquainted with all the Histories taken in the Hospitall" and its various sections; to identify their "most notable" elements, record these in writing, and "out of them, by the end of the terme of his service" to "collect a Systeme of Physick and the most approved Medicinall Aphorismes."[183] These aphorisms should then be compared with those of Hippocrates. Petty also discussed another compilation—a history of trades and manufactures, which he thought would take longer, possibly because it did not have the counterpart of the Hippocratic corpus as a baseline.[184] A few years later, Hartlib seemed to endorse Petty's vision, paraphrasing him as saying that "the only course in ordinarie is for the present to follow mainly the exampel of Hippocrates by way of industry and observations and by erecting a Nosocomium Academicum as hee hath described it."[185]

I need to say a little more on the relation between Petty's *Advice* and Pell's *Idea*. Although their subjects fell, respectively, under the distinctive categories of *historia* and *scientia*, there are affinities. Petty did insinuate that Pell's approach to mathematics might serve as a model for other subjects, and he specified, without explaining why, that the "Steward" of the nosocomium "shall be a Mathematician." There is also an apparent similarity between Pell's notion of the autarkic, or self-sufficient, mathematician and Petty's conceit that a person raised and educated in an institution furnished with maps, globes, botanical, and other specimens (such as in a "compleate *Theatrum Botanicum*") would "certainly prove a greater Schollar, then the Walking Libraries so called, although he could neither write nor read."[186] Both authors exploited the antibookish rhetoric so frequently found in some members of Hartlib's circle. But here the similarities end, because Petty's "Physician" and his "Compiler" of natural histories were not free of books; rather, the occupants of these roles were yoked to the burgeoning mass of notes collected from many sources and other people. The tasks of collating and refining this information, of working "to extract the Quintescence" of it, consumed their lifetimes. As Petty said, "the Compiler must be content to devote his whole life to this employment."[187] In contrast, in the case of mathematics, Pell proposed to do this himself, in quick time.

Nevertheless, Pell's *Idea* remained an inspiring model, even for nonmathematical domains. It captured the hopes of Comenius, Dury, Harrison, and others that once subjects had been collated and systematized it would be possible to crystallize the essentials, shedding all unnecessary information. However, there is a substantive difference between reducing already established knowledge (say, mathematics or music) into its key elements and gradually developing propositions and concepts from disaggregated "particulars" in new topics within *historia*. Hartlib seems to have appreciated this in so far as he underscored the different levels of knowledge, and championed the task of amassing particulars. This is why he doubted the empirical foundation of Harrison's project while rejoicing in the practical applications of his indexes. However, this left an unresolved tension within the Hartlib circle about the best way to deal with the information necessarily gathered over time in the Baconian natural histories and medical studies that interested them. As his diaries show, Hartlib was always looking for quick ways to condense the essentials of knowledge, often by way of schemata that aided memory. Pell's *Idea* relied on the feasibility of just such a publically shared set of pocket-books; but Petty's *Advice* implied that empirical subjects demanded deep *personal* memory of information gathered under various circumstances over many years. This experience grounded the comparisons, generalizations, and insights made by individuals.

In the next chapter I show how John Beale saw Robert Boyle as someone who could carry the spirit of Pell's *Idea* into chemistry and experimental natural history more generally. However, I think it better to view Boyle's challenge as closer to that facing Petty's physician—that of an individual seeking to collect detailed information from all relevant sources, possibly the quest of a lifetime. Boyle knew that it was foolhardy to attempt to avoid Hippocrates' aphorism by shortening the time devoted to science; however, he did believe that one might enrich one's own experiences by careful observation and increase them by incorporating those of others, recorded in notes and books.

[5]

Rival Memories

John Beale and Robert Boyle on Empirical Information

Robert Boyle, one of the leading scientific virtuosi, was regarded by some of his friends as a walking library. This is what Thomas Povey had to say after he missed the chance of conversation when Boyle had called to leave a copy of his *Certain Physiological Essays* (1661):

> I consider my-self to bee verie unhappy that I was not at home to receive
> you, aswell as your Booke. for, although I esteeme your Essaies . . . ; yet,
> you being in yourself a full and noble Librarie; and my waie of readinge
> and my improvements . . . being by Conversation and by the living Dis-
> courses of Such as are worthie to be Studyed: I have reason to preferr what
> I lost by my Absence, to what I found, when at my return, your excellent
> volumn was delivered to mee by my Servant.[1]

This amounts to an endorsement of Boyle's ability to deliver information from a copious memory, similar to that of renowned bookish polymaths. As such it appears to clash with Petty's negative allusion to "Walking Libraries" (in his *Advice*), by which he meant those who boasted a memory for textual passages but did not study the natural world. In fact, writing as a physician, Petty had counseled Boyle about the dangers of "your continual reading." In 1653 he joked that

> like a Quacksalver [an "empirick"] I might tell you how it weakens the
> brain, how that weakness causeth defluxions, and how those defluxions

hurt the lungs and the like. But I had rather tell you that although you read 12 hours *per diem* or more, that you shall really profit by no more of what you read, then by what you remember, nor by what you remember, but by so much as you understand & digest, nor by that, but by so much as is new unto you, and pertinently set down.[2]

In chiding Boyle for excessive reading, Petty was also asking him to give more weight to reason than memory. This is the point of his rhetorical questions: "what a stock of experience have you already in most things? What a faculty have you of making every thing you see an argument of some usefull conclusion or other? How much are you practiced in the method of cleere and scientifical reasoning?"[3] Petty implied that Boyle was trying to remember without appropriate abbreviation and rational ordering of ideas; he was overloading his memory with *copia* that had not been sufficiently reduced. Indeed, Boyle later admonished himself on this score, telling his prospective biographer, Bishop Gilbert Burnet, that he had "followed his studies rather reading every thing then choosing well."[4]

John Beale was even more explicit: he wanted Boyle to extend Pell's model of the self-sufficient mathematician into experimental natural philosophy.[5] In correspondence with Boyle between 1663 and 1666 he urged him to order his data more systematically, in the service of memory, hypotheses, and communication.[6] Their exchange shows how rival views of the role of memory were involved in disputes about the best ways of collecting and organizing empirical information for Baconian natural histories. Beale's promotion of mnemonic techniques that relied on highly structured arrangements of material seems to have reinforced Boyle's existing suspicion of premature systems. As early as 1657, in "A Proemial Essay," Boyle expressed distrust of "systems" and "superstructures" not founded on observation or experiment. He even included Bacon's *Novum Organum* among the works that he avoided reading (at least not closely) so that "I might not be prepossess'd with any Theory or Principles till I had spent some time in trying what Things themselves would incline me to think."[7] But one well-known virtue of systems was their ability to condense and order material. Boyle admitted as much in a suggestive passage from *The General History of the Air* (1692), when canvassing the "Peripatetick Doctrine about the Limits and Temperaments of the three Regions, into which they divide the Air." He cautioned that "it becomes a *Naturalist* to consider, not so much how easy a Doctrine is, by reason of its Concinnity [internal harmony], to be *remembred* or *supposed*, as how strongly 'tis to be *proved*."[8] Boyle thus conceded that certain features of leading doctrines might make them easy to remember, but insisted that this was not a voucher for their truth. Although harmony and order might be seductive *aide memoires*, he preferred an honest mass of empirical particulars, even if not yet methodized and unable to be remembered.[9]

The letters Boyle received from Petty and Beale can be regarded as advice from the Hartlib circle, but they should not be seen as letters to an outsider. Members of this correspondence network were among the young Boyle's earliest intellectual contacts. His surviving letters to Hartlib start in early 1647, and from early 1648 there are frequent references to him in Hartlib's Ephemerides as the source of reports, recipes (or "receipts"), and observations.[10] Boyle's first publication, in 1655, appeared in a volume orchestrated by Hartlib; its message was that "empiricks" should share their secret knowledge, such as cures for the stone.[11] Boyle was sympathetic to the general tenor of Hartlib's approach: in a letter to John Mallett he punned on "The Legacy of Husbandry" (a Hartlib publication), affirming that "for my part I make no doubt that the Husbandry of Knowledge will be dayly improv'd too: (though by throwing downe of Enclosures) & all Parts of Philosophy, be both better cultivated & more fruitfull."[12] Similarly, Beale had corresponded with Hartlib from the spring of 1656 and continued to write to him, almost on a weekly basis, until Hartlib's death in 1662.[13] When introducing himself to Hartlib, Beale had declared that he was "very willing to bee an incendiary to inflame with the Love of profitable knowledge. . . . You will finde mee a diligent collector, and fayre interpreter of other mens notions, but noe greater plagiary."[14] In 1658 Hartlib told Boyle about this Somerset virtuoso, saying that there was no one in the world to match his universal knowledge and Baconian enthusiasm.[15] He had already mentioned Beale in every letter to Boyle since September 8, 1657, giving news of his activities and summaries of his letters.[16] By the time Boyle received his first letter from Beale on February 23, 1663, he had read several by proxy, and may have known what to expect.

What was Boyle doing during the 1640s and 1650s, when Pell and Petty were communicating with Hartlib? In short, he was preoccupied with memory in a way that did not seem to match the approach of Pell, or what Beale would later put to him, but it did resemble some of Hartlib's own thoughts about building up individual experiences and memories in youth. In the 1640s Boyle began a lifelong preoccupation with the proper role of memory in the custody and organization of empirical information and ideas. He believed that a disciplined memory, combined with the use of notes, could assimilate multifarious experiences. The advice he was to receive from Beale in the 1660s was a different option within the range of views that Hartlib canvassed.

Improving Memory, Intensifying Experience: Boyle's Early Writings

Boyle began as an author of works on moral edification such as *Seraphic Love* (1659) and *Occasional Reflections* (1665).[17] Together with unpublished manuscripts such as "The Aretology" (1645–47), "The Dayly Reflection" (1646), and "The Doctrine of Thinking" (late 1640s), these were composed during his residence at Stalbridge.[18] Parts of the last two surfaced in *Occasional Reflections*. As Michael Hunter has put it, these writings deal with the "pursuit of moral balance, self control and piety."[19] They did so via a consideration of the reading of romances for moral lessons and the proper direction of thoughts in meditation.[20] Since Boyle was only about eighteen when the earliest of these was written, it might be objected that we are dealing with juvenilia.[21] However, some of the themes treated appeared in his early "scientific" publications, such as *New Experiments Physico-Mechanicall* (1660), *The Sceptical Chemist* (1661), and *Some Considerations touching the Usefulness of Natural Philosophy* (1663; written in late 1650s).[22]

I am interested in the strong role that Boyle assigned to memory in the cultivation of a virtuous self, and how this relates to his concern with intensifying individual experience, both moral and empirical. What kind of "experience" did Boyle recommend and seek? In these early writings he meant the incidents, behavior, and thoughts that touched upon one's moral self. Thus the purpose of *Occasional Reflections* was to "make the little Accidents of his Life, and the very Flowers of his Garden, read him Lectures of Ethicks or Divinity."[23] In his autobiography, "An account of Philaretus" (written during 1648 and 1649), Boyle said that as a young student he was "a passionate Friend to Reading" and confident in his own memory: "what time he [Boyle] could spare from a Schollar's tasks (which his retentive Memory made him not find uneasy) he would usually employ so greedily in Reading."[24] This assessment was backed by those who knew him at Eton: Robert Carew (a member of the staff reporting to Boyle's father) said that he possessed "the rarest memory that I ever knew."[25] Later, in dictating biographical notes to his amanuensis, Robin Bacon, Boyle recalled one teacher, "Mr or Doctor Harrison . . . observing, that as God had blest this Boy with a happy memory, he was thereby inabled to get his Lesson perfectly by heart in a short time."[26] In a wish list starting with "The Prolongation of Life," Boyle included "Potent Druggs to alter or Exalt Imagination, Waking, Memory."[27]

Boyle assumed that memory could be trained via practice, not necessarily by using classical memory techniques (sequences of places, or *loci*, combined with vivid images), but rather by drawing out the circumstances and consequences of observations and ideas through careful attention, repetition, and recollection. In the early moral writings, his emphasis was on choosing and selecting what to commit to memory. This was one basis for his attack on "those memorys now so much in Fashion, which are stuff't with almost nothing else then what deserves to be excluded." His target was the exercise of mnemonic techniques without proper regard to the content of what was stored. Thus Boyle derided those memories that keep "nought but Strawes, Dust, Feather, and such lighter Trash."[28]

Boyle discussed memory in relation to the practices of reading and thinking, or meditation, by which he meant a carefully directed train of thoughts. In "The Doctrine of Thinking" he explained how he "set my Thoughts awork . . . to recall to mind any thing I have almost forgotten, or repeate any thing I desire to retaine more firmly in my Memory."[29] He gave a list of six exhortations about how best to exercise his mind. The pervading theme was that active reflection was necessary to guard against "Intervening Fancys" and the proclivity of the mind to wander.[30] This watchfulness is also apparent in the "Dayly Reflection," addressed to "my Lady Ranalaugh," his sister. He advised that we should "recall orderly to mind, Whatever new observables in any kind of Knowledge, either your Study or Conversation has afforded or you owne Thoughts suggested yow, the foregoing Day." In this way, "the memory may be commanded to make this Restitution either in the order their Seniority/Priority . . . gives them, or that which their Considerablenesse assignes them in your Esteem."[31] Interestingly, given Boyle's scrupulous disposition, there was a potential downside to such memory training. In conversations recorded by Peter Pett, Boyle warned against the danger of jokingly mispronouncing passages from the Bible because

> our memories were not such table bookes as wherein with a spunge we could blot out what we would, and that perhapps as long as we lived, whenever we read or heard of that place of Scripture againe afterwards, the former ridiculeing of it would be apt to recurre upon our thoughts and enervate its majesty.[32]

Regular and careful reflection aided memory, guarded against the wandering of the mind, and improved thinking.[33]

Boyle claimed that such meditation could intensify observation and magnify individual experience. He contended that "what men Commonly stile Experience is nothing else but a certin Dexterity of conduct, resulting from the Remembrance and Consideration of [the] Occasions suitably circumstanced." But the retention of experiences in memory depended on active engagement with the world. Experiences must be examined and analyzed so that "our Reflections on what we have observ'd, improves it into consequences new Axioms and Uses." Such a practice helped memorization because "the Repetition of what we learn greatly contributes/conduces/ to secure our Acquists from (the Danger of) Oblivion." In this way, Boyle promised, an individual can gather more experience than his age might seem to allow, because "Experience Consists, not in the multitude of years but in that of Observations."[34] He elaborated on this theme in *Occasional Reflections* (1665), claiming that by such meditations "a man often comes to discover a multitude of particulars even in obvious things." He added that "this exercise of the mind must prove a compendious way to Experience, and make it attainable without grey-hairs, for that, we know, consists not in the multitude of years, but of observations, from Numbers and variety of which it results."[35] Boyle applied this conceit in a letter to Hartlib when describing John Hall, a poet and pamphleteer, as having "September in his judgment, whilst we can scarce find April upon his chin."[36]

Boyle's stance here owes something to Seneca, whose response to Hippocrates' complaint was that "It is not that we have a short time to live, but that we waste a lot of it. . . . We are not given a short life but we make it short." Seneca affirmed that dialogue with past thinkers afforded "a long period of time through which we can roam. We can argue with Socrates, express doubt with Carneades, [or] cultivate retirement with Epicurus."[37] Boyle's approach was slightly different: reading books was certainly a good use of time, but intensive observation and meditation were practices by which to enrich experience and thus extend it. In this way, "one man" could make "as great a number of Observations as less heedful Persons," even if his life was short. This habit was necessary because, in Boyle's view, knowledge is so long in generating, or purchasing, "that it seldome coms to the Enjoying. For we lern with labor, and by Peece-meal; and since, (as Hippocrates has it) Vita brevis, Ars longa, one half of a man's life is spent to instruct the other."[38]

There are good reasons to assume that Hartlib approved of Boyle's respect for accumulated experience. Hartlib's call for proper collection of "particulars" was often accompanied by a directive to cultivate both observation and memory. He believed that individuals should build up their experiences of particulars in memory, beginning in childhood and continuing until at least the age of twenty-five. Writing in his Ephemerides for 1639, he asserted that

> wisdome of Arts, Sciences and Inventions will never bee enlarged till Men furnish themselves in their yonger yeares with a World of all manner of particulars til 16–25, or 30, or more. Afterwards when they are come to maturity of judgment, they will bee able to some purpose to exercise their reason and philosophat upon them.[39]

Castigating traditional pedagogy, he alleged that children were taught "the most abstractive things and that in a most verbal way"; instead, there was no reason why children should not be encouraged "from their very infancy" to observe carefully, record, and remember. Hartlib's note-taking prescriptions matched this view: children should at first "write into their Ephemerides of whatever they shall see and heare," as these occurred, and only later should these entries be "reduced into Loci Communes when they are of judgment," a habit, he affirmed, that "would make them learned in the whole Encyclopaedia or Pansophia before they are aware of."[40] Both Hartlib and Boyle stressed the early cultivation of memory as a basis for later moral and intellectual development. When Boyle began to publish his scientific works he returned to what he had rehearsed with Hartlib, confessing that he was still "very young, not only in Years, but, what is much worse, in Experience," and therefore had much to learn.[41]

Boyle aimed to enrich his experiences by building up associations to the places, times, and circumstances in which these occurred. Cultivated in this fashion, memory supported thinking by providing a stock of experiences and by retaining previously forged links between them. The discipline he prescribed—careful selection of materials, repeated reflection on key themes, and rehearsal of a skeletal direction of meditation—promised to expand the experience stored in memory and facilitate its retrieval.[42] He argued that by taking "notice of the properties and circumstances of most things that Occur to him," and by relating them by "Resemblance or Dissimilitude" to each other, a person can not only "Revive the Memory" of good thoughts, but make "almost the whole World a great *Conclave Mnemonicum*, and a well furnished *Promptuary*, for the service of Piety and Vertue."[43] In this way, the individual was made self-sufficient, perhaps a walking library:

> Besides, whereas Men are wont, for the most part, when they would Study hard, to repair to their Libraries, or to Stationers Shops; the Occasional

Reflector [that is, a person following Boyle's method] has his Library always with him, and his Books lying always open before him, and the World it self, and the Actions of the Men that live in it, and an almost infinite Variety of other Occurrences being capable of proving Objects of his Contemplation; he can turn his eyes no whither, where he may not perceive somewhat or other to suggest him a Reflection.[44]

When his interest in natural philosophy developed, Boyle continued to find moral lessons in the study of nature and regarded the experimental life as a morally worthy one. Indeed, he forecast that the habits acquired in daily meditations might be applied not only to moral topics, but to "Oeconomical, Political, or Physical matters."[45] Boyle extrapolated what he said about the effects of gathering moral and devotional thoughts to scientific inquiry: "the Study of Nature is the Noblest Memoria Localis of a Christian, & that he may turne the whole world into a *Conclave Mnemonicum*."[46]

It is difficult to be precise about how Boyle envisaged this kind of memory. He explained the term "Mnemonicum" in a marginal note: "So they call a certain Room, Artificially furnish'd with Pictures or other Images of things, whereby to help the Memory."[47] Yet there is nothing in his account that suggests the use of any *ordered* sequence or chain (*catena*) of places (*loci*). In the art of memory it was this order that prevented confusion of images and guided recollection of material (both *verborum* and *rerum*) attached to particular images or symbols. It is more plausible that his habit of making vivid mental impressions of experiences created strong "episodic" memories of past observations and actions. Thus Boyle's reports of his memory of ideas were often instances of *remembering the experience* of learning something, of seeing, doing, or touching certain things.[48] In the terms suggested by Julia Annas, this can usefully be classed as "personal memory," as opposed to "nonpersonal memory" of a fact, a date, or a theorem—one that "lacks those features of [personal] memory which make the past acquiring of knowledge part of what is remembered."[49] Boyle's "Library" and "Mnemonicum" were personal, deriving from practices of concentrated observation and meditation that seem closer to the tradition of Protestant poetics than to the Baconian legacy.[50] Unlike several members of Hartlib's circle, he was not concerned in these early writings with the organization of specific bodies of knowledge. He showed no interest in putting his knowledge into the kind of condensed and codified form that Pell, for example, sought in the case of arithmetical facts and mathematical procedures.

There, is however, a plausible, if ambiguous, relationship between Boyle's attitude toward personal experience and Petty's *Advice* (1647). The humorous and encouraging tone of Petty's letter of 1653 (cited above) about excessive reading is appropriate, not just out of due respect, but because Boyle was an obvious candidate for Petty's own ideal person: that "one Man" (such as his physician or compiler) who may "see and comprehend all the Labour and Wit of our Ancestors." The basis for this hope was, as Petty explained, that a lifetime of experience in reviewing information, whether about medical diseases and cures or the history of trades, laid the groundwork for later discoveries.[51] Hartlib told Boyle about Petty's own plans for such a vast history, and Evelyn later recalled telling Boyle that he was also "intent on collections of notes in order to an History of Trades."[52] This emphasis on long-term effort is consonant with Petty's plea to Hartlib in his letter of early 1649 wishing that "the great wits of these times" would first "employ themselves in collecting & setting down in good order & Method all lucriferous Experiments & not bee too buisy in making inferences from them till some Volumes of that Nature are compiled."[53]

The unexpected addition in Petty's *Advice* is the notion that such a person might grasp connections "*uno intuitu*"—in, as it were, a single glance. Boyle did not think that such unifying insights were imminent, but it is not unreasonable to infer that he believed that an understanding of nature required *both* immediate experience gained through intense observation *and* the accumulated experience, held in memory, that accrues over a lifetime and improves interpretation.[54] In order to appreciate the expectations that Beale imposed on Boyle we have to recognize that, in the Hartlib circle, the emphasis on collective gathering of data did not preclude the anticipation that a single person, perhaps in a specific domain, might make a crucial advance. There was a precedent for this view in Bacon: "So far we have found no one with a mind steady and stern enough. . . . But if someone in the prime of life, with senses unimpaired and a mind washed clean, applies himself anew to experience and particulars we should hope for better things from him."[55]

Advice to Boyle

By 1663, when Beale began counseling Boyle about various ways of improving his memory, and his science, he was addressing someone who might listen. What transpired, I think, is that Beale urged Boyle to provide, in effect, a version of Pell's *Idea* for natural history, including experimental work. This required an arrangement of Boyle's ideas and results that would, in turn, serve as a mnemonic framework. Even though both men espoused the collection of empirical information, their exchange reveals a difference of opinion about the way natural history and natural philosophy should be done. Beale stressed early ordering of empirical information, even in artificial systems, because this aided memory. Boyle accepted the value of using Heads for the collection of particulars, but did not want to cast such Heads prematurely into a theoretical framework.[56]

Beale first wrote to Boyle on February 23, 1663, prefacing his letter with a long Latin ode to Boyle's accomplishments. He disclosed that he had acquired "the Nic Name of Erasmus Junior."[57] Writing again on September 28, he threatened weekly letters. The next day he sent a long paper titled, "The Mnemonicalls," and confided that before going to Eton he read and memorized key texts "in secrete corners, conceald from others eyes." Then "afterwards in Cambridge proceeding in the same order, & diligence with their Logicians, philosophers, & Schoolmen, I could at last learne them by hearte faster than I could read them."[58] The exchange of letters continued until Beale's death in April 1683 but, frustratingly, we no longer have Boyle's side of their correspondence.

There were two main preoccupations in Beale's letters from February 1663 to August 1666: the art of memory, and the best arrangement of Boyle's publications.[59] These were linked since, as all adepts of memory training agreed, orderly arrangement was a reliable aid. Beale announced his own scheme for an art of memory in the next extant letter of September 29, 1663 (it is hard to believe that he waited seven months without writing again). In what is actually an essay of about 8,000 words, Beale outlined a series of lessons about memory drawn from a wide range of sources: classical mnemonic techniques, the Cartesian theory of brain processes, and his own observations about the habits of individuals endowed with both naturally powerful and expertly trained memories. He offered Boyle a range of hints on memory improvement, some drawn from anecdotes about his mother's astonishing memory, others from his own experience. Some of these were proverbial, as seen in these extracts: "whatever is not offered to the Memory upon very easy Termes, is not duely tendred"; "The more we acquire, & the more often wee visite & imploy Memory, the firmer & stronger wee find it."[60]

What Beale said to Boyle about memory largely repeated his earlier, and ongoing, discussions with Hartlib. His main point was that natural memory was not harmed by mnemonic techniques and rules, and that a constitutionally weak memory could be enhanced by various methods.[61] Beale's own proposed memory art was a development of the classical technique. Instead of visual images, usually contrived and chosen by each individual, he proposed a grid of symbolic characters that had universal pretensions.[62] He announced to Hartlib that he set out "To devise Millions of Millions of Characters, each one soe apparently differing from each other throughout the whole immense variety, That the eye at first glance shall discerne, & distinguish the difference." Beale was confident that these characters would be "retained in the minde, & in a moment producible in fit order for any kind of occasion." Moreover, he promised: "All this, the reading, writing, use & practise, soe the Learner bee willing, & of ordinary capacity & skill in Clerkship, I undertake to teach with ease, due vacancyes & refreshments in one weeke."[63] This optimism was apparent in his assurance to Boyle that the characters could be taken in "as by one glance or blink of the eyes."[64]

Beale acknowledged, however, that the classical art of memory had fallen out of favor; he mentioned the "prejudice most learned men have against all discourses of Artificiall Memory." For this reason, he was not seeking, via Boyle, "to engage the Royal Society in it"—although a few years earlier he had suggested precisely this about his *own* scheme.[65] As well as promoting his own ideas, Beale concurred with Hartlib about the importance of childhood, affirming that that each person should, from an early age, build up a cognitive structure into which later, and novel, material could be more easily integrated. Beale asked an unidentified correspondent (presumably Hartlib), and also "Mr. Brereton," to consider how they came to have a strong memory for certain "Kinds of Knowledge," and suggested that this capacity was helped by the categories and themes internalized from "your Child-hoode, & particularly such things, as seemed leading heades, & conduct-pipes to the source of knowledge." These, he asserted, will be the "best & truest Topiques for our future improvement of Memory during life," especially if one can "string them in order" by some method.[66] Beale advocated this combination of early memory training and the adoption of an arbitrary classification and arrangement of categories. He appeared to believe that such artificial schemata could be learned quickly and used effectively, even if childhood memory had been undeveloped or structured along different lines.

Beale promised Boyle that the strength of memory could be prolonged beyond the usual decay occasioned by age. This was possible if an appropriate mental structure has been established, just as "the studdes or Nayles were engrafted in our childehood, as is the Alphabet generally to all that knowe letters." Each individual had to build up a framework of topics and the connections among them—otherwise, Beale forecast, each of us would live the fate of the "Merchant Whose wares are not placed in order. . . . His store confounds him, When the Fayre comes." The insinuation was that this was especially likely for someone trying to manage a large array of material. Beale added an incentive that must have resonated with Boyle, namely, that his method was better than those that produce merely "those quic returnes of Memory" for "Words & Sentences"; rather it was one "which heapes up, digests, & firmly retaines whole Libraryes."[67] This inducement was paired with a warning that an uncultivated memory would deteriorate: "And I know I could once boaste of a naturall memory beyond beliefe, but I found it true, that . . . the first grey hayre would signify the decay of it."[68] Beale later followed this up with a story intended to frighten Boyle:

> Joh Suisset Calculator, (As he was denominated) . . . that in the decay of his age, when he would have reviewed his owne Workes, he fell a weepeing excessively, because He was not able to understand, what himselfe had written. Sir, I do not threaten you this judgement. . . . But, if you doe not keepe your eye frequently upon your owne margins, you may live to find the perfidiousnes of grey hayres which doe sometimes steale our own labours & inventions out of our Memory.[69]

A year later, when discussing the instruction of "a young Quaker of 17yrs old" who had been introduced to Oldenburg, he made the point that the best support for memory was active mental activity — "Invention, reasone, Intelligence" — from youth onwards; but also, that any sudden "desisting" from such effort could see the "Memory utterly destroyd."[70] Boyle may have concluded that, at the age of fifty, this advice was too late for him.

What I have not said so far is that in advising Boyle about memory, Beale purported to do without notes. In one of his earliest letters he boasted that "when I read to the students of Kings College Cambridge, (which I did for 2 years togethr in all sorts of the current philosophy) I could provide my selfe without notes (by meere Meditation, or by glancing upon some booke) in lesse time than I spent in uttering it."[71] Indeed, he identified notebooks as among the "outward & false aydes" that are "Irrational." Thus after saying that "the Thread tyed at the finger, or ring" was an unreliable reminder if the impression at the time was not firm, he went on to attack one of the standard practices of the day:

> Topiques & Comon places in books, & not in the braine, may make a learned Dunce. And for this cause, Men should take heede of engrossing

in Table bookes what they ought to learne by hearte, except it be with purpose to take it off the Table booke into the Memory.[72]

It is striking that Beale should imply that string on a finger offered no better prompt to recollection than the well-known use of notes or passages under Heads in commonplace books. But citing Plato's suspicion of writing, he stressed the importance of internalizing what one wanted to remember: "He that trusts to his Table bookes, more then to the Tables of his Minde, hath tempted his phantsy to be treacherous or lazy." His general message was that "We should affect, & inwardly glory in the improvement of our owne Minds & lay up there evry valuable advertisement, as fitter to adorne our Insides, than Frontispices, Walls, and Paper-bookes."[73] This apparent rejection of external supports fits with Beale's comment to Hartlib about four years earlier that if some papers he had sent were lost it was

not in my power to repeate it, or to recover it. For tis the first draught, I as soone as I have engrossed, I am wont to discharge my Memory: And my Memory holds mee to that Covenant of Immunity; I as the beast will not feede on blowen fodder, so doe my spirite desire fresh aire, & free perambulations.[74]

By clearing the mind, by learning to forget, Beale claimed that his well-ordered memory could depend on its own resources. However, we get a slightly different picture of Beale's habits in a letter (probably addressed to Pell) of 1663, the year in which he began his exchange with Boyle:

I can recount unto you some of our Alliances of incredible Memoryes. And though for many yeares I have now beene grey; yet I am bound to blesse God for a kind of miraculous memory, that to my knowledge hath not hitherto beene guilty of the losse of any one Note, Notion, or Observation, in any Art or Science within the whole Encyclopedye, that at any time from my infancy hath beene committed to my memory. And having now for many yeares kept large Correspondence upon all promiscuous Arguments I could (before my papers were by carriage disordered) fetch any line upon occasions out of a Tun of loose papers.[75]

This revealing testament does not contradict Beale's attack on the use of notebooks, but it does describe a spatial memory that relied on physical locations of papers, perhaps themselves standing in for a schema of Heads committed to memory. It also suggests that Beale's supposedly self-sufficient memory could be disrupted by dislocation of its external prompts. As we shall see in the next chapter, Boyle had similar problems with his "loose notes," but rather more success in using them because they were, at least, deliberately assigned to paper as notes.

Publishing and Arranging Boyle's Works and Notes

From April 18, 1666, Beale turned his attention to the best publishing format and sequence for Boyle's works.[76] Much of what he suggested was influenced by his preoccupation with memory. Earlier in the correspondence Beale had said that printed material would benefit from special typography, colored ink, and visual symbols like those in medieval manuscripts. He referred to "the beautifying letters" which marked "the fronts of Chapters, & Sections" in such manuscripts, and stressed what a "vaste ayde might be to the Memory in the very printing of bookes that are worthy to be learned in the reading."[77] He now also considered the physical format of Boyle's works, suggesting that it was best that they "were abroad in Quarto, & rather in thinner Tomes, than in thicker." Thin volumes were desirable because "every man may sorte them in a Methode more agreeable to his owne humour & concernments."[78] He urged this strategy in a subsequent letter, reminding Boyle that the Jesuits had already pioneered it with success. These zealots had realized the power of short pamphlets, or even single pages, which could win hearts and minds more quickly than larger tomes. In this way, he observed, the "Jesuits, doe infatuate the world, as well by their shorte manualls, as by their endlesse volumnes . . . [for] by their single sheetes they catch him that runneth by." These brief pieces, like "shorte daggers" used against an enemy, did the job quickly, allowing one to move on to the next task. Beale urged that all proponents of the new philosophy, especially Boyle, should do likewise.[79]

Apart from assisting Boyle's readers, Beale wanted to make it easier for him to manage his own unpublished manuscripts. He may have known about Boyle's workdiaries containing entries of "Promiscuous Experiments," for he suggested that once these reached 100 experiments they should be published: "That assoone as they amounted to a Century, they deserv[d] to be abroad. For thus you may empty your deskes often; & be lesse overwhelmed with your owne abundance."[80] Nevertheless, in Beale's view, the density of empirical information that Boyle collected did not justify diminishing the role of memory. Having acknowledged the particularity and scale of empirical data, he reiterated the point about not publishing "your Pandects, or promiscuous Experiments . . . more than a Century at a Time," now adding that "they may easily overwhelme an ordinary Industry, & confound Memory." In the same letter it is clear that this stress on brevity was meant to facilitate the rehearsal of information: "Tis impossible that we should keepe *our Memoryes firme for our owne improvements*, if by thiese strong impressions, renewalls, ruminations, & inculcations they should not be fortifyed."[81]

What might Boyle have made of all this? Was he spooked by Beale's tales of decaying memory capacity? Did he feel that his own memory was becoming impotent in the face of empirical particulars, as Bacon had warned? Unfortunately, Boyle's replies are not extant, but we can detect something of what he might have said from the conciliatory mood of two letters from Beale in 1666. There is a sign in a letter of July 13, 1666, that Beale recognized (perhaps because Boyle raised it) that this preoccupation with brevity, order, and memory might not be appropriate for the new empirical Baconian sciences. Beale acknowledged that some heavily systematized subjects (he referred to these as "loosely notionall" but did not cite examples), might be "skim'd over, as with the glance of the eye" and understood, and remembered. This was not the case, he conceded, with the material of "Experimentall Philosophy" in which the various observations and experiments needed to be continually revisited and revised: "But thiese [new sciences] doe require a frequent & assiduous reviewe, & a kind in incubation, as for innumerable applications, for remoter discoveryes, & for seasonable inventions upon all imaginable occasions."[82] Nevertheless, his compulsion to look for brevity and compression is seen in a marginal scribble suggesting "Howe you may make 100 experiments serve for 1000 uses, & escape all oppositions."[83]

In a letter (about 5,000 words) of August 10, 1666, Beale acknowledged Boyle's suspicion of premature systems: "I do not forget that you have renderd sufficient reasons against the praesumptuous affectation of Methodes, & hasty Systemes." He did not relent, however, drawing attention to Boyle's

> Concessions in *your Proœmiall Essay pag. 5. & 6* That there is a Usefullnes & a Seasone for Systems. And certainly when store off good materialls are collected, . . . it will have more usefullnesse, ornament, & strength, if skillfully Ordered into a fit building, than in a confused heape.[84]

His plea to Boyle was "*to drawe foorth your Experiments & Observations into Hypothesis, such as they doe fayrely beare, & unite them as far as they give mutuall strength, & light, & assistance.*"[85] Beale maintained that Boyle owed it to the world to publish a systematic presentation of the new philosophy, one that might compete with the traditional doctrines.[86]

Beale's advocacy of a nomenclature of plants offers an example of his conviction that multifarious data could be reduced to a simple, and memorizable, form. The background is a letter from Cyprian Kinner (an associate of Comenius) to Hartlib of June 27, 1647. Addressing the difficulty of remembering the huge number of details about the qualities of plants, Kinner asked:

> For how few of the most experienced Botanists are there who know all the virtues and names of every plant, when the Authors are so at variance with one another about every single one of them; and how few of them can by the common way of learning impress them on the memory and retain them in it?[87]

Kinner thought this was impossible, and called for a new botanical terminology that captured the qualities and powers of plants, herbs, and so on. Combinations of consonants and vowels in syllables would indicate these common and differentiated natures, so that

> if someone can at least memorise that kind of technical word in this way, he will have fully in his grasp all the virtue of the whole plant denoted by that word, its usage and its common nomenclature, in a wondrous, easy and pleasant small compass.[88]

On October 11, 1665, Beale mentioned to Boyle a manuscript he received from Hartlib called *De Herbis sine Duce cognoscendis* (A way of recognizing plants without a guide). He gave this to William Brereton to pass on to Boyle.[89] Beale referred to it again on April 28, 1666, expanding the title to say that it enabled one to distinguish "all plants by their affinityes & differences, in their rootes, stemms, branches, blades, stature, color, leaves blossoms, fruites, seedes &c." He wanted it made more public because he saw it as a model "allso for the collections of those infinite varietyes into fewe heades"; we might then "cleare our apprehensions of the nature of gravity, & levity, It may give us some satisfaction concerning the Systeme of the World." Beale called it the "*Cribrum divinum*" (divine sieve) by which the essential natures of things in the world are sorted. We do not know whether Boyle commented on this, having received it from Brereton; in any case, Beale had to remind him of it in a letter July 13, 1666. Now he asked Boyle to have the pamphlet translated into English and disseminated. He thought this way of reducing and codifying information could work just as well for Boyle's own experimental natural history: "to Exemplify what you have written of colors, & other qualityes in that Generall Physiology."[90] Beale seemed to endorse the view that there are simple natures, or primitive forms, which are few in number and underlie all complex phenomena.[91] One implication of this outlook was that once these forms had been identified, results would follow quickly.[92] This is certainly what Beale professed in a letter to Hartlib: "Tis my great joy that Mr. B[oyle] is so far engaged to give us the rest of his notes and following experiments. In these he hath obliged all the intelligent inhabitants of this world, and hath given us hope, that we shall shortly complete humane sciences."[93]

Here again is the hope that Hippocrates' aphorism could be avoided. But this is not what Boyle thought. Indeed, by 1665 when he was composing *The Excellency of Theology, compared with Natural Philosophy* (published in 1674), he contrasted the feasibility of an individual mastering the subject of "divinity" with the impossibility of a similar achievement in natural philosophy or natural history (here he refers to the individual as "a naturalist"). Regarding theology or divinity, Boyle contended that "a man who values and inquires into the mysteries of religion" may attain "an eminent degree" of knowledge — under the guidance, of course, of Scripture. In contrast, he considered the study of nature far more daunting, "being so vast and pregnant a subject that . . . almost every day discovers some new thing or other." Premature systems were likely to be shipwrecked by the latest "novelty." Instead, Boyle promoted his own style of writing "loose tracts" that could deal with new information and conjectures as they appeared. As I discuss in the next chapter, he saw note-keeping as the means by which one man could build up a lifetime of experience that might lead to some discovery in natural knowledge, or even to that limited "fame" or "reputation" that scientific inquiry allowed.[94]

Summary

Boyle's early writings show some affinity with Hartlib's emphasis on individual memory as a store of embodied experiences. Boyle wanted to thicken and deepen his own experiences so that these would be more securely impressed in his memory. He aimed to select experiences (from observations and experiments) and to draw out inferences, but he was not interested in placing the facts and ideas thus acquired in a sequence within some mnemonic grid. What we accept as Boyle's "empirical" attitude was not so much a greater commitment to gathering matters of fact—also professed by Hartlib and Beale—as a refusal to condense and arrange material in the way they demanded. Beale's promotion of memory techniques that relied on highly structured arrangements of units only reinforced Boyle's antipathy toward premature systems. To various degrees Hartlib and Beale believed that essentials, radicals, or simple natures would be discerned after past and current information was sifted and arranged, and that this ordering would provide a sound basis for the integration of new discoveries. They aimed for an agreed set of abbreviations in various subjects and, potentially, a classification of the world that supported a shared mnemonic system.

Among Hartlib's correspondents, there were interestingly *different* approaches to the question of how far collaborative Baconian natural histories (collections of medical, chemical, and other data) should rely on individual memory, either natural or trained. Beale was an avid contributor of empirical information (especially, but not only, about fruit trees) to the Royal Society. He insisted that such information must be arranged to assist memory and thinking. Boyle tried to extend his own empirical collections by pooling data gathered by others, and thus confronted the sheer mass of information entailed by Bacon's project. More so than Beale, he stressed the time and collective effort required to reach the stage at which general theories might be supportable. In the next chapter I show how a large part of his individual contribution to this effort consisted in making and collecting notes that recorded reading, conversations, observations, and experiments. This activity is not incompatible with the cultivation of a personal episodic memory of thick observations and reflections, as expressed in his notion of a mnemonicum. However, Boyle feared that memory aids might sanction premature systematizing of empirical information, and that this would lead the mind away from the world. Thus when he warned about the attraction of systems or doctrines in *The General History of the Air* (1692), it is quite likely that Boyle was thinking about his earlier exchanges with John Beale.

[6]

Robert Boyle's Loose Notes

At the time of his death, Robert Boyle was compared with Nicolas Fabri de Peiresc, the great French collector and virtuoso. Pepys wrote to Evelyn: "Pray let Dr Gale, Mr Newton, and my selfe have the favour of your company to day, forasmuch as (Mr Boyle being gone) wee shall want your helpe in thinking of a man in England fitt to bee sett up after him for our Peireskius."[1] On the score of polymathic learning and European reputation, this comparison made sense; but there is also a nice irony. Whereas Peiresc was seen as a person in command of his books, notes, and papers, Boyle came to be regarded as an exemplar of mismanagement. In the estimate of his biographer, Pierre Gassendi, Peiresc distrusted his memory and therefore took care to order his notes, so that although "his House" might appear to be "a confused and undigested Masse, or heap; yet was he never long in seeking any thing."[2] After Boyle's death, some friends encountered a very different scene. Evelyn told William Wotton that Boyle's bedchamber was crowded with "Boxes, Glasses, Potts, Chymicall & Mathematical Instruments; Bookes & Bundles of Papers."[3] After making his own inspection, Wotton agreed that "His Papers were truly, what he calls many Bundles of them himself a Chaos, rude & indigested many times God know's."[4]

Whereas Peiresc was aware that the limits of his memory called for scrupulous arrangement of his papers, Boyle's "chaos" was in part due to a habitual recourse to memory. Of course, the disarray of his papers also resulted from his use of loose sheets for notes and drafts, and from inade-

quate organization. As Michael Hunter, Harriet Knight, and Charles Littleton have shown, Boyle recognized this and sought to order and index his
papers, although not with much effort until the 1680s.[5] Some time after 1665
he composed mnemonic verses as a way of remembering the order in which
he sought to arrange his treatises.[6] Later, he tried colored string and various combinations of letters and numerals as codes, but he never settled on
whether these index markers were to be applied to general subject classes,
to the kind of writing (for example, preamble, advertisement, or appendix)
in which salient material was contained, or to the order or titles of his own
works.[7] I think that this failure to develop an effective indexing system resulted from years of trusting in memory in tandem with notes. Almost per-
versely, the disorder of Boyle's papers necessitated a reliance on memory by
way of compensation. Indeed, he professed an ability to recover information
from memory, either with or without the prompt of a note or fragment of the
missing material. Why this confidence?

In this chapter I explore Boyle's views on the use of memory and notes,
taking account of the precepts and options of his day. Like many other early
modern virtuosi, Boyle made copious notes comprising both textual extracts
and empirical information — the latter deriving from testimony, observation, or experiment. Unlike his friend Evelyn, he did not maintain large
commonplace books of the kind recommended by Renaissance humanists;
nor did he publicize an account of his note-taking methods, as John Locke
did.[8] However, through his early contacts with Samuel Hartlib, Boyle was
exposed to the ways in which diverse information could be noted, stored,
and used. His friends believed that he had made such note-taking habitual:
after engaging in conversation, "what was remarkable in Experiment or oc-
currence, he note[d] down Every day when the company parted."[9] Furthermore, his practice exemplified the well-known dual function of notes as both
prompting memory and relieving it. Boyle did not pause to write a sustained
essay on these issues; nevertheless, scattered throughout his prefaces, ad-
vertisements, works, notebooks, and manuscripts there are significant
comments on his practice of making what he called "loose notes." Boyle was
aware of the kind of notes he kept and his reliance on both memory and
notes as prompts to reflection and thought. These remarks might be seen
to constitute a self-conscious and positive rationale for his personal style
of managing information. I draw examples of Boyle's note-taking mainly
from his workdiaries, but do not pretend to offer a full account of the ways
in which he collected and used notes, or how this may have changed over
time.[10] I consider in more detail his comments about the use of memory and

the combination of memory and notes, which involved what is best called "recollection."

During the 1670s and 1680s, Boyle gave several reasons for his decision to write on loose sheets. He told Oldenburg that in order to "secure my selfe against the like losses of a whole Treatise at a time, [I] resolv'd to write in loose and unpag'd sheets: and sometimes, (when I had only short memorials and other Notes to set downe) even in lesser papers."[11] Additionally, in a manuscript (c. 1680) justifying this choice, he said that he was discouraged from the usual practice of lodging "thoughts & Observations in Papers bound up into Bookes" because loose sheets were less tempting to thieves, since they did not promise "Coherence."[12] In *The Excellency of Theology, compar'd with Natural Philosophy* (1674), composed about 1665, he had already put this last point about coherence in another context—the form of writing best suited to empirical science. Here he contrasted "Methodical Treatises" and "the Systematical way" of writing with "a more loose and unconfin'd way" of presenting novel, yet often inchoate, particulars in "loose Tracts and Discourses." Boyle declined the former systematic style even though he agreed that orderly presentation of material served as an *aide memoire*. The problem was that by imposing such order, "for the most part they prove greater helps to the Memory, than the Understanding."[13] Thus Boyle's resistance to premature systems deprived him of one of the standard aids to memory.[14] Consequently, his preference for short pieces of writing (including notes) on loose sheets placed extreme demands on his ability to remember and recollect. Of course, loose sheets are more likely to be mislaid or lost and, indeed, in "An Advertisement" of 1688 he declared that many of his papers had been lost or damaged.[15] Here we have the conundrum that Boyle created for himself.

By the 1660s, it is likely that Boyle was using what he described as "loose and unpag'd sheets" in the main subjects that concerned him—theology, philosophy, and empirical scientific inquiry, especially in chemistry and medicine.[16] It is important to say that the content of such loose sheets was not uniform: it ranged from short discursive prose passages (many being drafts of future works) to numbered notes of various lengths in the workdiaries. The common feature, as Boyle seems to have conceived it, was that of breaking down both information and thought into small units. These could then be added or interpolated into works being prepared for publication, or stored as notes of observations and experiments for future use.[17] In an "Appendix" to the work on "Final Causes" (1688 or later) it is evident that this practice applied to the full range of subjects—hence Boyle's indication that "I have to each distinct Particular, whether Observation, Experiment, Reflection &c. prefixt an ordinal number according to the usual Arithmetical series."[18] Similarly, the broadsheet of 1688 about missing papers made it clear that the notes belonging to lost sets of "centuries" (groups of 100 entries) included empirical and more discursive ones. Boyle mentioned "four or five Centuries of Experiments of my Own, and other Matters of Fact, which from time to time I had committed to Paper, as they were made and observ'd" as well as "seven or eight Centuries of Notions, Remarks, Explications and Illustrations of divers things in Philosophy, which I had committed to Writing as they had chanc'd to occur to my Thoughts."[19] In a sequel to this announcement, he reported "loosing six Centuries of matters of fact in one parcel."[20]

However, in another irony perhaps unnoticed by his contemporaries, Boyle was often not worried about being able to *find* something because he was confident that he could *remember* it. In his justification for using single sheets, he explained that he was prepared to take the consequent risk of having material lost or stolen because these were "inconveniences" which "I could sometimes easily repair out of my Memory." For this reason, too, he was prepared to "leave Competent Blanks or Intervalls between the distinct Observations, Notions, &c."[21] In the letter to Oldenburg (cited above), he said that "I presumed that by the help of the remaining Papers and my memory, I should quickly be able to supply the Loss of a sheet or two, in case it should happen."[22] These are manifestations of Boyle's strong expectation that he could recall (without notes) or recollect (with the aid of a note or fragment) material of various kinds. On many occasions he claimed to revive memories of past experiments, even of the circumstantial details surrounding them. We find this statement in the publisher's notice to *New Experiments and Observations touching Cold* (1665): "not taking care to leave all the first Copy, the Author found, (besides several Blanks, that he filled up out of his Memory, or by repeating the Experiments, they belonged to)."[23] Consider also this more extensive account, by proxy, in *Hydrostatical Paradoxes* (1666):

> But part of the Appendix consisting of Experiments, which the Authour
> has several times made, but trusting to his memory, did not think it neces-
> sary to Record, when he came to recollect particulars, he found that some
> years which had pass'd since divers of them were try'd, and variety of
> intervening occurrents, had made it unsafe for him to rely absolutely upon
> his Memory for all the circumstances fit to be set down in the Hystorical
> [historical] part of the design'd Appendix.[24]

This admission that he did not trust his memory with circumstantial details after a considerable lapse of time reinforces the fact that Boyle *did* believe that on many occasions he could remember various pieces of empirical information. It is likely that his ability to do this diminished over time: writing in 1682 about a work started in 1666 he said that, confronted with relevant papers, he had tried to "make up the Gaps that remain'd between their Parts, by retrieving, as well as, after so many Years, my bad Memory was able to do, the Thoughts I sometimes had, pertinent to those purposes."[25] Often, in those instances in which he acknowledged that his unaided memory was untrustworthy, Boyle implied that notes assisted him to recover relevant details:

> And I speak less promisingly of what I am to say in the remaining part of this paper, because I have not by me any Notes to assist my Memory, (which I dare not trust alone) concerning the Issues of the not numerous Trials, I had the Opportunity to attempt in pursuance of these Thoughts.[26]

This recourse to both memory and notes was a regular part of Boyle's intellectual practice.

Lessons about Loose Notes

The great German polymath Gottfried Wilhelm Leibniz confessed that he was not careful in keeping track of his ideas, notes, and papers. In March 1693 he revealed that "after having done something, I forget it almost entirely within a few months, and rather than searching for it amid a chaos of jottings that I do not have the leisure to arrange and mark with headings, I am obliged to do the work all over again."[27] Leibniz knew that he should have been following the precepts about keeping notes under appropriate Heads in notebooks, as advised by the Renaissance humanists and practiced by many leading scholars of his own time. Leibniz endorsed this discipline, even referring (in another letter) to Placcius' *De arte excerpendi* (1689), which summarized several generations of advice about note-taking techniques and methods of arranging entries in notebooks.[28] The consensus was that one should make notes under Heads: that is, under categories, themes, topics, and so on. This in turn encouraged recollection of, and improvisation on, cognate material rather than verbatim recitation of things learned by rote. Consequently, commonplace books were regarded as supports, not substitutes, for memory; notes kept under Heads were prompts to recollection, considered as part of memory.[29]

As Leibniz's admission implies, the standard advice cautioned against the use of notes on loose slips. The danger was that these would never be assigned to a Head, and thus not deliver all the advantages that commonplacing was meant to achieve for memory and thought. There was also the standard tip that scholars (especially when traveling) should use pocket-sized paperbooks to collect information or quotations, later copying these into the notebooks normally prescribed for the collection of "commonplaces." However, deviations from such recommendations were common. When Montaigne acknowledged that he relied on notes rather than his memory, he mentioned "flipping through my notes (which are as loose as the leaves of the Sybils)."[30] Placcius remarked that those who disliked loose slips called them "foliis Sibyllinis," suggesting that these can be easily lost.[31] Nevertheless, the appeal of such loose notes was on the rise. As mentioned in chapter 2, Richard Holdsworth reported in about 1637 that he had been told about "a box" divided into partitions for various Heads into which notes on pieces of paper could be thrown. In principle, this box sounds like a limited version of what we now know as Thomas Harrison's "invention," first discussed in Hartlib's circle and outlined in a manuscript in his possession.[32] As part of Harrison's attempt to secure Parliamentary finance for his invention, it was described as "most aptly disposing of every observation taken out of any Booke whatsoever by vertue of Notes all of them loose."[33] Placcius later printed Harrison's anonymous Latin manuscript in his *De arte excerpendi*.[34] As we have seen in chapter 4, this explained how extracts from books were kept on small loose pieces of paper, tied or folded together, and then placed on hooks fixed to a specially designed cupboard (*Arca Studiorum*). Placcius regarded Harrison's design as a safer way of managing loose slips. However, it is a nice thought that Leibniz's predicament might have followed from the fact that he was indeed using pieces of loose paper (*Zetteln*). We know that he had an excerpt-cabinet constructed according to Placcius' description.[35]

Physically loose sheets need not be conceptually loose notes. The slips of paper described and promoted by Placcius were assigned to Heads of some kind. Indeed, prior to Harrison's invention, Hartlib had been entering information and ideas in his Ephemerides as they occurred to him rather than assigning them to separate pages on the basis of topics. However, in the margins adjacent to the entries he usually wrote keywords or Heads, thereby tying a particular entry to a subject. He was able to cite figures such as Keckerman, Hübner, and Jungius who devised similar techniques for assembling data that did not follow strict commonplace method. There was, therefore, an alternative practice that preferred short, loose notes, taken as the moment invited or required, and perhaps assigned to a temporary heading.[36] There was also a view that these loose notes could more easily be re-sorted into new combinations, thus enabling exploration of complex relationships among data. As we shall see in chapter 8, Robert Hooke recommended this procedure. In discussing Boyle's reflections on his note-taking, I bear these alternatives in mind.

Boyle's Notes

In the Boyle archive in the Royal Society of London there are some twenty bound notebooks, some of which probably exist today in the form in which Boyle used them, whereas others were rebound in the nineteenth century.[37] Boyle usually called them "Note-book(s)" or "Note booke (s)."[38] There are also many loose folio sheets stitched together and folded to make small, pocket-sized, booklets. Boyle referred to these as "Memorials" or "Adversaria" or loose notes, but usually not as notebooks, presumably because they were unbound. From the late 1640s until his death in 1691, Boyle used forty of these "workdiaries" (as we now call them).[39] In describing some of the earliest of these, Hunter and Littleton remark that they are "virtually indistinguishable from the commonplace books familiar to students of Renaissance literature."[40] Although this is true of the content, which comprises mainly aphorisms drawn from English and French authors, the format already ignores any methodical use of headings (see, for example, workdiaries 1–5). The entries do not seem to be grouped in any way, except that some sequences may have come from the one text. Those devoted to medical recipes (workdiaries 6–11) are similar. The titles of many of these diaries—such as "Diurnall Observations" (workdiary 1) and "Miscellaneous excerpts" (workdiary 5)—are more suggestive of the journal or diary form than that of the commonplace book. But in any case, as Hunter has observed, by the 1650s the workdiaries became more focused on science, and the entries then included not just extracts from texts, but observations and experiments.[41]

The titles Boyle gave to individual diaries imply loose collections: for example, "Promiscuous Observations begun the 24th of September 1655"; "Promiscuous Experiments, Observations, & Notes"; and "Loose Experiments, Observations & Notes about The Preservation of Bodyes."[42] The title of workdiary 19 (used in the early 1660s) suggests Boyle's procedure: "Philosophicall Entrys & Memorials (of all sorts), Here confusedly throwne together; to be hence transferr'd to the Severall Treatises whereto they belong."[43] Perhaps as good resolutions, several start on the first day of the year: "Memorialls Philosophicall Beginning this Newyears day 1649/50 & to End with the Year. And so, by God's Permission, to be annually continu'd during my Life."[44] At least one starts on his birthday: "Physiologicall Notes, Begun the 25th of January."[45] Yet despite Boyle's own descriptions, it would be misleading to characterize these workdiaries as wholly chaotic; indeed, several announce themselves as "Philosophical." They comprise long sequences of short entries, often numbered in "Centuries," thus emulating Bacon's *Sylva Sylvarum* (1626) with its ten centuries, or 1,000 observations or experiments (see fig. 6.1).[46] It is interesting that the young Boyle displayed a penchant for numbered entries: some of his student notebooks boast "centuries" that clearly involved a considerable forcing of the material into the requisite sets.[47] In some publications of unfinished material, Boyle was also content to use "pentades" of experiments and observations, numbered 1–5.[48]

Boyle did not attend university and hence did not experience the almost-daily discipline of assigning material to topical Heads. It is possible, however, that Isaac Marcombes, his tutor during his time in Geneva, advised the young aristocrat on note-taking practices.[49] It is clear that Boyle was aware of the status of the commonplace method, occasionally mentioning a decision not to use it. In *Some Considerations touching the Style of Holy Scriptures* (1661) he said that he would not follow authors who "allow themselves to be lavish in Ornaments, to expatiate into Amplifications, and to drein Commonplaces." In the *Usefulness of Natural Philosophy* (1663), when stressing the importance of both the "History of Diseases" and "the Materia Medica," he explained that although "these particulars, . . . might easily be enlarged on" he did not have "the leisure nor designe to handle them commonplace-like."[50] Moreover, in rejecting the standard commonplace book, Boyle identified the kind of notes he was taking. In the *Advertisement* of 1688 about the lost papers, he mentioned

> four or five Centuries of Experiments of my Own, and other Matters of Fact, which from time to time I had committed to Paper, as they were made and observ'd, and had been by way partly of a *Diary*, and partly of *Adversaria*, register'd and set down one Century after another, that I might have them in readiness to be made use of in my design'd Treatises.[51]

The deliberate distinction between diary and adversaria is significant: we may reasonably infer that the "diary" format was chronological, which im-

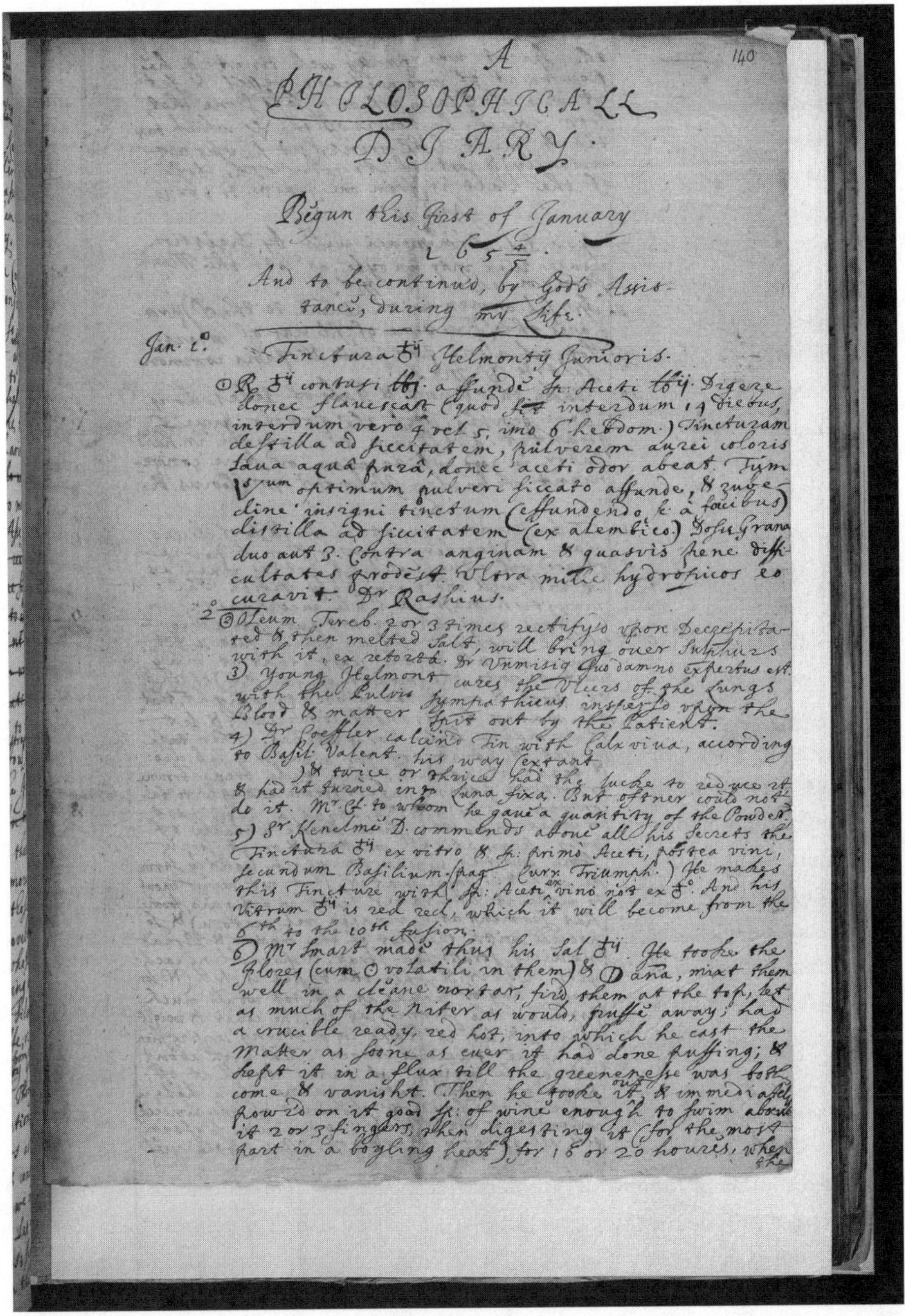

Figure 6.1. Robert Boyle's "A Philosophical Diary" starting on January 1, 1654/5, showing numbers 1–6 of 100 entries (a "century") in workdiary 12. BP 8, fol. 140ʳ.
©The Royal Society of London.

plies, by contrast, that the "adversaria" were notes assigned to topics, albeit perhaps general ones able to accommodate as many as 100 items. Such a
combination resembles the format of Hartlib's Ephemerides and, indeed, he
described this procedure in an entry of 1635.[52] In another entry of 1648 about
the need "to keepe diaries exactly of all whatever wee heare or see," Hartlib
added that "Mr Boyle also is very much for Adversaria."[53] About this time,
Boyle mentioned his plan "to keepe a kind of written Diary," telling his sister that he once drew up a "Draught Modell" and "for a time, made use of."[54]
Even though Hartlib kept a diary, not a commonplace book, his "adversaria"
were notes marked in some way by a topic or Head word.

For Boyle, the "loose notes" in his workdiaries are his "adversaria," even
though they are not always accompanied by a marginal Head. Take, for example, the preface to his tract of 1674 about the "Preservation of Bodies":

> My willingness to make the bulk of the Papers about the *Hidden Qualities
> of the Air* less inconsiderable, by things that were of affinity to the Subject,
> inducing me to tumble over some of my *Adversaria*, I met among them with
> divers loose Notes, or short Memorials of some Experiments I made several
> years ago (and some of a fresher date) about the *Preservation of Bodies* by
> excluding the *Air*.[55]

In his published works there are eighteen reports about finding something
"among my *Adversaria*": for example, "About which Experiment I find this
short memoriall among my *Adversaria*."[56] In a set of *Tracts* (1670), Boyle
stated that he "will here annex 2 or 3 testimonies, the first of which I find
thus set down among my Adversaria." There follows an almost verbatim copy
of entry 204 from workdiary 21 (late 1660s).[57] Elsewhere he pointed to the
"following Observation transcrib'd *verbatim* out of one of my *Adversaria*."[58]
Sometimes he reported more limited success: "Of my Observations about
things of this kind, I can at present find but few among my *Adversaria*; but in
Them I find enough for my present turn. For They and my Memory inform
me."[59] On other occasions he regarded his notes as a secure record, suitable to
be communicated in print. When discussing the effect of different combinations of salts on the temperature of fluids, he said that "The clearest Instance
I found of this Observation was afforded me by an Experiment made with the
Solutions of Album and Nitre; a Relation of which I find among my *Adversaria*."[60] So it appears that the combination of short numbered entries, occasional marginal Heads (usually added post facto), and perhaps the title of the
workdiary, sustained Boyle's confidence in the help his notes would supply.[61]

Most of Boyle's allusions to his "adversaria" or, more often, simply to his "notes," are avowals of his reliance either on memory or notes, or on both. Such declarations would be unremarkable were it not that Boyle broke with traditional precepts in at least two ways. First, his loose notes transgressed the advice that only careful commonplacing under Heads guarded against the natural weakness of memory. Second, the declining state of his eyes, a condition that led him to consult the physician, William Harvey, in 1656–57, meant that Boyle could not make his own notes.[62] In *New Experiments, Physico-Mechanical* (1660) he complained about "the distemper in my eyes forbidding me not onely to write my self so much as one Experiment, but even to read over my self what I dictated to others."[63] From the mid-1650s he relied mainly on amanuenses, as we can see from the handwriting of the workdiaries.[64] In 1696 Evelyn told Wotton that when Boyle "had quite given-over reading by Candle, which Impair'd his sight: This was supply'd by his Amanuensis, who sometimes Read to him, and wrote-out such passages as he noted, and that so often in Loose Papers, pack'd-up without Method."[65] According to the contemporary manuals, such a disjunction between reading and note-taking removed a crucial aid, because attention and, hence, memory, were better served by writing and noting than by repeated readings. When Peiresc conceded that he did not trust his memory, he observed that he "had found by experience, that the very labour of writing did fix things more deeply in his mind." He also ensured that he "wrote on the top of the leafe, or the upper part of the margent, the Subject or Title of what he was to note down."[66] In not writing all of his own notes, was Boyle deprived of this aid to memory? We need to consider what he believed his memory could retain, and what his notes might help him to recollect.

Memory and Recollection

Before further consideration of Boyle as a note-taker, it is salutary to appreciate his optimism about the capacity of memory to retain and recall information and ideas *without* the aid of external records. As seen in the previous chapter, this attitude can be traced to his early writings on moral and mental discipline, which professed that memory could be improved by drawing out the circumstances and consequences of observations and ideas through meditation. The striking point is that Boyle maintained that this method allowed the individual to recall experiences and trains of thought without any assistance from notes. He anticipated an objection: "That to practise this kind of thinking, one is oblig'd to the trouble of writing down every Occasional Reflection that employs his thoughts; and they conclude it far easier to forbear making any, than to write down all." However, Boyle assured these readers that such recording was "no less unnecessary than tedious." Naturally, if one wanted to communicate some thoughts to others, these should be committed to paper, but "for the rest of our Occasional Reflections, though they fill our heads, they need not employ our hands, as having perform'd all the service that need be expected from them within the mind already."[67] Here Boyle seemed to align himself with the tradition of mnemonic arts that relied on internalized images and associations to cue recollection of things stored in memory. In contrast, as seen in chapter 2, the proponents of methodical note-taking, such as Sacchini and Drexel, argued that writing, including short notes, offered more reliable support to recollection. Boyle's attitude therefore needs to be examined in the light of early modern understanding of how the retrieval of information might be distributed across memory and notes of various kinds. This issue was discussed under the aegis of the *ars memoriae* tradition, and also with reference to the function of commonplace books —by humanists and Jesuit scholars, and by Francis Bacon.

In the *Advancement of Learning* (1605), and later in the extended Latin version, *De augmentis* (1623), Bacon contended that the classical mnemonic arts might be improved so as to assist in what he called the "Custodie or retayning of Knowledge."[68] This task belonged to the office of "Memorie," but Bacon affirmed that "Writinge" was a crucial auxilliary: "The great help to the memory is *writing*; and it must be taken as a rule that memory without this aid is unequal to matters of much length and accuracy; and that its unwritten evidence ought by no means to be allowed."[69] There was much at stake because Bacon acknowledged that his program demanded the intensive collection of empirical particulars, far beyond the capacity of memory to retain:

> But even after such a store of natural history and experience as is required for the work of the understanding, or of philosophy, shall be ready at hand, still the understanding is by no means competent to deal with it off hand and by memory alone; no more that if a man should hope by force of memory to retain and make himself master of the computation of an ephemeris.[70]

In this connection, Bacon emphasized the function of writing—not only in the transmission of knowledge but also in its discovery. In the passage just cited, he went on to urge that "Now no course of invention can be satisfactory unless it be carried on in writing." On this point it is helpful to view note-taking as a special form of writing that fits Bacon's concept of "writing in discontinuous sections"; this was, in his opinion, one of the ways to create "tags," cues, or prompts that "help the memory." He explained that the intellect was unable to do anything when confronted with an "army of particulars" unless these were brought into some order, such as that provided by "appropriate tables of discovery." With information thus arranged and stored, it would then be possible for the mind to "buckle down to the organised assistance made ready by these tables." He regarded this as part of "the ministration" to memory and reason.[71]

Sophie Weeks and Rhodri Lewis have recently underlined the importance of the concept of "*experientia literata*" (literate experience), which encompassed the summary and arrangement of information in written form, including notes.[72] In the *Novum Organum* (1620), Bacon referred to this notion in discussing the role of natural history collections as a preparatory stage of the inductive method.[73] In principle he did not see this process as a simple accumulation of an "abundance of particulars" in an "Indigested Heape"— although, according to William Rawley, Bacon conceded that this might well be the appearance of his *Sylva Sylvarum*.[74] He distinguished two stages of inquiry, each using somewhat different written formats. Before the intellect could begin to produce general laws or axioms by induction, Bacon insisted on the collection of adequate natural histories of particular topics, ranging from celestial phenomena to technology. In sketching a list of "Titles" in his "Parasceve," or "Preparative to a Natural History," he stressed that the information must be "set down" in writing: "For I want this primary history to be written up with the most religious care."[75] He urged the use of Titles, or Heads, for the lists of observations and experiments in each natural history. In this way, the data was recorded in permanent form, and the Titles aided recollection and retrieval of the material, which was too diverse and detailed to be retained in memory. However, Bacon contended that although written records relieved memory, these were not sufficient in themselves. He explained that "even after the abundance and matter of natural history and experience" is collected, "the intellect is still quite incapable of working on that matter unprompted and by memory."[76] This called for a new step in the process of inquiry and a specially devised written display. The aim was, as he said, to make experience "literate" in a way that brought "men to particulars," and "especially to ones digested and laid out in my tables of discovery."[77] Bacon's famous "*Tables*, and *Structured Sets of Instances*" sifted the data in the preliminary histories and arranged it in new ways, allowing the "understanding" to discern possible patterns indicative of the underlying "form" (or cause) of, say, Heat.[78]

How do Boyle's views compare with Bacon's advice? The use of "General Heads," as outlined in his paper of 1666, does accord with Bacon's stress on the organizing value of "Titles" for the preliminary (or "primary") natural histories.[79] It is important to recognize that for Boyle these Heads had a prospective as well as a mnemonic function: they served to guide the collection of additional material and to suggest new lines of inquiry.[80] However, Boyle eschewed Bacon's tables of instances, which were supposed to allow re-sorting and comparison of information in combinations that could not be contemplated by unassisted memory. Boyle's Heads continued to function mainly as data-collection points rather than as a means of evincing the presence or absence of qualities inhering in the phenomena. One reason for this different approach is that he was mainly concerned with the early stage in which observations and experiments were still being collated. He also thought that Bacon's preliminary natural histories were too brief and insufficiently copious.[81] He sought to correct this by taking more careful account of "the Immensity and variety of the Particulars that pertain to Natural History" and the most appropriate "way of writing" and "compiling a Natural History."[82] In fact, Boyle's conception of written data was more extensive than Bacon's, since it encompassed all the notes in his workdiaries and other notebooks, prior to any extraction into lists, or allocation to Titles, or arrangement in Tables. Indeed, we might say that Boyle was more interested in what Bacon said about the role of short notes, that "host of circumstantial details or tags [which] help the memory, as writing in discontinuous sections."[83] Boyle wanted to revisit, reread, and cross-reference the raw material stored in his loose notes. Rather than seeking distillation of this information as a way of reducing the load on memory, Boyle regarded his notes as a way of accessing memories of related experiences, experiments, and associated circumstances. We therefore need to examine more closely how he envisaged his notes as doing this.

Recollecting with Notes

In the second part of his *Christian Virtuoso* (written in 1691–92), Boyle pondered the storage capacity of memory, having accepted the "received opinion" that "the seat of memory" lay in "one part or region" of the brain. Boyle asked how it was that "so vast a multitude of things," especially in the case of a "learned man," could be "crouded into and yet commodiously lodged in a portion of the brain, and that so durably, that after forty or fifty years they may be at pleasure brought to appear before the mind."[84] Although Boyle expressed wonder, this picture of an individual in command of a lifetime of experience and learning describes his own case. Moreover, the notion of bringing ideas to mind invites the distinction between remembering and recollecting: Boyle continued to trust his memory in the absence of external aids, but his notes served as crucial prompts to recollection, including the rich episodic memories built up in his youth.

Despite Boyle's early interest in memory training, he did acknowledge that writing was a necessary support.[85] Throughout his life as an experimental philosopher, he often conceded that he did not trust his memory with *details* in the absence of notes. In the preface to *New Experiments Physico-Mechanical* (1660) he apologized for "being somewhat prolix in many of my Experiments," saying that "divers Circumstances I did here and there set down for fear of forgetting them, when I may hereafter have occasion to make use of them in my other Writings."[86] In 1672 in his work about gems, he referred to some "purposely contriv'd Experiments" but said that "not having the Minutes of them by me, and not daring to trust my single memory in Experiments so nice, and so long since made, as those were, I shall here put an end to your trouble."[87] Boyle believed that not only the passing of time, but also the intervention of other subjects, could confuse or weaken memory. In *Memoirs for the Natural History of Humane Blood* (1684), he explained that he had drawn up "a Set of Enquiries, . . . *yet* those Papers being since lost, and a long Tract of Time, and Studies of a quite other nature, having made me lose the Memory of most of the Particulars."[88] However, when Boyle speaks of being assisted by his notes, we need to be careful about the kind of recall involved.

One complication is that a significant number of entries in the workdiaries are actually records of what Boyle reported orally to an amanuensis — as memories. Boyle used the phrase "I remember" (or variants such as "if I misremember not") in twenty-one of his forty workdiaries; and within some diaries he repeated the phrase: thus workdiary 21 has twenty-nine instances of this kind. Take, for example, the records of experiments and observations about the colors produced by various chemicals in combination with metals and other substances:

The same spirit being put upon a small Quantity of Salt of Potashes & soe likewise upon Salt of [tartar] did upon the former (&, as I remember, upon the latter,) colour itselfe in the Cold, & that in few howers with a reddish colour somewhat muddy like that of claret-wine.[89]

Or this remark: "I tooke strong spirit of salt & with it made of a solution of filings of Copper. Solutions thus made I remember I have observd to have a Blewish, wrinkled filme." In some of these instances Boyle purported to recall facts from observations and experiments made a decade earlier: an entry from the early 1670s about the making of "butyrum Antimonii" (butter of antimony) records that he first looked into this "as I remember about the year 1660."[90] Some other entries were notes of what he had once meant to do: thus, in a workdiary 19, used in the early 1660s, he noted that

> I remember that once I had a mind to try whether the coldnes producd
> upon the solution of beaten Sal [sal ammoniac] in water might not be more
> probably referrd to some change of Texture, or motion resulting from the
> Action of the Liquor upon the salt then to any infrigidation of the water.[91]

We can see that some of these entries contained quite specific empirical details, either relating to Boyle's own experiments, or to what people had reported to him about events, animals, and natural phenomena. Boyle's recovery of these from memory did not seem to be assisted in any obvious or reliable way by the general Heads he suggested in 1666.[92] Rather, his memories of these particulars were often connected with temporal and spatial circumstances, as discussed in *Occasional Reflections*; some, perhaps, were triggered by his train of thought during dictations to amanuenses. In any case, Boyle made a note of an existing memory, turning it into a permanent record. This did not mean that he trusted all of his notes as accurate records. Thus when discussing measurements made with thermometers in *New Experiments and Observations touching Cold* (1665), he advised that

> Among the several notes, I find among my loose papers, and in a Diary
> I kept for a while of these observations, I shall content my self to transcribe
> the following two, because, though divers others were made by my Amanu-
> ensis, whose care is not to be distrusted, yet by reason of my absence
> I could not take notice of them my self.[93]

In the same work he treated some readings from a "Weather-glass" with caution, saying that "Thus far the notes, I have yet been able to recover: and though, as I said, I dare not build very much upon them."[94]

Although most of Boyle's notes were entered chronologically, as they occurred to him, and without being assigned to topics, there is no doubt that he recognized the importance of Heads as prompts to recollection. Indeed,

in this way he expected to be able to recollect additional ideas and information not in the notes themselves. One recurring theme in his apologies and
reflections is the belief that he could recover at least part of lost materials if
they were originally entered under a *memorable* Head. In a manuscript draft
from 1660–80 entitled "Of the several degrees or kinds of Natural knowledge," Boyle explained that he had lost the original version "but yet, since I
retain some memory of the chief heads it consisted of, I shall here present
you with a summary of them."[95] There is a hint of how such memories might
have been sustained by a habit of reading over notes. In 1665 he told Oldenburg that he had come across

> some rough copys of my notes about some subjects . . . & some of them
> I have not yet, that I remember read over this 5 or 7 years, the cheife heads
> are about sensation in generall, about the pores of greater & figures of
> smaller Bodys; & about Occult Qualitys.[96]

He even inscribed this conviction about recollection into a notebook of 1689–
90, describing some entries as "A Continuation of loose notes &c most of
them set down to recall to minde fuller passages referable to them from January 25th."[97] This preoccupation continued to the end of his life and is en-
shrined in an "Advertisement" (c. 1690) about his unwillingness to receive
visitors because of some "unlucky accidents" that had corroded or maimed
"many of his Writings." He therefore needed time "*both* to recruit his spirits, to range his Papers & fill up the *Lacunae* of them." Such gaps he intended
to repair "out of his memory or Invention."[98] We have seen this assurance in
the passages I have cited from some of his experimental works. What is not
clear is whether the impediments to memory—Boyle cited excessive detail,
lapse of time, interference from other subjects—would also weaken the recollective search prompted by a note.

If topical Heads aided recollection of information, why did Boyle feel confident about notes made in other ways, such as those entered in chronological
sequence in his workdiaries? One possibility is that the marginal heads he
assigned to some numbered entries (albeit often retrospectively) may have
provided equivalent help.[99] Another alternative is that in the choice between
taking notes quickly and pausing to allocate Heads, he gave priority to the
former. Here it is worth emphasizing that when Boyle talked about his notes
he was often referring not to textual excerpts (although he did make these)
but rather to notes of observations, experiments, and the thoughts arising
from these.[100] In one seemingly casual remark he indicated that this in itself

showed that traditional methods were not suitable for his purposes. In the
Greatness of Mind, he began by addressing a "friend," saying that "you seem
to desire but my own Thoughts, so I know not, whether common Place-books
would afford me any great Assistance on so uncommon a Theme."[101] When
Boyle referred to "my own Thoughts," it was invariably in the sense of a rejection of authority, and here it carried the additional insinuation that commonplace books did not usually include the thoughts of the compiler.

Boyle evinced a compulsion to take down thoughts on the run, as they
came to him, without worrying about thematic arrangement. In contrast
with the standard instruction to ponder and digest before making a note,
he assumed that material not entered rapidly would be forgotten. Thus in
Occasional Reflections, he tells his sister that after dinner with others he re-
tired "into another Room, and set down in short hand, what I have hitherto
been relating, least either delay should make the particulars vanish out of
my Memory, or they should be confounded there by the accession of such
new Reflections."[102] The assumption was that thoughts, more so than passages found in texts, were quite likely to escape unless recorded. Boyle extended this point to thoughts about experimental observations, explaining
that he tried to "set down their Phaenomena whil'st they were fresh in his
Memory, if not objects of his sense."[103] In the early 1670s he noted in one
of his workdiaries that "because the last tryall is freshest in my memory
I shall set down that instead of all."[104] When it came to using such notes in
publications, Boyle maintained that expeditious recording of experimental
results was more important than their systematic or discursive arrangement: "I shall venture to subjoyn the naked Transcripts of my Experiments,
as I had in an artless manner set them down with many others for my own
remembrance among my *Adversaria*."[105] He requested some indulgence:
"Tis hop'd that neither Connection nor style will be expected in loose Notes
hastily set down at several times, to secure the Matters of fact, then fresh
in Memory, from being, as to any necessary Circumstances, forgotten."[106]
Boyle even convinced himself that his mode of presentation was not too exceptional, claiming that

> the Method of Writings that treat about Experimental Philosophy, is not
> much minded and remembered by the Reader, at least after the first perusal;
> the Notions and Experiments themselves, abstracting from the Order they
> were deliver'd in, being the things that Philosophers use to take Notice of,
> and permanently retain in their Memories.[107]

Boyle gave an explicit analysis of his note-taking in a manuscript entitled "Introduction to my Loose Notes Theological," written in the 1670s or 1680s. Here he admitted that a large part of his papers might seem "a Chaos of Promiscuous Notes confused'dly thrown together."[108] Similarly, in "Cogitationes Physicae" (1670s or early 1680s), he admitted that he was sending some material "without any other order then that wherein they chanc'd to occurr to me, they being parts of a chaos of thoughts confusedly set down in short notes." However, he asserted that short notes, entered quickly, helped his thinking: they functioned as "hints to excite my own thoughts."[109] It is also evident that Boyle had given his practice some consideration. He alluded to "Notes of various sorts, unequal Lengths and differing degrees of Moment," distinguishing, for example, between short and long notes, between those that registered matters of fact, or encapsulated oppositions, or fixed a series of thoughts.[110] On this last point it is worth underscoring the way he singled out notes that

> might do my self the service of recalling, when need should require, into my memory, the *Series* of thoughts which came into my mind at the same time, upon the same occasion, with those short notes, some of which by being but simple, or little more than bare Expressions or References, you will easily guess, need not be rendred insufficient for the intended purpose, since this was not to informe Other Mens understandings, but only to recall things into my own memory.[111]

In this passage we hear Boyle's typical accent on the personal value of his notes—they helped him (not others) to recollect certain related ideas and, as we have seen, they prompted recollections of other notes, some of which were lost or misplaced.

Additionally, there is the implication that the utility of such a note consisted in its preservation of a series of thoughts or arguments in the original order. It has long been recognized that notes, diagrams, and images may perform at least two functions: as Mary Carruthers puts it with respect to their role in medieval books, these "can serve as 'fixes' for memory storage, or as cues to start the recollective process."[112] In his *Rules for the Direction of the Mind* (c. 1628), Descartes stressed the former benefit in explaining how "a continuous movement of thought" was necessary in making "a long chain of inferences." He admitted that "weakness of memory" made this difficult and revealed his own attempts to rehearse the intermediate steps until I have learnt to pass from the first to the last so swiftly that memory is left with practically no role to play, and I seem to intuit the whole thing at once. In this way our memory is relieved, the sluggishness of our intelligence redressed, and its capacity in some way enlarged.

However, in rule 16, he emphasized reliance on written reminders:

But because memory is often unreliable, and in order not to have to squander one jot of our attention on refreshing it while engaged with other thoughts, human ingenuity has given us that happy invention—the practice of writing. Relying on this as an aid, we shall leave absolutely nothing to memory but put down on paper whatever we have to retain.[113]

Descartes later reinforced his point in a letter to Cornelis van Hogelande, asserting that "the fewer items we fill our memory with, the sharper we will keep our native intelligence for increasing our knowledge."[114] Boyle acknowledged this when he remarked on how difficult, if not impossible, it was "to go through with a long Geometrical Demonstration, without the help of a visible Scheme, to assist both the Fancy and the Memory.[115] From this perspective, notes (and pictures or diagrams) worked as external aids for thinking, precisely because their content did not have to be memorized. But Boyle also often stressed that notes played the second function: they led back via recollection to his memory or to other notes stored in his papers. Reliance on memory and recourse to notes were two sides of the one coin.

Communicating by Notes

What, however, did Boyle think about the use of notes as carriers of information to others? Here Bacon's warning was salient: some short notes worked well as memory prompts for the individual who made them, but for collaborative purposes notes need to be detailed records that could be understood by others. Yet although Boyle's notes were indeed closely involved with his personal memory and recollection, he also imagined a public use for them. In 1671 an article in the *Philosophical Transactions* about the recent observation of sunspots by Giovanni Cassini in Paris suggested that "the last observation in *England* of any *Solar Spots*" was made by Boyle. It printed the account as Boyle now "found it registred in his *Notes*": namely on "Friday, April 27 1660 about 8 of the clock in the Morning, there appear'd a Spot in the lower limb of the Sun a little towards the South of its Aequator, which was entred about ¼₀ of the Diameter of the Sun."[116] In some cases, Boyle published such notes, especially if he considered the material useful, albeit unfinished. We can glimpse one moment of decision when he wrote "Magnetical Notes" in pencil at the top of a manuscript later published with related matter in *Experimenta et Observationes Physicae* (1691).[117] And to complete the circle, there are examples of individuals making notes from Boyle's works.[118] One physician made an "Index rerum" of medical topics from *Usefulnesse* (1663) containing approximately 1,000 terms arranged alphabetically from "Ambusta" to "Vulnera interna."[119]

Boyle probably accepted Bacon's notion of "*experientia literata*" as an important rationale for communication of information, especially his emphasis on bringing diverse observations and preliminary data together for close inspection. In the preface to his work on *Cold* (1665), Boyle explained the reasons that "induc'd me to resolve to draw together the Notes I had on several occasions set down."[120] In 1671 he mentioned a similar consideration for the publication of what he called "several scatterd Experiments and Remarks":

> It may serve to beget a Confederacy and an Union between parts of Learning, whose possessors have hitherto kept their respective Skills strangers to one another; and by that means may bring great Variety of Observations and Experiments of differing kinds into the Notice of one man, or of the same persons; which how advantagious it may prove towards the Increase of knowledge, our Illustrious Verulam has somewhere taught us.[121]

But in contrast with Bacon's stress on reduction and methodical arrangement, Boyle thought that raw notes could be included in material that might benefit later generations. He defended the value of "heaps of Fragments" for the reason that "the Particulars they mainly consist of are Matters of Fact; their being huddled together without Method (though not always without Order) may not hinder them from being fit, if well dispos'd of, to have places some where or other in the History of Nature."[122] Indeed, he announced that he did not mind taking "Notice ev'n of Ludicrous Experiments," and he justified this as part of a projected sequel to Bacon's *Sylva Sylvarum*.[123] In his *Tracts* (1674), Boyle highlighted the decision to communicate some experimental notes about the effects of the exclusion of air on the preservation of bodies:

> And though these be only such as come now to hand, and were most of them set down rather as Notes than Relations, yet being faithfully

register'd, and most of them having been made in Vacuo Boyliano (as they call it) they will probably be New, and so perhaps not altogether useless to Naturalists, who may vary them, and requite me for them, by trying the same Experiments.[124]

Shortly after Boyle's death, his mode of scientific inquiry was criticized for being excessively preoccupied with empirical and experimental detail. In his *New Essays* (written in about 1704), Leibniz agreed with Benedict Spinoza in saying that Boyle "does spend rather too long on drawing from countless fine experiments no conclusion except one which he could have adopted as a principle."[125] As we have seen, one factor in Boyle's style was his possession of a vast number of notes, taken by himself or, more often, by an amanuensis. Thomas Kuhn put this in a more positive light when he commented that Boyle not only recorded data that fitted a theory, but almost all information pertaining to the circumstantial details of his observations and experiments: he, and others, "began for the first time to record their quantitative data, *whether or not they perfectly fit the law*, rather than simply stating the law itself."[126] More recently, Lorraine Daston has indicated that by the early 1700s some experimenters regarded this "strenuous heed to particulars" as distracting. She rightly surmises that Boyle would have been shocked by the routine of the French chemist Charles Dufay, who regularly edited his laboratory notes "to delete individuating particulars and variability."[127] For Boyle, such notes, many of which he described as "loose," were the essential raw material for his own ability to recollect ideas and information; if suitably presented, they were also potentially useful data for others. The study of nature, in his estimate, was "a subject so vast & comprehensive, [it] will afford exercise to the Curiosity & Industry of more than one Writer, perhaps more than one Age." In the meantime, he insisted that the focus must be on the collection of copious particulars, complete with their circumstances, even to the point of including some that "seem mean trivial and as to immediate use barren."[128]

From an early age, Boyle's measured trust in his memory was grounded in habits of close observation and meditation. The practices he outlined in early manuscripts and in *Occasional Reflections* favored disciplined trains of thought on various topics, forming intricate associations that supported retention in memory. Boyle contended that this process did not always require the aid of writing. However, when he began to gather large stocks of empirical particulars, he entered these in numbered entries in his work-diaries. These notes were not only backups for confused or lost memories, but prompts to the recollection of associated material stored in memory. Boyle acknowledged that the use of Heads, as prescribed by traditional commonplace techniques, reliably delivered the material allocated to them, but he also felt reasonably confident of his ability to recollect the essentials even when the content was not assigned to a topic. In using notes in this way, he exploited his youthful habits of meditation: the notes set off a chain of associations or reactivated previous trains of thought. The paradox in Boyle's case is that even though his notes might resemble an unregimented "army of particulars," as Bacon warned, he believed that he could use them productively, both because of the information and ideas they recorded, and through the recollections they stimulated.[129]

Boyle's use of loose notes flouted the standard recommendation to keep entries under topical Heads in bound notebooks. However, to some extent his practice represented a more general deviation from such precepts, especially among scholars and collectors of multifarious information such as Hartlib, one of his early mentors. Putting aside the caveats about topical arrangement sustaining memory, there were strong grounds for taking notes on the run. Such notes, entered quickly, secured empirical details, captured fleeting ideas, and anchored a stage in an argument or a chain of deduction. Boyle's sometime-assistant, Robert Hooke, suggested that the flexibility of such notes was crucial: information on loose sheets could be moved around physically and tried under various Heads in a conceptual scheme, such as one of the Baconian natural histories outlined in his "General Scheme."[130] This reshuffling could help to generate hypotheses.[131] However, although Boyle's numbering of discrete entries may have permitted such conceptual juggling, he was more concerned (belatedly) with assigning his notes to larger classes or Titles, so that he could find them.[132]

Boyle kept personal notes, arranged in ways that suited him. In 1662 he referred to some experimental observations on "cold" in this way: "I find them, in a kind of Note-book, wherein I had thrown them for my own private use."[133] He did believe that some of his raw and fragmentary material might be of public value. However, although Boyle's notes could provide additional information or cues to further inquiries, most of them must necessarily have lost their original powers when passed to others. Detached from their maker's mind, memory, and papers, such notes were no longer avenues to the associations and recollections developed over a lifetime.

[7]

John Locke, Master Note-taker

Mr Locke above all things, loved Order; and he had got the way of
observing it in every thing, with wonderful exactness.

Pierre Coste, "The Character of Mr. Locke," xvi

:: :

This character appraisal invites a consideration of John Locke's note-taking.
From his early student days at Christ Church, Oxford, Locke began to keep
notebooks, and by 1660 he had developed the central features of a method of
entering and retrieving information that he followed for the rest of his life.
At the request of some friends, he composed an account of his way of making and indexing commonplace books that was published in Amsterdam in
the *Bibliothèque Universelle* of July 1686.[1] In 1706 this article about his "Methode Nouvelle" appeared in two different English translations. What became
known as Locke's "New Method" (as I will call it) made him, at least for most
English readers, the default authority on note-taking. Tellingly, in their
entries on commonplacing both Chambers' *Cyclopaedia* (1728) and the *Encyclopédie* (1751–65) singled out this "New Method." Neither of these entries
cited earlier humanist or Jesuit literature on note-taking, or even Locke's
original French article. In introducing the use of commonplace books, Chambers said, "Several Persons have their several Methods of ordering them: but

that which comes best recommended, and which many learned Men have now given in to, is the Method of that great Master of Order, Mr. Locke."[2]

This association of Locke with commonplace books is unsettling. By this time, the practice of commonplacing was firmly linked with an intellectual *ancien régime* characterized by deference to textual authority; in contrast, the "modern" approach, often identified with Locke's own *Essay* (1690), evaluated concepts, arguments, and evidence without reference to the authority or status of their sources. This is what Enlightenment *philosophes* such as d'Alembert were thinking when they remarked of Locke that "he did not study books, because they would have instructed him badly."[3] Nevertheless, it would be a mistake to depict Locke as antibookish: he owned a personal library that included, at its peak in the 1690s, some 3,641 titles, comprising up to perhaps 4,000 volumes, a very large collection for an individual in the late seventeenth century.[4] He did not acquire books for show but rather for reading and study, as witnessed by the copious notes he made. The huge quantity of these makes Locke a successor to prolific note-takers such as Conrad Gessner, Sir Julius Caesar, Gabriel Harvey, William Drake, Nicolas Fabri de Peiresc, Joachim Jungius, Samuel Hartlib, and Vincent Placcius.[5] Yet Locke rejected the role of commonplacing in university pedagogy, in part because he believed that it instilled bad habits of thinking and failed to appreciate the nature and scope of empirical knowledge. Nevertheless, we do not have to accept Ann Moss' severe assessment that the commonplace book was "by Locke's time a rather lowly form of life, adapted to fairly simple tasks, and confined to the backwaters of intellectual activity." At least for the English scene, we have to confront various counter-instances, not least of which is Locke himself, whom Moss rightly describes as an "indefatigable maker of commonplace-books."[6] But before looking at Locke's actual notebooks, it is worth recapitulating the experience of most of his contemporaries, whose first exposure to his note-taking came from the description he gave of it in 1686.

The "New Method"

Apart from a few friends and amanuenses (especially Sylvester Brounouver), Locke's contemporaries knew nothing of his note-taking until he outlined its key features in the "New Method." The readers of the *Bibliothèque Universelle* saw the result of a correspondence between Locke and Nicolas Toinard that began about seven years earlier, in August 1679.[7] Toinard repeatedly asked Locke to produce an explanation of his note-taking practices, but Locke probably did not do this until 1685 when he composed a description in English, sending it to Toinard on February 14/24, 1685.[8] He subsequently made a Latin version (not a direct translation from the English), sending this on March 30/April 9, 1685.[9] In March 1685 they discussed publication options, including the *Journal des Savants*, but nothing eventuated.[10] Later that year Locke must have sent another copy, most likely in Latin, to Jean Le Clerc, who translated it into French.[11]

Before explaining Locke's method, a caveat is necessary: he did not employ the usual English terminology of the day. Invariably, he wrote about his *adversaria*, rather than "commonplace books," and referred to "Titles" rather than to "Heads." When the "New Method" was published in English the translator altered these terms and called the account itself *A New Method of a Commonplace Book*—a title Locke never used.[12] Even in his original English draft of this piece in early 1685, Locke titled it "Adversariorum Methodus." Similarly, Boyle rarely mentioned commonplace books, and he also used the term *adversaria* and often referred to Titles (as well as Heads) as the label for topics, categories, or subjects under which entries were made. In preferring "Titles," Boyle and Locke may have been deliberately following Bacon, which itself might signal their determination to adapt humanist techniques to new ends. However, many of Locke's correspondents did use the more-conventional terms, although occasionally laced with jibes against bookish attitudes: Benjamin Furly, an English Quaker living in the Netherlands, wrote in 1694 about a work he had translated as being "altogether new, and out of the Road of the common place makers."[13]

Locke was very definite about the format in which this account of note-taking should appear, stipulating that "I want these pages to contain not only a description but an example of my method; and I should advise that this description be printed in the same manner; for examples are more informative than instructions."[14] Accordingly, the published French version is set out as a mockup of one of Locke's own notebooks; by building a miniature commonplace book his explanation of the method models what it describes. It opens with the index on two facing pages (see fig. 7.1). An alphabet of twenty letters is arranged in four columns. Each column is divided horizontally into twenty-five cells, allowing five letters to a column with each letter allocated five cells, one for each vowel—thus making 100 cells in the index.[15] As Locke explained in the English draft, "When any thing occurs that I thinke convenient to write in my Adversaria I first consider under what title I thinke I shall be apt to looke for it."[16] Readers of the published English version saw this rendition:

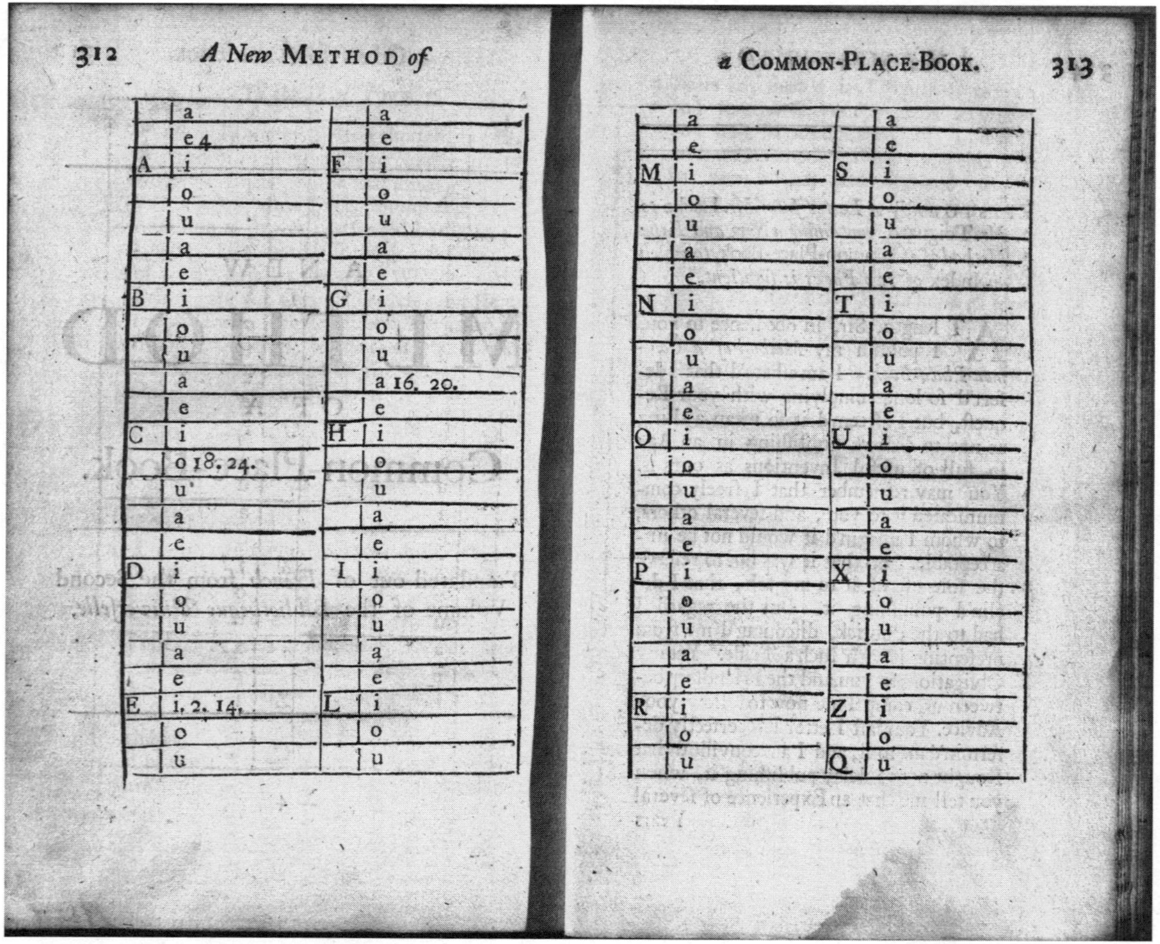

	a			a	
	e	4		e	
A	i		F	i	
	o			o	
	u			u	
	a			a	
	e			e	
B	i		G	i	
	o			o	
	u			u	
	a			a	16. 20.
	e			e	
C	i		H	i	
	o	18. 24.		o	
	u			u	
	a			a	
	e			e	
D	i		I	i	
	o			o	
	u			u	
	a			a	
	e			e	
E	i	2. 14.	L	i	
	o			o	
	u			u	

	a		a
	e		e
M	i	S	i
	o		o
	u		u
	a		a
	e		e
N	i	T	i
	o		o
	u		u
	a		a
	e		e
O	i	U	i
	o		o
	u		u
	a		a
	e		e
P	i	X	i
	o		o
	u		u
	a		a
	e		e
R	i	Z	i
	o		o
	u	Q	u

Figure 7.1. Locke's two-page index in his "New Method." Horizontal lines within each letter/vowel set are ruled in red, and sets are separated by black lines. John Locke, "A New Method of a Common-Place-Book," in *Posthumous Works* (1706), 312–13. Kindly supplied with permission by Rare Books and Special Collections, the University of Sydney Library.

If I would put any thing in my Common-Place-Book, I find out a Head to which I may refer it. Each Head ought to be some important and essential Word to the matter in hand, and in that Word regard is to be had to the first Letter, and the Vowel that follows it; for upon these two Letters depend all the use of the *Index*.[17]

To illustrate this procedure, Locke placed what he had said so far under two Heads — "Epistola" (for his opening letter to Toinard) and "Adversariorum Methodus" (for the explanation of his method). To make these entries in his commonplace book, this is what Locke did. He turned to the first double opening and wrote "Epistola" in the left margin ("in large letters in the margent"), thus ensuring that it would be immediately visible.[18] He made the entry itself next to the Head, keeping the left margin free. He then entered the number of that page in the index, in the cell for *Ei*. To record the material for "Adversariorum Methodus," he first checked the *Ae* cell of the index. If, as he said, he found "noe figure at all I then turne to the first cleare page," namely, the first *new* double opening; here he made the new entry, and all subsequent ones bearing a Head with the *Ae* letter/vowel combination.[19] Similarly, the remaining space under the entry for "Epistola" was reserved for the next Head of the *Ei* class, such as "Ebrietas, Elixir, Epilepsia, Ebionita, Eclipsis &c," to use Locke's examples.[20] He mentioned the prospect of further subdividing by considering the second vowel of each word: as he put it, "if any one thinke these 100 classes too few, soe that some of them will have too many various titles in them," the additional vowel would increase the number of cells in the index from 100 to 500.[21] This would prevent overcrowding of page references in the index because, for example, "Eclipsis" (*Eii*) and "Epistola" (*Eio*) would now be indexed in separate cells. It also meant that entries under these Heads would appear on separate double openings in the notebook (see fig. 7.2).

In the great majority of his commonplace books, Locke followed the steps so far outlined. Thus a new entry either is added to a page already containing other entries with the appropriate letter/vowel combination or, if that page is full, is entered on the next empty double opening. Consequently, the sequence of the entries in these notebooks is not alphabetical; rather, it reflects the order in which the entries were made, so that the first page could just as easily contain "Zeus" as "America." This is why Locke needed an index, of the kind described in the "New Method," to locate the pages on which similar letter/vowel Heads (or titles) would be found. (See fig. 7.5 below showing entries in one of his medical commonplace books.) However, in two of his

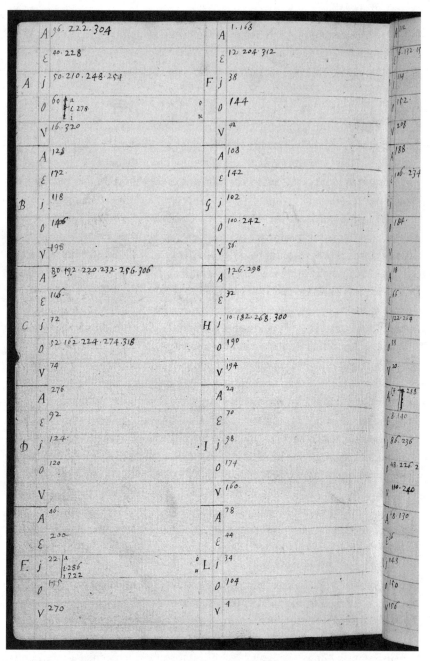

Figure 7.2. The left-hand side of the Index from Locke's "Adversaria Physica," showing page numbers for the Heads in this large commonplace book, and in the cells *Ao* and *Ei* his use of a second vowel. MS Locke, d. 9, p. 534. By permission of the Bodleian Libraries, The University of Oxford.

early notebooks, which he called "lemmata," Locke tried a different approach. He preformatted the pages, assigning to each double opening a first letter/first vowel combination (e.g., *Aa*, *Ae*, *Ai*). Then he subdivided each left-hand side of the double opening into five columns, labeled *a*, *e*, *i*, *o*, *u*, thus incorporating a second vowel. In this case, entries are essentially in alphabetical order and could easily be found without an index.[22] Although Locke continued to make entries in these two "lemmata" he did not ever construct another notebook in this fashion.

Here we have the essence of Locke's method for entering notes in notebooks and finding them again on subsequent occasions. It can be considered innovative in technical terms because each entry was indexed as it was made; thus the index grew *with* the commonplace book rather than being drawn up at a later date, or when the notebook was full. It also avoided the waste of space often seen in commonplace books in which Heads were preallocated to pages that were either subsequently not used, or used so frequently that the entries spilled over onto pages assigned to other Heads — as noticed, respectively, in some notebooks of Southwell and Newton. Furthermore, the distinctive features of the "New Method" had the effect of concealing the topics it covered, disaggregating subjects, and juxtaposing unrelated passages. Locke offered a way of retrieving items, provided that the maker of the note remembered the Head under which it was entered. As seen in earlier chapters, Locke was certainly not the first maker of commonplace books to flout traditional philosophical hierarchies of the university curriculum or the canonical moral topics of the humanists. Furthermore, among his contemporaries and friends, Hartlib, Petty, Evelyn, Aubrey, and Boyle all amassed considerable stockpiles of notes.[23] But only Locke published a description of the method he employed in his own practice. By so doing, he allowed this account to be taken as a source of precepts.

What Locke presented in 1686 was a *formalized* description of the method that he had been using for over twenty-five years.[24] It is therefore unhelpful to view the "New Method" as an ideal model against which to compare Locke's actual practice, or to speak, as Guy Meynell did, of inconsistent application.[25] Apart from the method of indexing and entering, the salient aspects of his note-taking are covered in the "New Method" and, when Locke mentioned choices that might be made, or ways in which his method might seem awkward, his comments were always grounded in experience. However, he did not set out to catalog the trials and variations that are evident in his notebooks from the late 1650s.[26] We can now consider these actual notebooks, none of which was seen by any but a tiny number of people until the mid-twentieth century.

Locke's Extant Notes

Locke died on October 28, 1704, in a room full of his own books in the house ("Oates") of Damaris and Francis Masham in Essex.[27] He had moved there in January 1691, and during this time he completed a catalog of his library, a document of over 1,500 pages.[28] However, the full extent of his notes was only apparent to his cousin and executor, Peter King, to whom Locke bequeathed half of his books and "all my Manuscripts."[29] King thus inherited the bound notebooks and all the loose manuscript notes. Locke stored his books in shelves or wooden boxes that suited the size of the volume, from duodecimos to quartos (ranging from five to eleven inches), or up to twenty-two inches for the largest folio. Six large folio notebooks, titled "Adversaria" and "Lemmata," were kept with the books and listed in his final catalog.[30] The smaller notebooks were not catalogued, and some of them may have been squeezed, together with papers, into the pigeon-holes of Locke's desk, which served as a filing cabinet. True to form, Locke had sketched the dimensions for such a desk in one of his folio commonplace books, specifying that "the pigeon holes on top" be four inches high.[31] His drawing shows these holes as eleven inches deep, which would accommodate at least the octavo-size notebooks. In 1829, when the seventh Lord King published some extracts from the manuscripts, he stated that they had been "in the same scrutoir in which they had been deposited by their author."[32]

Locke's vast storehouse of notes and the range of topics covered deserve comment. On J. R. Milton's estimate, the extant notebooks comprise "thousands of pages and tens of thousands of entries."[33] The notebook titled "Adversaria Physica" (MS Locke d. 9) has about 1,800 separate entries.[34] Locke possessed at least forty-five notebooks, comprising commonplace books (thirteen or fourteen if the two parts of MS Locke c. 42 are separated), journals (eleven), other notebooks (ten), memoranda (seven), and ledgers and account books (four).[35] In size and format these ranged from large folios to small octavos. Some notebooks may have been lost; among these are the quarto-sized "Adversaria 59" and "Adversaria 62" that Locke listed in his journal on July 14, 1681, but did not mention in his final catalog.[36] There were also loose notes of various kinds among his unbound papers, although once in the Bodleian these were sorted under rough subject headings and sewn or pasted into guard books, such as the one titled "Memorandum of books read" (MS Locke c. 33).

The great majority of Locke's notebooks were not available to scholars until the fourth Earl of Lovelace presented the manuscripts bequeathed to him to the Bodleian Library in 1942.[37] A further donation from Paul Mellon, who purchased the Lovelace collection in 1960, added the eight large commonplace books that had been among the books from Peter King's moiety.[38] Locke's nineteenth-century biographer, H. R. Fox-Bourne, was not given access to this collection, so the only notebooks he was able to use were those "escapes" that had settled in the British Library.[39] Nevertheless, he guessed that what he saw belonged to "the sort of notes" that Locke was "in the habit" of making.[40] Fox-Bourne assumed these notes were mainly "philosophical," such as those entries on space, species, and memory from the journals of 1677 and 1678, which King had published in 1829. However, as we shall see, Locke's note-taking extended far beyond what later came to be considered as belonging to "philosophy."[41]

It is important to appreciate that Locke's methodical note-taking pervaded most areas of his life. It was, of course, primarily evident in connection with his reading and study, but it also extended into nonscholarly tasks such as recording expenses and financial transactions and filing letters. In one of his account books he applied the "New Method" style of entering and indexing, making the name of the creditor, debtor, or item the keyword for the index; he did likewise in some of the small memoranda, which mixed reading notes, book purchases, and exchanges of money.[42] In saying that Locke was "a great lover of Order, and Oeconomy, and an Exact Keeper of Accounts," Damaris Masham knew her friend very well.[43] Almost each time he traveled, Locke made a record of the distance of the coach trip in his journal: thus on both August 18, 1681, and May 30, 1682, he noted "From Oxford to London 47^m."[44] This is the distance given in a printed almanac for 1669, which he owned.[45] Indeed, Locke was in a position to supply, and correct, such information: in December 1681 and January 1682 he noted the outward trip from London to Bexwells in Norfolk as twenty-four miles and the return as twenty-three.[46] Here, if not already manifest in his other note-keeping, we may be witnessing a compulsion. Indeed, because we now know more about the force and extent of Locke's habit, the remark of the antiquarian Anthony Wood, who attended a "course of chimistry" in Oxford in April 1663, sounds amusingly off target. Wood recalled Locke as

a man of a turbulent spirit, clamorous and never contented. The club wrot
and took notes from the mouth of their master . . . but the said J. Lock
scorn'd to do it; so that while every man besides, of the club, were writing,
he would be prating and troblesome.[47]

But in fact Locke *did* make notes on these chemical lectures given, at Boyle's
invitation, by the German chemist Peter Stahl. We can see the evidence in
one of his small notebooks titled "Adversaria 4 Pharmacopea."[48] The variety
of other chemical sources in this densely stocked notebook shows that Locke
was by no means reliant on these lectures.

Although Locke's note-taking was fastidious, and perhaps obsessive, it
was not without rationale. This can be demonstrated by examining and con-
textualizing a selection of notebooks from his student days until the years in
which he made his European reputation as a philosopher.

A Lifetime of Note-taking

The young Locke was a pupil at Westminster School in London from 1647,
under the famous headmaster Richard Busby. In May 1652, while still at
school, he was elected to a "studentship" at Christ Church, Oxford. This studentship, equivalent to a fellowship in other colleges, was tenable for life.
Locke took the BA on February 14, 1656, and an MA on June 29, 1658. In May
1661 he was appointed tutor, a position he held for the next six years, in conjunction with formal college positions as praelector in Greek (1661–62) and
in rhetoric (1663), and censor of moral philosophy (1664).[49] By this time he
had already embarked on a mission to take notes on a range of subjects. One
of his earliest account books records payment (in 1661–62) of four shillings
for a "paperbook."[50] As discussed in chapter 2, the keeping of notes was part
of the pedagogical practice of the day. At Oxford, Robert Sanderson's *Logicae Artis Compendium* (1615) was still in use, and it included a section on "*Loci
Communes*" for the treatment of topics set for disputations and, more generally, as a storehouse for copious speech and writing.[51] Students were advised to avail themselves of reading lists, such as those that accompanied
the hints on study and note-taking supplied by influential tutors like Duport or Holdsworth at Cambridge. A more comprehensive bibliography of
this type was issued by Thomas Barlow, the Bodleian librarian between 1652
and 1660. In one of Locke's early notebooks, under the Head of "Barlo Biblio-
theca," there are transcriptions from one of the copies of this list, which circulated in Oxford.[52] The list of books is not in Locke's hand, but he supplied

marginal Heads, giving subjects that largely follow the division that Barlow used, starting with logic, ethics, natural philosophy, and metaphysics.[53] On one page the word "Note" is written next to Barlow's advice about reading authors of logic and metaphysics, such as Christoph Scheibler and Suarez, in order to attain "a sufficient measure of knowledge in that Generall Learneing wee call Metaphysicks."[54] It is plausible that Locke used this list as a guide to his reading and, if so, may have attended to Barlow's recommendation about how to arrange the fruits of reading in a "Methodicall Common Place booke," or his more specific advice about creating two of these, one being "onely for References."[55] Or Locke may have simply absorbed the oral instruction about notebooks, soon improvising on this to create his own method.

Of Locke's surviving notebooks, the two earliest were inherited from his family. The first, with entries from the 1640s, belonged to his father and contains material relating to his work as a magistrate, and some concerning the family property in Somerset.[56] There are definitions of philosophy and ways of teaching "morall philosophie."[57] The second notebook (MS Locke e. 4) is a compilation of medical recipes probably used by various family members; the first page carries the title "Farrago" and the names of "John Locke" (either Locke senior or junior) and "Agnis Locke" (presumably his mother, Agnes). The inside cover bears a date of 1652 but this may have been added later by Locke; most of the entries are from 1657 onward.[58] When the young Locke took over these notebooks he made his own entries, especially in the second one, which he continued to use for medical topics. Here the entries up to page 24 are medical remedies placed mainly under the name of the complaint, such as "Ague," "Gout," "Wounds," "Cough," and "Measles," or a part of the body, such as "Stomach," "Back," or "Teeth." These were standard for the time, as we have seen in Hartlib's diary. On page 23 there is a recipe "for sore Eyes" followed by initials "AL," indicating that it came from Locke's mother. In these pages (and also pp. 165–58 reversed at the back) all entries are in English, although some of the marginal Heads are Latin, many of which have been added by Locke in a different ink (pp. 232–33). A definite change is apparent from page 24 where Latin entries, usually drawn from medical texts, begin to predominate. There are also a small number of entries that point to Locke's involvement in the experimental inquiries conducted by Richard Lower, the physician, whom he may have known at Westminster School.[59]

The great majority of entries in Locke's early notebooks do not contain new ideas or experiments. He assembled extracts, quotations, references, and wish lists that accrued information for possible future use.[60] A high proportion of the notes he made were from books he did not own and perhaps could not be sure of obtaining when required; other notes concern books he had heard about but not yet seen or read.[61] In a small notebook, dated 1667 on the inside front cover, but used between c. 1659 and c. 1667, he gathered commendations (as well as some negative comments) about many famous ancient and modern authors. The latter included Renaissance figures such as Guicciardini, Castiglione, Montaigne, and Scaliger, and near contemporaries such as Campanella, Descartes, Gassendi, and Boyle. The opinions he recorded are usually not his own, because in many cases when he made the entries he had not read the books; rather, he noted the judgments of others, whose works he had read—such as Robert Filmer on Hobbes, Robert Sanderson on Calvin, Richard Hooker (negatively) on Petrus Ramus, Boyle on Bacon, Mersenne, Descartes, Pascal, and Gassendi.[62] The purpose of these entries differed from that of the longer passages copied from the medical works he was currently studying.[63] They are best seen as an attempt to build up a body of extant opinion from reputable authorities. Locke noted what he would need to know in the future, and seemed to anticipate that the sheer volume of entries might overwhelm him. From an early stage, his mode of entering material was also a mode of finding and retrieving it.[64]

From at least 1660, Locke forecast the kinds of material he would be collecting and made a decision to divide his notes into two large categories— "Physica" and "Ethica." The term "Physica" covered medical illnesses, diagnoses, and treatments, pertinent chemical and botanical facts, and experiments and speculations in natural philosophy; "Ethica" included entries on religious doctrines, beliefs and rituals, social customs, and political systems.[65] Initially, this division reflected Locke's two main academic preoccupations: his own keen medical reading, especially from 1658, and his teaching of Greek, rhetoric, and moral philosophy from 1661.[66] Locke named two of his largest commonplace books accordingly: "Adversaria Physica" and "Adversaria Ethica." On the first page of each of these notebooks he also wrote dates next to the titles, underlining the last two digits: thus 1660 for "Adversaria Physica"[67] and 1661 for "Adversaria Ethica."[68] Both notebooks are large-folio paperbooks comprising 544 and 321 pages respectively. In cross-references to these, Locke referred to them as "Adversaria 60" and "Adversaria 61."[69] In spite of these dates, these two notebooks may not have been used, respectively, until about 1666 and 1670.[70] The allocation of these titles exemplifies Locke's advice in the "New Method" that his mode of entering worked best if one used "severall books for severall sciences or at least if he make two different repositorys for those two great branches of knowledg morall & naturall."[71] Accordingly, "Adversaria Ethica" contained what we now class as Locke's philosophical writing, incorporating the first draft of the *Essay* and one copy of *Essay concerning Toleration*.[72] The companion notebook, "Adversaria Physica," covered medical topics, but also more general scientific ones, and it included his long-running "Register" of the weather. I will focus on this notebook, which Locke used between 1666 and the 1690s, to illustrate some key features of his note-taking.

"Adversaria Physica" functioned as a notebook for the full range of Locke's scientific interests. Although the medical content of this large notebook is similar to that of his smaller medical commonplace books, Locke also entered (especially after 1679) material pertaining to some other scientific subjects, such as horticulture, weather observations, and weights and measurements. It is set out according to the "New Method" in which Heads sharing the same alphabetical code are entered on the same double opening. The index at the back of the notebook made it possible to locate entries entered at various times. Most of the early entries are medical, reflecting Locke's reading in the 1660s. We find groupings of topics on the same page, such as Menstrua, Melancholia (p. 6); Hypochondria, Histerica, Hydrophobia (p. 10); Anorexia, Apoplexis, Atrophia, and Abdominis (p. 60). Although this method scatters related topics, such as Febris and Quartana (a fever occurring every fourth day), Delerium and Phrenesis, and Respiration and Sanguis, it also brought together sources and commentaries on a single topic. Thus extracts under "sanguis" from various authorities, old and recent, might be on the one page.[73]

Locke's commonplace books conform in some respects with the standard practice of the day: they mainly comprise excerpts from books, and the entries are not usually dated.[74] However, the exceptions are instructive. In "Adversaria Physica," after the earliest strata of entries, there are some that derive from observation, experiment, or testimony. Such entries normally carry a date. For example, under "Aeris gravitas," Locke recorded barometric measurements (dated April 23, 1666). About a month later he made a chilling entry under "Rhackitis": "June 4. 66 we cut up a child of Dr. Aylworths dead of the Rhickets. It was a boy of about 1½ years old."[75] Similarly in a small commonplace book devoted largely to chemical inquiries made during 1666–67, he regularly dated the observations and experiments, sometimes even including the time of day.[76] Thus if the source was not a book, Locke treated the information as a specific event to be recorded as if it were a journal entry. Another deviation from the genre of commonplacing is Locke's inclusion of his own comments, observations, thoughts, or queries (often marked as "Q").[77] In a small memorandum book (MS Locke f. 27) used mainly between 1664 and 1666 there is a run of pages devoted to such queries.[78] For example, under the marginal word "Sal," the entry asks "Whether volatile or urinous salts acid & alcalizat may by any art of chymistry be changed into another & what difference is to be found amongst the particulars of each of these 3 species."[79] Many of these entries carry his signature, thereby distinguishing them from the excerpts from books in the commonplace style.[80]

Weather Notes

"Adversaria Physica" also incorporates another form of note-taking quite distinct from textual commonplacing. These notes are weather observations in the "Register of the Air" that Locke kept at Oxford from June 24, 1666, to March 28, 1667, and then, with some interruptions, until June 30, 1683.[81] The gathering of this information demanded new habits, such as routine observation and measurement at various times of the day and night. He also made some entries regarding the weather in his journals while in France during the mid-1670s and occasionally after he returned home in May 1679. However these tend to favor more dramatic weather events, such as "a storme of haile the bigest that I ever saw," recorded on May 18, 1680.[82] He did not resume weather observations on a regular basis, or with detailed records, until he took up residence in Essex. His final register, made at Oates "in the great Chamber," runs from December 9, 1691, until May 22, 1703 (the year before his death).[83] Locke's registers from 1666–1683 were printed in Boyle's posthumous *General History of the Air* (1692), probably at Boyle's request. They appear under the section covering the "Weight of the Air" (title XVII) even though barometric measurements were only one component.[84] As well as making a private register public, the publication of this material gave Locke the chance to add some explanatory comments. Thus both the original version in his notebook and the printed one need to be considered. (See fig. 1.1 above for a page from Locke's register.)

Locke's decision to start a register in the autumn of 1666 was influenced by Boyle, and it was a step in their developing personal and intellectual relationship. Boyle had taken up residence in Oxford in early 1656, and Locke probably met him soon after via medical friends such as Richard Lower.[85] In November 1665 Locke left for Cleves on a short diplomatic mission headed by Sir Walter Vane to meet the elector of Brandenburg, Frederic William of Hohenzollern.[86] While there he wrote to Boyle apologizing that "little worth your notice" had cropped up, but then tried his best to send news about local chemical practitioners and the effects of the plague. He justified this small offering by praising Boyle's capacity, grounded in "all manner of knowledge," to put even "slight and barren" information to good use.[87] Not long after returning home, Locke made what may have been his first formal recording of weather, sending it to Boyle in a letter of May 5, 1666:

> Near the House where I sometimes abode, was a pretty steep and high Hill. *April. 3. hora inter 8 et 9. Matutin.* the Wind West, and pretty high, the Day warm, the Mercury was at 29 Inches and ⅛, being carried up to the Top of the Hill, it fell to 28 Inches ¾, (or thereabouts, for I think it was a little above 28 Inches ¾).[88]

The inclusion of air pressure was compensation for Locke's failed attempt to make barometric readings, as requested by Boyle, at the bottom of a lead mine in the "Mine-deep" ("Mendip") hills in Somerset.[89] In June, when he started his own register, he set out to collect the kind information that Boyle, two months earlier, had urged others "to set down in their Diarys."[90] Once Locke embarked on this task, the combination of compulsion and exactitude already on show in his other note-taking was applied with good effect to the observation and presentation of empirical information. In a delayed response to Locke's account of his efforts in Somerset, Boyle made a further request about "the nature of Mineralls," concluding that he looked upon him "as a Virtuoso."[91]

The keeping of personal weather diaries was, of course, not novel, but in England during this period *daily* recordings were more scant.[92] However, not long before Locke began his register, a Scottish gentleman, Andrew Hay, kept a diary from May 1659 to January 1660 in which he made daily remarks on the weather in Lanarkshire, as well as notes on books and sermons. These remarks have been described as one of "the earliest sources of continuous meteorological data extant in the British Isles."[93] Yet, as the authors of this comment admit, Hay's entries, although regular, included no quantitative measurements and his descriptions were colloquial rather than technical: thus, during 1659, for Sunday, May 1: "A very filthie raine all day"; Wednesday, June 1: "A fair, drying day"; and Saturday, July 16: "Fair in the morning, and foul after."[94] Apparently Hay always made these appraisals at the end of the day, after his other notes on reading and a daily assessment of the state of his soul.[95] These are judgments about the dominant feature of each day's weather rather than records of its several components, such as heat, wind, and rain at particular times of the day.[96] Hay's diary confirms what is implicit in Boyle's recommendation in the *Philosophical Transactions* of 1666: namely, that careful measurement and recording were not common. It is important to review this situation in order to appreciate what Locke was undertaking.

The young Christopher Wren argued that useful observation of changes in the weather demanded collective labor and careful use of instruments.[97] Among his desiderata was a "History of the Seasons"; this, he said, was "not the Work of any one Person, and therefore fit for a Society."[98] For the registration of the necessary meteorological data he called for separate "punctual" diaries of significant phenomena including wind; temperature; moisture of the air; the "State of the Air" as seen, for example, in clouds and rain; and a "Register" of unusual events such as "accidental Meteors." He warned that the proper keeping of a "Diary of the Winds and Air" would be arduous because it required "constant Attendance." He proposed that the task be delegated to "four or five Men" who "have Weather-Cocks in view" from their homes, and who should "sometimes compare Notes" in order to compensate for discrepancies and omissions.[99] However, Wren believed that this arrangement was a poor substitute for instruments that automatically recorded the various phenomena, and he offered his own inventions, such as a self-emptying rain gauge.[100] This was one component of the "Weather Clock" that Hooke devised on the basis of Wren's initial design. As Sprat explained, the mechanism moved "a black-lead-pencil" so that its traces on paper showed "what Winds had blown" in the absence of the "Observer."[101] However, even with the possibility of mechanical recording of data, there were problems in making comparisons of readings taken with different instruments, both thermometers and barometers (or baroscopes).[102] These remained unresolved well after Wren's suggestions.[103] In 1684 the issue arose in exchanges between the philosophical societies of Dublin and Oxford: Petty argued that "keeping a Diary of the Weather" was difficult not only because of the apparatus involved but because instruments currently available were not governed by "constant Standards": and if "we make new ones every year, we can make no estimate of the Weather by them, in relation to what was observed last year by others."[104] Such doubts confirmed Wren's assessment of the challenge posed by a natural history of the weather, but also reinforced his call for some group to muster the "Patience to pursue it."[105] Locke was familiar with the ongoing debate and was capable of the vigilant note-taking this task required.

Locke's first "Register" featured a carefully formatted "table," incorporating both quantitative and qualitative information.[106] In contrast with Hay's diary (unknown to others), various features of Locke's register indicate an attempt to improve the quality of information, thus allowing for more plausible comparisons over time and, potentially, with other contemporary registers. The obvious model to invoke here is Hooke's "Scheme" of the weather included in Sprat's *History* in late 1667 with the aim of achieving some uniformity in both the collection and presentation of weather observations.[107] A year or so earlier, Locke's table was meant for his own use, but he was aware of the need to enter information in a way that could be understood by others. In general terms, the tables of Locke and Hooke included the kind of information that Boyle requested in April 1666: they both implicitly accepted his insistence that barometric measurements be taken in conjunction with those of the strength and direction of the wind.[108] This call was part of an important shift in which the barometer came to be seen as not only an instrument for measuring air pressure but as one also capable of detecting, and predicting, changes in the weather.[109] In keeping with Hooke's "Scheme," Locke incorporated columns for temperature, as measured by a thermometer, and for humidity (from July 1666) as indicated by a hygrometer. The latter device was based on the disposition of a beard of wild oat to unwind in humid air, as described in Hooke's *Micrographia*.[110] Locke used letters to indicate the direction of the Wind, placing the dominant direction first, so that, as he said, "WN signifies more West that North." He estimated its strength on a scale of 0 to 4 where "0 signifies not so much Wind that mov'd any leaf that I could see in a Garden I look'd into out of my Window"; and 4 marked "a very violent Storm."[111] Locke wrote these definitions for the "Explication of the foregoing Register" added to the printed version of 1692, but he had also used these letters and numbers in the first register. The final column on the right covered descriptive remarks about the "Weather," covering rain (but without an instrument), cloud, hail, thunder, lightning; and the aspect of the air, as in "fair," "close," snow, mist, and fog.[112] As mentioned in chapter 1, this is the column in which Locke noted the effects in the night sky of the great fire blazing some fifty miles away in London.

Locke embraced his new task with enthusiasm. On the third day he entered six sets of observations at various hours between nine in the morning and ten at night, possibly excited by rain, hail, and thunder around lunchtime.[113] Usually he made only one set of entries each day, most often after breakfast and chapel. There are some late-night entries in the first register: for example, there is one at 11 p.m. on July 7, 1666, when Locke recorded "Lightning," possibly because he was woken by the storm. Even at this hour he also entered temperature and air pressure, but nothing about the wind.[114] Late-night records were more frequent at Oates. Locke does not seem to have aimed to make his measurements at the same times each day, but he was very particular about recording the times they were made.[115] As with the style of note-taking summarized in the "New Method," he made only minor variations to his initial design over the several separate registers, retaining it in the final, long-running, one from December 1691 to May 1703. One of these changes was an improvement in the way the hour of the day was shown. In the first register, he gave the hours as 1–12, placing a dot above or below the numeral to indicate, respectively, morning and afternoon. On April 9, 1682, in a resumption of his register, Locke adopted a twenty-four-hour clock; in the published version of these registers he converted all the original entries made in "Adversaria Physica" into the new system, without mentioning that he had not used it in his first register.[116] He also marked the dates on which he introduced a different thermometer or barometer.[117]

Locke also took care about the qualitative descriptions in the column for "Weather." At first he employed only a few basic terms, such as "Fair," "Cloud," and "Rain," but by the end of 1666 his comments became more specific: thus "fog, hoare Frost," "small misty Rain," "hard Frost," "Little Snow last night."[118] There is a similar effort to paint word-pictures of the weather in a register which Boyle kept between December 1684 and January 1686: his right-hand column includes "Frost, but not very hard, somewhat foggy Evening"; "foggy dark day, moonlight but cloudy."[119] Always more methodical than Boyle, Locke wrote an "Explication of my Register of the Air" at the back of "Adversaria Physica" in which he defined the terms used to depict both clouds and rain.[120] For example, he explained that "Cloudy from hence forward signifies more of the skie (as far as I could see it out of my Chamber window) covered with clouds than was clear from them."[121] He made similar attempts for "Fair," "Mist," and "Close." These were all terms that occur in his early registers, but their careful characterization here suggests that Locke sought to improve, and standardize, them for the register he kept at Oates throughout the 1690s. He also made discriminations within "Rain," at least at the minimal threshold, deciding (on April 19, 1695) that "A few drops signifies rain but not soe much as makes the houses drop" [drip?]. No quantification was involved in these designations, but Locke's wording is both precise and vivid, allowing for comparison of observations made by anyone who adopted his calibration of terms.

On the basis of his daily observations, Locke occasionally reviewed weather events and patterns, entering longer-term summaries in the right-hand column. There are at least six substantive comments of this kind (and other shorter ones) in the early registers, later published in the *General History of the Air*. Locke conveyed exceptional weather with pithy observational reports, such as that of March 8, 1667: "Thames/Charwell frozen, Carts went over."[122] More frequently, he made comparative observations about the lengths of wet and dry spells, adding above the entries for July 1673 that there had been six weeks with "scarce one dry Day, but so great Rains, that produced greater Flouds than were known in the Memory of Man." Or, in contrast, recording that "the driest Spring that hath been known" occurred in 1681, with no rain "from the End of *March* to the End of *June*."[123] He probably made these synopses by skimming through the descriptive terms in the "Weather" column, rather than by checking the measurements—although it was the presence of the latter, backing these generalizations, which give Locke's remarks credibility. In the register for Oates, he also began to note other seasonal events, such as the arrival and departure of swallows, etching this word in large, heavily inked letters, usually in the right-hand column.[124] For September 23, 1692, he wrote: "this 23[d]: was the last day SWALLOWS were seen" (see fig. 7.3). If he missed these moments, Locke entered them retrospectively: thus for March 30–31, 1693, he wrote "SWALLOWS As they told me." And on September 21 of that year, he sheepishly jotted "SWALLOWS gon what day they went I forgot to observe."[125] These summaries show that Locke not only made his regular entries but reviewed them in order to infer patterns, mainly on the assumption that seasonal changes were worth noticing.

What was the point of all this? When Sprat summarized Wren's "History of the Seasons" he distinguished between meteorological data and the collection of information about those things likely to be affected by changes in weather, such as agriculture, the breeding and migration of animals, and, crucially, in Wren's words, "a good Account of the epidemical Diseases of the year."[126] This correlation was expected under the Hippocratic doctrine about the effects of both seasonal and extreme weather.[127] Boyle, Locke, and Sydenham attributed the onset of various diseases, such as influenza and smallpox, to the action of "morbific" particles carried in the air.[128] With adequate data, accrued only with "Patience for some Years," as Wren cautioned, it would be feasible to discern "the Difference of Operations in Medicine according to the Weather and the Seasons."[129] Locke's interest in the role of climatic factors is evident in the way in which he sometimes entered both weather measurements and medical facts or speculations under the Head of "Aer" in his notebooks.[130] He also theorized about the effects of the constitution of the air on respiration.[131] Although Locke did not keep such medical

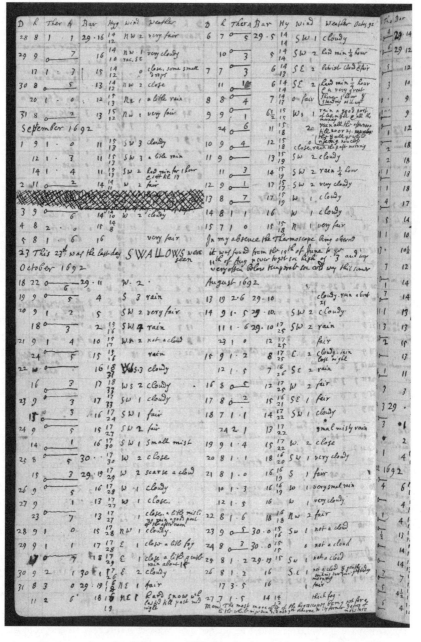

Figure 7.3. One of Locke's notices of "Swallows" in his weather Register in Essex, September 23, 1692. To the right of this comment, Locke mentions his Thermoscope and, in the bottom-right corner, he records that a hygroscope (using a beard of wild oat) left in the sun did not return to "its former degree of moisture." MS Locke, d. 9, p. 520. By permission of the Bodleian Libraries, the University of Oxford.

observations in his weather registers, he was enthusiastic about accessing pertinent data, such as Bills of Mortality (another of Wren's suggestions) and willing to marshal his own. In the early 1690s he collaborated with the physician Charles Goodall to draw up a questionnaire seeking information about the frequency of diseases, mortality rates, and quality of the air in different climates throughout the world.[132] By the 1730s physicians such as Francis Clifton were urging the Royal Society to collate medical and weather data in tables that allowed both easier collection and analysis.[133] Locke's registers, with their columns and routine entering, anticipated at least one portion of this tabular form.

In March 1704 Locke sent Hans Sloane his register of the weather for 1692 for publication in the *Philosophical Transactions*, and reflected on his past efforts: "The pains is [sic] so little that I indulged my curiosity when it cost me the writeing not so much as a line a day."[134] Having now seen the discipline he applied to these registers we can appreciate that this sanguine assessment came not only from an addicted note-taker, but also from a person encouraged by involvement in a long-term Baconian project. For some years, while living in France, Locke suspended his weather registers, but the daily routine of note-taking continued in a new way, and in a type of notebook which he had not yet used: the travel journal.

Locke's Journals

Locke lived and traveled in France between November 1675 and April 1679, and during these years there was a significant development in his note-taking.[135] For the first time (apart from memoranda of accounts), he used the journal, or diary, format. His first journal starts on November 12, 1675, in a notebook consisting of loose quires subsequently stitched together; the next three journals, made in similar fashion, are joined with a French almanac for the current year. He used a separate notebook for each year, compiling four of these during his sojourn and taking some 1,500 pages of notes. The entries are all dated because they are in journal format, thus distinguishing them from most of the notes Locke had previously made in his commonplace books. In the latter, he usually followed the convention by which excerpts under Heads carried no date (bar the exceptions already discussed). He did not take any of his commonplace books to France, certainly not the folio-sized ones that were too large and too precious.[136] However, when he returned home in May 1679 he began to include dates in his commonplace books, the first of which were transfers of entries from the journals.[137]

When Edward Gibbon pitched into the long-running debate on the value of "foreign travel," he voiced some high expectations. In specifying the desirable aptitudes of the *learned* traveler, he contended that "He should be a Chymist, a botanist, and a master of mechanics."[138] To some extent, Locke met the first two criteria, and his interests were in fact even more extensive. Some of the journals' content derives from his pre-existing compulsion to record various details about prices and distances, and the large total number of entries is partly accounted for by his quasi-professional involvement in medicine and his more general concern with ethical and religious subjects. Thus among the recurring subjects are weights and measures, coinage, currency exchange rates, wine making, farming, horticulture, cooking, inventions, medical treatments, taxation, and the legal situation of religious groups, especially the Huguenots.[139] The kind of note-taking ranged from tourist-like observations on gardens, buildings, city plans, food, wine, weather, and the hardships of travel to finely detailed accounts of agricultural and technological practices, and meticulous medical reports on his own health and that of his patients. However, Locke was also keen to gather new information on a variety of topics that might be of interest to others. From Paris, he wrote to Boyle asking for a recommendation to "any one of the virtuosi you shall think fit here," and offering to make any inquiries on Boyle's behalf. About a year later he reiterated this offer, mentioning a number of things which he had seen: a new microscope, a watch motion moved by air, a new hygrometer, an unnamed medicine, a report on a fish from the Bahamas, and a boy with nails growing to five inches on his fingers and toes.[140]

Due to this breadth of information, Locke's journals might seem to resemble a miscellany fuelled by unbridled curiosity.[141] However, what appears to be indiscriminate note-taking assumes another guise when considered in terms of the information demands of Baconian natural histories.[142] As discussed in chapter 3, Bacon's call for such collections envisaged a vast bank of particulars grouped under convenient, but usually temporary, Heads. From this it followed that information had to be sought from many observers, ideally in a collaborative spirit, and with the recognition that short-term results were unlikely. By the late 1660s, especially through his involvement with Boyle and other scientific figures in Oxford, and with Thomas Sydenham in London, Locke had signed up to this Baconian program.[143] He accepted the importance of detailed empirical particulars as the foundation for any future causal understanding of natural processes; and he believed that useful generalization required comparative cases. Both these convictions informed the weather register he began in Oxford in June 1666.

Although Locke was not directly involved in the Royal Society's attempts to create questionnaires for travelers, his own reading of travel books showed an eagerness for information from other places. By the time he set off for France in late 1675, Locke was the ideal Baconian collector: he possessed the appropriate philosophical orientation and motivation and, crucially, he had demonstrated the rigor and persistence necessary for making numerous and precise notes. The extent to which this promise was borne out can be considered by discussing the kind of information he accumulated in his journals, how he conceived its purposes, and how he managed to do this without his commonplace books.

In the absence of these commonplace books, the diary keeper in Locke was released. While in France he was no longer able to distribute topics into separate notebooks, such as those for medical and nonmedical information. Instead, his journal became the main receptacle for most of the material recorded each year. The result was that Locke, like Hartlib and some others before him, created a hybrid form of notebook in which entries made in chronological order were given marginal Heads that served as keywords for an index to each journal. Such keywords were essential because of the range and diversity of the subjects he noticed. As travelers were advised, Locke used his journal to respond, on the spot, to things of interest. Thus in some pages, an assortment of topics follow each other in quick succession, often with jarring effect, such as with "Diarrhoea" and "Cartesian philosophie," in March 1678.[144] However, his note-taking was also directed by some definite projects. While in France, Locke continued to regard himself as part of the earl of Shaftesbury's household, and he accumulated information on agriculture and wine-making techniques at his patron's request.[145] He was at work on this task in the first months of his journey via Paris to Montpellier, making entries under the Heads of Olives, Oil, Vines, and Grapes.[146] These incorporated his first-hand observations and the testimony gathered in conversation with local farmers.[147] A different kind of entry was that devoted to reflection on philosophical topics. Although Locke had already written the first draft of the *Essay* in 1671 (and carried a copy with him), he made some important notes about several topics not yet fully developed in that work.[148] There is a burst of these in his first journal, all entered in *one* day (July 16, 1676): Passions, Love, Desire, Hope, Hatred, Pain, Pleasure, Wearinesse, Vexation, Sorrow, Grief, Torment, Melancholy, Anxiety, Anguish, Misery, Mirth, Delight, Joy, Comfort, Happynesse, Misery, Bonum, Pleasure, Desire, Power, Will.[149]

Locke's journals continued to include material stimulated by his reading but, compared with those in his commonplace books, the notes taken in

France comprise a higher percentage of entries relying on direct observation or which contain his own thoughts and comments.[150] Locke the traveler also drew on observations and conversations as new sources of ideas and information about various empirical matters, especially those relating to medicine and natural history, and also to local customs, crafts, and religious beliefs. He relied on introductions and tips from English residents, such as William Charleton (actually Courten) in Lyon and the countess of Northumberland, the wife of the English ambassador Ralph Montagu, in Paris. In Montpellier he tapped into medical circles, meeting Charles Barbeyrac, Pierre Magnol, and Pierre Jolly. But it is also evident that he tried to talk directly with strangers encountered on the way to Montpellier (via Paris), while traveling in Provence and Languedoc, and on his return trip to Paris, begun in early 1677.

Damaris Masham said that Locke believed "he could learn something which was usefull, of every body"; and that he accomplished this by "suteing his Discourse to the Understanding, and proper skill of every one he convers'd with."[151] In Lyon he wanted to ask about the workings of various contraptions in a famous cabinet of curiosities initiated by Jean Grolier de Servières in the mid-sixteenth century. Early in his journey, however, Locke's lack of French was a drawback, as admitted in this account of his conversation with the person in attendance: "he haveing not Latin nor I French . . . I could not particularly enquire into."[152] However, the journals show that he later spoke with peasants in the fields, gardeners, fruit cultivators, wine makers, artisans, craftsmen, clerics, apothecaries, and physicians.[153] These people told him what to look for, what to see, who to ask. He usually recorded the names of informants, especially when the content included factual details about plants, animals, or medicinal remedies and treatments.[154] Thus in May 1676 he noted that "Mr Upton tried Balaruck waters with galls & found them not change colour at all."[155] In many other instances, he took notice only of trade or profession, and when summarizing his conversations with groups of people he used referents such as "they," "some," or "a few," thereby capturing the range of views on a particular topic within a community. For example, in summarizing the opinions in Montpellier about pressing olives, he noted that "All confesse that oyle is better which is made of Olives fresh gatherd. . . . Others say the reason why they are not pressed sooner is because every bodys grist cannot be ground at once." In recording local estimates about the number of small county houses (bastides) near Marseilles, he wrote that "some tell you, 22,000 they that speake lowest say 16 or 17,000."[156]

In the great majority of cases, Locke sought local knowledge about natural processes and causal properties. Depending on the subject matter, he seems to have been more concerned with the skill and expertise, rather than with the social status, of the informants.[157] In the *Essay*, he gave "Integrity" of witnesses as one of the six criteria for the evaluation of testimony.[158] Indeed, there were people whose knowledge and judgment Locke certainly valued, such as François Bernier, a physician and traveler. In October 1677 he made a detailed note of what Bernier told him about "the Heathens of Indostan," treating it as first-hand and expert testimony.[159] But this criterion could not feasibly be applied to all people, and so in many exchanges Locke mainly recorded what he was told. When visiting a chateau in Castries, near Montpellier, he made measurements of the section of an aqueduct near the house but had to be content with reports of its scale near its mountain source: he added, regretfully, "I never could speak with any body that had seen these 50 arches." He entered a detailed note on silkworms as explained to him by the wife of his landlord (Monsieur Fesquet) in Montpellier.[160] Making notes about how plums, peaches, and pears were dried, he recorded that "This was taught me by Madame de Superville."[161] Locke recognized that there were good and bad reporters; he sometimes signaled a doubtful testimony with a "Query," often marked simply as "Q."[162] At other times he was less reticent: in Paris he recorded a conversation about the transporting of orange trees (without branches or roots) from Italy, saying, "I am afraid in this later part of the story the gardiner made bold with truth." While traveling to Paris he made an entry under the marginal heading of "Vines": "To make vines beare in a barren ground put a sheeps horne to the root & it will doe wonders." Later, when he wrote up the material he gathered on wine making and other agricultural practices at Shaftesbury's request, Locke reactivated this note, adding "I have no great faith in it, but mention it because it may so easily be tried."[163] This practice of recording empirical information from available sources without overly stringent assessment made sense: Locke held that strong theoretical judgments must be postponed until information on particular topics was more fully collated. But first it had to be gathered, if not by personal observation, then by examining the right sources. He sometimes contrasted published reports with his (or others') first-hand observations. Thus when visiting the Canal du Languedoc (commissioned by Louis XIV in 1666 and finished in 1680), he remarked that "the dam of the Reservoir is a great work, but seems not to be as big as Froidour describes it."[164]

Another of Locke's abiding concerns, the diversity of customs and beliefs, is also apparent in the journals. Of course, as an avid reader of travel books, he came with expectations, but was now able to augment this book knowledge with his own observations.[165] Whereas, Locke's conversation *with* strangers was a major source of new empirical information, the conversations *of* strangers provided comparative illustrations of what he already thought about the influence of customs on opinions and beliefs.[166] He took many notes about the religious practices and moral beliefs in various parts of the country. There are some scathing comments, some in shorthand, about the Catholic rituals he witnessed.[167] Locke showed the appetite of an anti-Papist for scurrilous stories about priests, bishops, nuns, and the theater of the liturgy. On February 11, 1679, he recorded the account of a French visitor and his wife attending "a Masse [in Rome] where the pope was present." After the elevation of the host a "very considerable cardinall" asked the visitor: "*Che dice vostra signioria di tutta questa fanfanteria* [furfanteria?]?" (What does your wife say about all this trickery?).[168] Yet even when Locke showed his annoyance with local beliefs about relics and miracles, the descriptive detail of his accounts made them potential data for a natural history of religion and, more generally, of opinion.[169] Alongside these observations of Catholic practices, Locke collected information about the culture of the Islamic world, the Indies, and the Americas, both from books and travelers.[170] By recording illustrations of different customs and beliefs, Locke contributed toward a baseline of information that supported potential comparative analysis.

The large range of topics in a *single* journal each year defeated Locke's attempt to apply his usual method of indexing. The fundamental problem was that in his "New Method" there was a nexus between entering and indexing: the page on which entries were to be made was determined by the initial letter/vowel codes of the Heads. Once a double opening in a commonplace book had been assigned to a certain alphabetical code, say *Ae* for "Aer" or *Me* for "Memoria," it was reserved for all subsequent instances of this code—until the two pages were full. This method prevented Heads of the same code from being scattered across many different pages; it therefore also reduced the number of separate listings required in the Index. The case of journal entries was, of course, quite different, since these were simply made as they occurred, in chronological order. Despite this, at the back of the first journal, after it was completed, Locke tried to apply the "New Method" system of indexing keywords under initial letter/vowel codes. However, due to the great number of marginal keywords (522) *and* their distribution across many pages, he found that even with the use of a second vowel the number of index terms was excessive, causing the various cells of the two-page index to become crowded. After indexing the entries up to page 254 of the journal he abandoned the attempt and reverted to an alphabetical list of the Heads (see fig. 7.4).[171] This experience must have confirmed his perception that the "New Method" index worked best if separate notebooks were allocated to various subjects, thereby reducing the number of Heads to be indexed. When Locke remarked, in the *Bibliothèque Universelle* of 1686, that he had rarely needed to use a second vowel, he did not say that the time in France, a decade earlier, had produced this failure.[172]

Locke's Reflections on Note-taking

It is a nice irony that Locke wrote most *about* note-taking while in France and, later, in the Netherlands — in the absence of his commonplace books. While in Amsterdam he acceded to requests and published the description of his note-taking as the "New Method." But he had also thought about this while traveling in France. We need to appreciate that his attitude toward commonplacing was part of a deeper concern about proper intellectual disposition. This is revealed in a long entry, titled "Of Study," started on March 26, 1677.[173] In what is effectively an essay of some 8,000 words, Locke attacked key facets of the tradition of commonplaces while continuing to affirm the importance of note-taking. He did so via a wide-ranging medita-tion on reading, study, the balance of bodily and intellectual effort, the fea-sible scope of individual learning, and the tools and practices by which study was pursued.[174] One likely stimulus was the exchange of letters with Denis Grenville, an English clergyman of royalist persuasions residing in France at the time. Grenville first wrote to Locke in March 1677 while they were both in Montpellier, sending three letters; he expressed pessimism about the time and effort needed for proper study in the context of other duties, including religious ones.[175] At this time, Locke also encountered more sub-tly argued doubts about the capacities of human reason in his translation of three essays from Pierre Nicole's *Essais de morale* (1671–79).[176] In one of these ("Discourse on the Weakness of Man"), Nicole exposed the vanity of intellectual pretensions and alleged "that philosophy is a vain amusement, & that men know almost noething."[177] In a journal entry alluding to Nicole's position, Locke concluded that such weaknesses required that "all needless difficulties should be removed out of the way."[178] As we know from the *Essay*, Locke did not succumb to extreme pessimism, maintaining that we have a duty to seek knowledge and a fair chance of achieving it, providing that our aims match our limited capacities.[179] Accordingly, his criticism of the com-

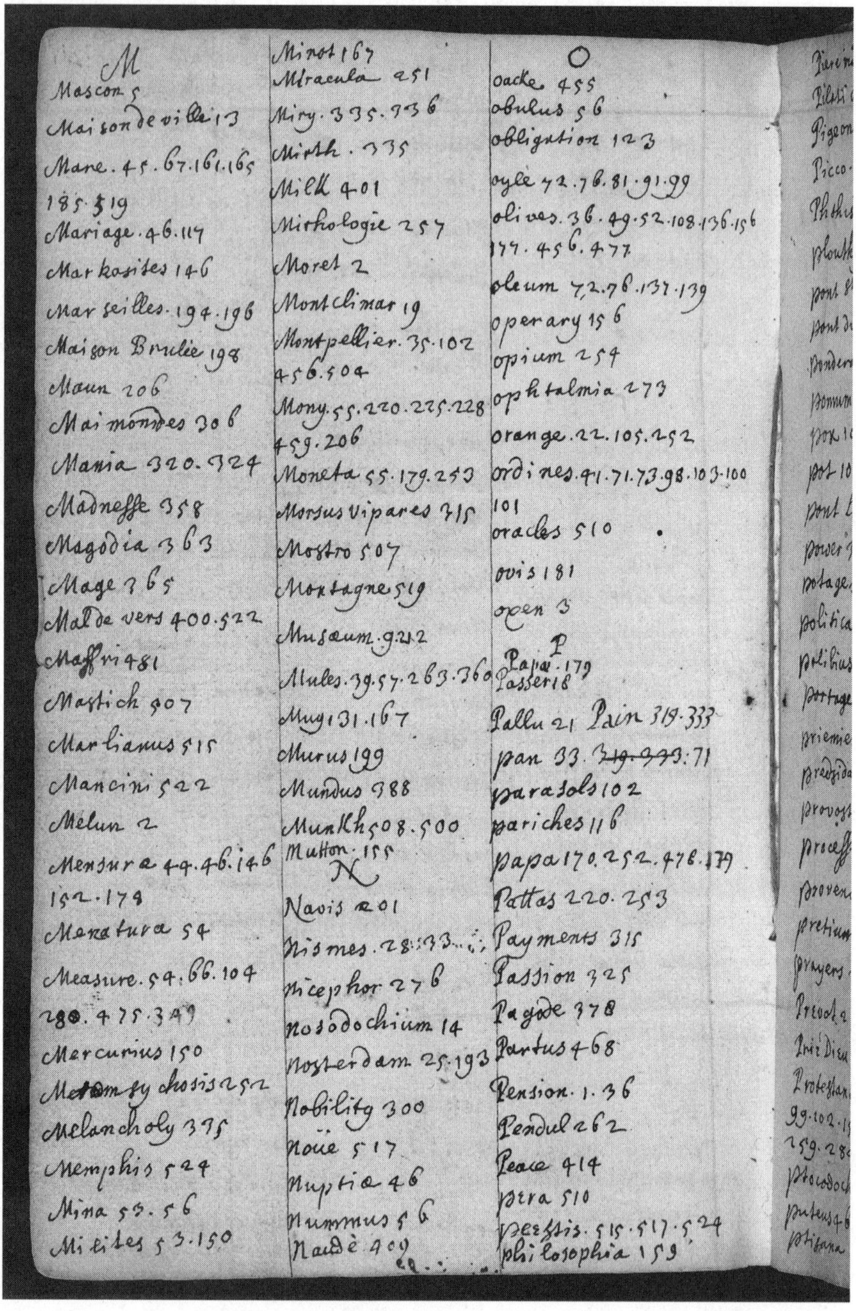

M

Mascon 5
Maison de ville 13
Mare. 45. 67. 161. 165
185. 519
Mariage . 46. 114
Markasites 146
Marseilles. 194. 196
Maison Brulie 198
Maun 206
Maimondes 306
Mania 320. 324
Madnesse 358
Magodia 363
Mage 365
Mal de vers 400. 522
Massri 481
Mastich 507
Marlianus 515
Mancini 522
Melun 2
Mensura 44. 46. 146
152. 174
Meratura 54
Measure. 54. 66. 104
280. 475. 349
Mercurius 150
Metempsychosis 252
Melancholy 375
Memphis 524
Mina 55. 56
Milites 53. 150

Minot 167
Miracula 251
Miry. 335. 536
Mirth . 335
Milk 401
Mithologie 257
Moret 2
Mont Climar 19
Montpellier. 35. 102
456. 504
Mony. 55. 220. 225. 228
459. 206
Moneta 55. 179. 253
Morsus vipares 715
Mostro 507
Montagne 519
Musaeum. 242
Mules. 39. 57. 263. 760
Musg. 31. 167
Murus 199
Mundus 388
Munkhs 508. 500
Mutton. 155

N

Navis 201
Nismes. 28. 33...
Nicephor 276
Nosodochium 14
Nosterdam. 25. 193
Nobility 300
Noue 517
Nuptia 46
Nummus 56
Naude 409

O

Oacke 455
Obulus 56
Obligation 123
Oyle 72. 76. 81. 91. 99
Olives. 36. 49. 52. 108. 136. 156
177. 456. 477
Oleum 72. 78. 137. 139
Operary 156
Opium 254
Ophtalmia 273
Orange. 22. 105. 252
Ordines. 41. 71. 73. 98. 103. 100
101
Oracles 510
Ovis 181
Open 3

P

Papa. 179
Passer 18
Pallu 21 Pain 319. 333
Pan 33. 319. 333. 71
Parasols 102
Pariches 116
Papa 170. 252. 476. 479
Pattas 220. 253
Payments 315
Passion 325
Pagode 378
Partus 468
Pension. 1. 36
Pendul 262
Peace 414
Petra 510
Perssis. 515. 517. 524
Philosophia 159

Figure 7.4. A page of the alphabetical index in Locke's Journal of 1676. The terms collated here appear as Head words in the margins of the journal. MS Locke f. 1, p. 530. By permission of the Bodleian Libraries, the University of Oxford.

monplace method, as usually deployed in his day, was that it did nothing to instill the necessary care about thinking and judgment.

In "Of Study," Locke accepted that the commonplace book remained the notebook of choice for scholars, but questioned the value of the kind of memory training that it was said to assist. At its worst, he regarded this method as a lazy way of assembling a toolbag of quotations, such as those for and against standard topics in moral or natural philosophy. As discussed in chapter 2, this procedure was well suited to the public disputations in which undergraduates were required to propose or counter a thesis. But Locke objected that this pedagogy failed to demand "a true & cleare notion of things as they are in them selves."[180] Instead, students memorized arguments that they had not properly looked into, and so did not come to form and trust their own judgments. This kind of education produced the worst possible outcome — "an ignorant man with a good memory" or, as Locke dubbed this person, the "topical man, with his great stock of borrowed & collected arguments," constantly in danger of contradicting himself. This scenario could be avoided through deliberate judgment in the selection of material and cognizance of its relationships to other parts of knowledge. Furthermore, once the grounds of an argument had been mastered, a person would be able to debate with consistency because the relevant ideas were "in less danger to be lost," having been "placed in the judgment" rather than burdening the memory.[181]

Having made these criticisms, Locke addressed the question of how best to record and retrieve information. He declared that "Reading methinks is but collecting the rough materialls, amongst which a great deale must be laid aside as useless." He confessed that his own reading was promiscuous, manifesting a tendency to "change often the subject I have been studying read books by patches & as they have accidentally come in my way, & observe no method nor order in my studys."[182] Given this, Locke urged the importance of explicit classification: "A great help to the memory & meanes to avoid confusion in our thoughts is to draw out & have frequently before us a scheme of those sciences we imploy our studys in, a map as it were of the mundus intelligibilis."[183] He believed that such a "map" was even more necessary to sustain varied and diverse reading outside the canonical topics. He reassured himself, in this private journal, that "I have avoided confusion in my thoughts. The scheme I had made serving like a regular chest of drawers to lodge those things orderly & in their proper places which came to hand confusedly & without any method at all." The implication here is that a departure from standard commonplace Heads sanctioned by tradition and pedagogy required awareness of the kind of knowledge being collected. When mentioning the assignment of materials to their "proper places" (or Heads), Locke declared that this "perhaps will be best done by every one himself for his owne use as best agreeable to his owne notions; though the nearer it comes to the nature & order of things it is still the better." He advised that each person should "endeavour to get a true & cleare notion of things as they are in them selves; this being fixd in the mind will (without trusting to or troubleing the memory which often fails us)." As he said, this called for "now & then some little reflection upon the order of things as they are or at least as I have phantasied them to have in them selves."[184] This last point was crucial: Locke thought that individuals should be free to choose whatever Heads worked best as organizing devices or memory aids. A decade later he included this licence in the English draft of the "New Method," saying with regard to his use of separate notebooks for natural and moral knowledge that "I have also upon different occasions varied it into other formes, which perhaps would better please some mens phansys or be more proper for some particular purposes."[185]

After more than a year in France, the range of material Locke was collecting in his journals may have led him to think about the "chest of drawers" best suited to contain it. In an entry of September 4, 1677, under "Adversaria," he classified "the principall parts or heads of things to be taken notice of" in making notes, delineating these four general categories: Philosophica, Historica, Immitanda, and Acquirenda.[186] In fact, this entry constitutes one of twelve schemata Locke produced that show the main divisions of knowledge. Close inspection reveals that these schemata are of two kinds: one in which the classification is based on major disciplines and subjects (type A), and another based on *ways* of knowing (type B).[187] The first examples of type A occur in "Adversaria 1661" and were probably done in about 1670. One is at the start of the notebook and uses a fourfold division of Theologia, Politia, Prudentia, Physica, with each division further subdivided; another, later in the notebook, adds Metaphysica, Historia, and Semiotica.[188] In another set of Locke's papers, there is a very neatly composed version, in Brounover's hand, which may have been meant as a final schema.[189]

Those of type B were all done between August and November 1677.[190] They depend on three main divisions: "Cognoscendorum," "Reminiscendorum," and "Agendorum," thus indicating the different ways in which knowledge can be acquired and retained—by thinking (or meditation), by remembering (largely through historical records), and by making practical or technical interventions.[191] There is a strong link between these schemata and Locke's reflections on the practice of note-taking—for example, these divisions are laid out in the journal entry about "Adversaria" of September 4, 1677. There are three additional reasons for suggesting the connection with his consideration of note-taking. First, Locke used the term "Adversaria" (or notes) as the marginal Head for all these versions, and never applied it to those in type A. Second, in a manuscript of one schema from type B, dated November 12, 1677, the annotation "Studia 77" is written in the margin next to the subdivisions of "Agendorum," thus associating it with the entry "Of Study" in which the rationale of note-taking is examined.[192] Third, the subdivisions of "Agendorum" (of things to be done) and its two subheadings—"Acquirenda" and "Immitanda"—are new, as well as being pertinent to Locke's experience as a traveler taking in a wider range of sources. As he defined these terms, "Acquirenda" referred to "the naturall products of the country fit to be transplanted into ours & there propagated, or else brought thither for some very useful quality they have." "Immitanda" denoted practices that applied to "a mans private self or any beneficial arts employed on natural bodies."[193] Thus "Acquirenda" encompassed the entries made about the samples of vines and fruit requested by Shaftesbury, whereas "Immitanda" covered his patron's directive to gather intelligence on the techniques for wine making and extraction of oil from olives. The latter category also included many of Locke's observations about the beliefs, customs, and rituals of both Catholic and Protestant communities. In many cases, this kind of information was drawn from observation and conversation rather than from books. Although Locke sometimes used these two terms as marginal keywords, in both the journals and in the commonplace books into which he copied them, he did not develop them further; that is, he did not make the schemata of type B the basis of a new division of notebooks. Rather, the entries under "Acquirenda" and "Immitanda" were absorbed into his pre-existing, main division between "Physica" and "Ethica." The category he did suggest for a third class of notebook was "Semiotica," which belonged to the maps of knowledge in type A.[194]

Moving Notes

Due in part to Locke's interest in travel books, various editions of his collected works attributed to him the introductory chapter of *A Collection of Voyages and Travels* (1704). The final paragraphs of this introduction stress the importance of taking, and transferring, notes: "Let them, therefore, always have a table-book at hand to set down every thing worth remembering, and then at night more methodically transcribe the notes they have taken in the day."[195] This advice describes Locke's own practice upon his return from France in May 1679. He began to transfer some of the material in his journals into appropriate commonplace books. He copied selected entries into "Adversaria Physica," "Lemmata Ethica," and the two companion notebooks bound in MS Locke c. 42.[196] In moving notes in this way, Locke used the marginal keywords of the journals as Heads; these usually identified the topics as either scientific (including medical) or ethical (in his wide definition of that category). Some of this material came from the notes he made under Heads on loose sheets folded together, thus serving as traveling adversaria.[197] The standard alphabetical index he had made at the back of each journal allowed him to locate the entries he wished to transfer. It is significant that after returning to England, Locke did not bother to index his journals—because they no longer functioned as the single, temporary, repository of his information and ideas. Once back home he was able to reassign selected material to his set of adversaria and lemmata, that is, to his commonplace books.

When he made these transfers, Locke did not always simply make a verbatim copy. Thus in 1681 when moving an entry on "Hysterica" of July 8, 1676, he expanded the shorthand of the journal into longhand, also adding a cross reference in the margin of the journal to show that the material was now in "Adversaria Physica."[198] The collation of entries, often from different years in France, gave him the chance to order material in new ways, sometimes taking account of extant entries in the notebook, as is evident in the entries on hysteria and related psychosomatic complaints.[199] There are also instances in which the transfer of entries served as an opportunity for Locke to review both his information and his thoughts. Thus in MS Locke d. 1, he copied short entries from the journal for 1679 but greatly expanded them: a short notice of Samuel Clos' book on mineral waters activates several entries, some under his name, but also under the topics of "acqua mineralis," "acqua medicinalis," and "terra," which are signed by Locke as his queries and speculations.[200] These entries (many marked with "Q" for query) exemplify what Locke described as the best way to make hypotheses about causal factors—by joining analogous instances and cognate material. Such hypotheses could then serve, as he later wrote, as "great helps to the Memory" and, by extension, as aids to thinking. In considering how best to theorize about diseases, he acknowledged that hypotheses contracted key features of phenomena and hence could act "as distinct arts of memory" provided they did not hinder observation.[201] At the very least, by transferring notes Locke was able to consolidate cognate material, to benefit from the stimulus of seeing older notes, and to revise or amplify his thoughts from earlier occasions.[202]

As well as moving notes between his own notebooks, Locke imparted some information, based on these notes, to the Royal Society. When he returned from France, there was a letter waiting for him from Richard Lilburne, an informant in the Bahamas, containing an account of a poisonous fish.[203] Locke sent an extract from this letter to Oldenburg who read it to the Society on May 27, 1675.[204] However, despite the wealth of documented observations in his journals, this is the only matter that Locke thought worth sending, at least at this time. It may be that he set high expectations about such reports. Indeed, Locke told Oldenburg that the person had not provided "so perfect an account . . . as one could wish," and in his "Adversaria Physica" he drew up a list of thirteen queries designed to elicit more precise details and sent these to Lilburne.[205] He received a reply but did nothing more with it.[206]

Another communication from Locke to the Society occurred almost twenty years after his original note about it. In 1696, in his role as secretary, Sloane wrote to Locke, saying, "I hope you will think of the Society now you are in the Country and lay by any observations proper for them, I remember one about long nailes you halfe promis'd."[207] Locke must have spoken about his experience at the hospital, La Charité, in Paris in May 1678. His journal carries accounts of two visits in which he met "a young lad of between 19 & 20 years old" with nails growing to five inches on his fingers and toes. On May 24 he made an entry under "Cornua digitorum," describing the nails growing like "horns" out of "all his fingers," but noting that this happened by a "thickening of the nail" rather than an increase in length, and that the angle of the extension resembled the "shape of a birds claw."[208] Historians of medicine have remarked on this as the first close description of a case of onychogryphosis.[209] But Locke did nothing with this information except mention it among other news to Boyle in 1678, saying, "I have a large piece of one of them, which was broke off in my presence, and the whole history of the case amongst my things at Paris."[210] Almost a year later, he was able to recover this "whole history" by returning to his journal, largely copying the original entries into this letter. Locke seriously discounted the value of this report, or at least Boyle's likely interest, concluding that "I know you think the works of nature worth taking notice of, and recording too, even when she seems extravagant and out of her way. This must be my excuse for sending you this long, and (as it would be to another) tedious story."[211] However, much later, at Sloane's urging, Locke sent him both the "peices of horny substance" and his account of the boy with these monstrous nails.[212] Locke indicated that he was supplying this "just as I set it down for my owne memory and therefor [you] must excuse the way it is expressed in."[213] However, in a subsequent letter he reconsidered this, suggesting that "I thinke what I writ for my owne private memory ought to be revised and corrected a little before it appear in the world. To which possibly I may finde something to adde, for if I misremember not I visited this yonge man a second time, and tooke some farther notes, these I will looke out amongst my papers."[214] Locke's article appeared in the *Philosophical Transactions* in 1697, preserving the form of the two original journal entries, together with an engraved illustration of the nails.[215]

Locke's final communication with the Royal Society was initiated by him, again with Sloane as intermediary. In contrast with the case of long nails, Locke was keen to have his weather records made more public. It seems that his publisher, Churchill, had plans to update the registers already printed in Boyle's *History of the Air*, but Locke was keen to see something "published in my lifetime," probably because he wished to supervise it.[216] His presentation of the 1692 register for publication shows the care he must have taken in instructing his amanuensis to transcribe the records from his notebook.[217] In his accompanying letter of March 15, 1704, Locke was modest about these records: "If you find it worth your acceptance (for it is bare matter of fact)." Yet he went on to put a forceful case for the function of such observations and measurements in the natural history of both weather and diseases, that is, the program he had committed to four decades earlier. He admitted that "this solitary one" was insufficient, but affirmed that "if such a register" were "kept in every county in *England* and so constantly published" then much useful knowledge might be gained, especially by "a sagatious man."[218] By this time, in his *Essay*, Locke had made the strong claim that documented observations in "concurrent Reports" were beyond reasonable doubt: as he said, if many creditable observers attested that "it froze in *England* the last Winter, or that there were Swallows seen there in the Summer, I think a Man could almost as little doubt of it, as that Seven and Four are Eleven" (IV.xvi.6). Locke himself was an observer able to add to such reports, especially those about weather and swallows.

Methodizing Boyle's Notes

Given the notorious disarray of Boyle's papers, it is not surprising that Locke, when he had the chance, would seek to put them in some order. This opportunity came when Locke acted as an informal executor after his men-

tor's death in December 1691. But well before that he had made copies of
some of Boyle's papers, thereby, as it happened, compensating for the loss
of originals. As Michael Hunter has noticed, these included Boyle's sets of
Heads for inquiries into cold, human blood, and the history of diseases — all
topics covered in Locke's own notebooks.[219] In the preface to his *Memoirs for
the Natural History of Humane Blood* (1684), Boyle admitted that "a set of Enquiries" crucial to this work were among papers "since lost." He addressed
this preface to Locke and explained that he was fulfilling Locke's request to
"set down what I can retrieve."[220] Presumably, he was helped by the copy of
one of the lost lists that Locke had entered into a medical commonplace book
used to record some of the inquiries pursued in Boyle's Oxford circle, including "Tryall about the blood especially humane" (see fig. 7.5).[221]

Locke's retention and organization of information contributed to Boyle's
General History of the Air, published in 1692, after his death. This project
began in the early 1660s when Boyle sent out a set of Heads or Titles as a
guideline for the collaborative gathering of information. No copy of this
printed list survives, but there are likely responses to it among Boyle's
papers, some from the mid-1660s. Locke's letter of May 5, 1666, may have
been one of these, since as well as including a barometric measurement,
it recounts the effects of water in mineral streams mixed with air from
the lead mines in Somerset.[222] From the published work it is evident that
Boyle's request had stimulated the participation of a range of observers —
friends, travelers, virtuosi, and even royalty (the duke of York, in titles XX
and XLVI) — thus marking it as a collaborative project of the Baconian kind,
but perhaps also one that posed Bacon's own warning about delegated note-
taking. In his preface, Boyle said that to guide the work of the participants
"I drew up a set of Heads and Inquiries of that sort, which . . . , tho' purposely set down without any anxious Method, were comprehensive enough
to have a good Number and Variety of Particulars conveniently referr'd to
them."[223] However, the incoming papers do not seem to have been assigned
to such Heads, and several manuscripts from the early 1680s reveal attempts
at settling on a set of "Titles" under which the various letters and papers
could be published.[224] One of the Boyle manuscripts Locke copied was "The
Titles of the Naturall and Experimentall History of the Air."[225] This entry
is dated 1682, and it was probably about this time that Locke began to help
Boyle assemble the material. In preparing the text, he also added information from his own sources, including some gathered in Holland.[226]

Locke saw the *History of the Air* through the press and did not disguise
his frustration in editing Boyle's messy papers.[227] This reaction flowed over

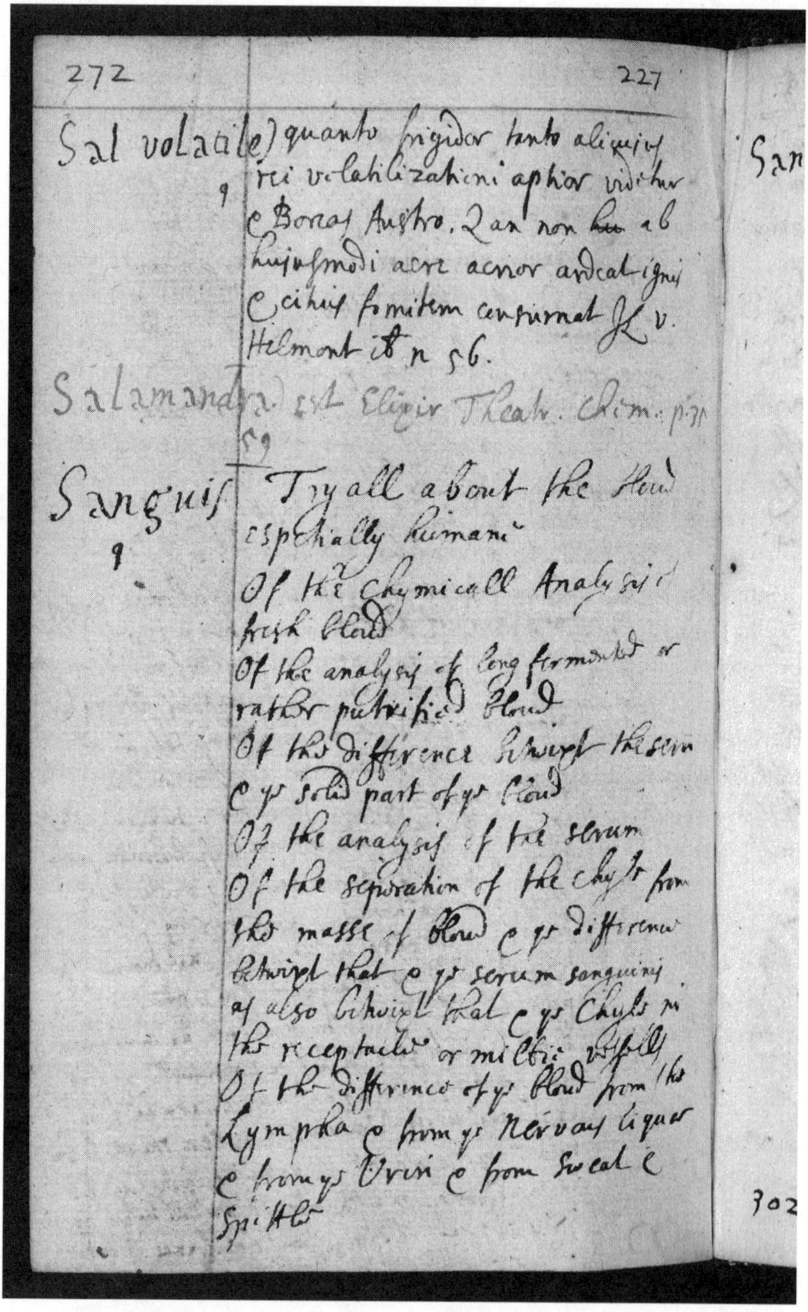

Sal volatile) quanto frigidior tanto alicujus
rei volatilizationi aptior videtur
e Boreas Austro. Qua non ab
hujusmodi acre acrior ardeat igni
e cibus fomitem consumat ℔ v.
Helmont it n 56.

Salamandra) est Elixir Theak. Chim. p. 71
59

Sanguis | Try all about the blood
espetially humane
Of the Chymicall Analysis
fresh blood
Of the analysis of long fermented or
rather putrified blood
Of the difference betwixt the serum
e ye solid part of ye blood
Of the analysis of the serum
Of the separation of the chyle from
the masse of blood e ye difference
betwixt that e ye serum sanguiny
as also betwixt that e ye Chyle in
the receptacle or milkie vessels
Of the difference of ye blood from the
Lympha e from ye Nervous liquor
e from ye Urin e from Sweat e
Spittle

Figure 7.5. Locke's list of inquiries concerning blood, under "Sanguis," in a medical
commonplace book. MS Locke f. 19, p. 272. By permission of the Bodleian Libraries,
the University of Oxford.

into his "Advertisement" in which he indicated that the work was imperfect on several scores: Boyle's "Expectation of Assistance" had not been fulfilled, and some details of the provenance of the various contributions had been mislaid. Indeed, Boyle admitted in his preface that he had lacked the "leisure to methodize my incoherent Notes." Nevertheless, Boyle regarded the "Titles" of the collection as "a kind of Common Places, what my Memory, or some old Notes about divers things relating to the Air, and especially to the *Causes and Effects of its Changes*, supply me with in reference to that Body."[228] Locke seized on this confession as an opportunity to underline a moral lesson about note-taking:

> Nor could it be hoped that the Authors own Memory (were he in a State of Health fit to be troubled with it) should after so long a time as this Collection has been making, and in that Variety of Men and Books he has had to do with, be able to retrieve them.[229]

He explained that Boyle had set out guidelines, albeit rather general ones: "In the first Draught he [Boyle] followed my Lord *Bacon's* Advice, not to be over-curious or nice in making the first Set of Heads, but to take them as they occur." Locke cautioned that for the collection to proceed with the involvement of many assistants, these Heads needed to be a "little more increased or methodized"; only then might Boyle's work "serve to some Men as a common Place for the History of the Air." Locke also made a significant revelation: the material as he now *collated* it did not fully match the manner in which it had been *collected*: "The Scheme of Titles under which these Materials for a History of *Air* are ranged, is somewhat different from that printed by him several Years since, and distributed amongst his Friends."[230] He was even more specific in a letter written just before Boyle's death, saying, in a postscript, "I had forgot to mention above, that I have a little altered some of your titles, the better, as I think, to accommodate them to the papers [which] are to be ranged under them."[231] Locke had reorganized the various reports, assigning the material to Heads he thought most appropriate. There are forty-eight of these. Several Titles have no observations: they wait as blank spaces to be filled in by subsequent observers. This is true not only of the last two, which are in effect calls for future additions.[232]

Locke had not finished with Boyle's notes and papers. In the last weeks of his life, Boyle sent his *Medicinal Experiments* (1692) to the press; it took the form of remedies and recipes arranged under "decades," that is, groups of ten items. However, the components shared no strong resemblance: the first decade includes remedies for coughs, bladder stones, and toothache; occasionally there are two or three that relate to each other but most appear to remain in the order in which Boyle gathered them.[233] Together with the physicians Edmund Dickinson and Daniel Cox, Locke was charged with producing two further volumes.[234] His hand is evident in volume 2 (1693) in which approximately 300 remedies are arranged according to the complaint or disease they treat, and then ordered alphabetically so that A includes aches and agues, P covers piles, pains, palsie, and so on. An index at the front allows location of material by both complaint and remedy.[235]

There is some irony in the fact that having uncovered haphazard features of Boyle's *History of the Air*, Locke heard its praises sung by his friends. James Tyrrell, one of the contributors, said that he was "highly pleased with it not onely for the matter but the method; it being as a perpetual common place booke to which wee may still adde fresh observations." Although someone had remarked to him that "there is little new in it, more then what hath bin publisht already in his discourses about air, or other subjects," Tyrrell defended the project as a way of making a collective commonplace book that invited many new contributions, and he professed himself willing to offer "any thing too [sic] so usefull a designe, or which the Authour could think worthy the inserting."[236] A couple of months earlier, Locke made a similar point to William Molyneux, saying that he had sent a copy of Boyle's work:

> it is cast into a method that anyone who pleases may add to it, under any of the several titles, as his reading and observation shall furnish him with matter of fact. If such men as you are, curious and knowing, would join to what Mr. Boyle had collected . . . we might hope in some time to have a considerable history of the air . . . but it is a subject too large for the attempts of any one man, and will require the assistance of many hands to make it a history very short of compleat.[237]

Locke also knew from the task of making sense of Boyle's material that such a collaborative project required conventions for note-taking and presentation of information.

Memory, Recollection, and Retrieval

In the *Essay*, Locke averred that memory "is of so great moment, that where it is wanting, all the rest of our Faculties are in a great measure useless."[238] He called the memory "the Store-house of our *Ideas*," thus adopting a metaphor long associated with the tradition of mnemonic arts. However, he construed this storehouse role in a distinctive way: "For the narrow Mind of Man, not being capable of having many *Ideas* under View and Consideration at once, it was necessary to have a Repository, to lay up those *Ideas*, which at another time it might have use of." This appraisal rested on his supposition about the mental capacities of "some superiour created intellectual Beings," such as angels, whose memories held "constantly in view the whole Scene of all their former actions, wherein no one of the thoughts they have ever had, may slip out of their sight." However, in Locke's view, human memory fell short of far less demanding criteria. The message in his published works was that natural memory, however it functioned, was constitutionally a weak instrument, suffering the "defects" of slowness in retrieval and decay over time. Underlying all Locke's remarks about memory is the stark confrontation of the fact that without special attention, or other precautions, our ideas fade rapidly, "leaving no more footsteps or remaining Characters of themselves, than Shadows do flying over Fields of Corn; and the Mind is as void of them, as if they never had been there."[239]

Locke partially absented himself from the contemporary debates on memory. He did not say anything about classical mnemonic arts, nor did he contribute to the discussions about the physical capacity and workings of memory.[240] But as a physician he was interested in the effect of disease and physical trauma; and he agreed with the familiar comment that "we oftentimes find a Disease quite strip the Mind of all its *Ideas*" (*Essay* II.x.5). Some of his early notes contain the usual stock of anecdotes and quotations on this topic, such as Seneca on prodigious memory and the story about Corvinus' loss of memory.[241] Additionally, he was concerned about the proper office of memory in both study and everyday life. While gathering information in France from books, testimony, and observation he urged the importance of selection and organization—"it being both impossible in itself, and useless also to us to remember every particular."[242] Locke regarded note-taking as a significant way of aiding memory in its proper role of retaining and disposing information for the exercise of judgment and reason.[243] However, the "grand miscariage in our studys" was the effort spent on lodging material in memory rather than acknowledging that comprehension of ideas and arguments was the proper work of judgment, not of memory.[244] In any case, Locke questioned the efficacy of the usual attempts to improve memory: his most acerbic criticism was directed at various exercises to encourage rote learning. In *Some Thoughts Concerning Education* (1693) he rejected the opinion "that Children should be imploy'd in getting things by heart, to exercise and improve their Memories." If this were so, he argued, then "Players [actors] of all other People must needs have the best Memories." Locke submitted that this was not the case. The simple point was that "strength of Memory is owing to an happy Constitution, and not to any habitual Improvement got By Exercise." He concluded that "what the Mind is intent upon, and careful of, that it remembers best," adding that "if Method and Order be joyn'd, all is done, I think, that can be, for the help of a weak Memory."[245]

Apart from suggesting methods for taking notes from books and other sources, Locke speculated about habits that assist thinking. In 1704 Pierre Coste reported Locke's advice that

> whenever we have meditated any thing new, we should throw it as soon as possible upon paper, in order to be the better able to judge of it by seeing it altogether; because the mind of man is not capable of retaining clearly a long chain of consequences, and of seeing, without confusion, the relation of a great number of different ideas.[246]

In the *Essay*, Locke defined "Contemplation" as a way of keeping an "Idea" in the mind "for some time actually in view."[247] He analyzed two relevant problems. The first was that knowledge produced "by *Demonstration*," that is by what he called "*intervening Proofs*," did not sustain "that evident lustre and full assurance" of "*intuitive*" perceptions (*Essay* IV.ii.4–6). Second, it was almost impossible, even for those with "admirable Memories, to retain all the Proofs" (IV.xvi.1), especially not "in the same order, and regular deduction of Consequences, in which they have formerly placed or seen them; which sometimes is enough to fill a large Volume upon one single Question" (IV.xvi.2). Like Descartes and Hobbes, he doubted that attention could be sustained over "a long train" of deduction (IV.ii.6), and that such chains of thought would not be retained in memory. The only dependable response was reliance on writing, even if only in the form of short notes.[248]

Toward the end of his life, Locke gave a clear indication that writing, including brief notes, was essential for the retention and clarification of our best thoughts. He did so in response to a letter from Samuel Bold, an admirer and defender of the *Essay*.[249] In a perfect illustration of the problem Descartes, Hobbes, and Locke had described, Bold admitted his inability to keep track of his thoughts: "I loose a great many things, . . . Several things I am fain to let slip, because my own thoughts are not steddy and strong enough to follow and pursue them to a just Issue."[250] Locke's ample reply acknowledged the habit of note-taking, which he identified with Bacon:

You say you lose many Things because they slip from you: I have had Experience of that myself, but for that my Lord Bacon has provided a sure Remedy. For as I remember, he advises somewhere, never to go without Pen and Ink, or something to write with.[251]

More specifically, Locke urged Bold

to be sure not to neglect to write down all Thoughts of Moment that come into the Mind. I must own I have omitted it often, and have often repented it. The Thoughts that come unsought, and as it were dropt into the Mind, are commonly the most valuable of any we have, and therefore should be secured, because they seldom return again.

This echoes what Petty and Aubrey said about precious "flying thoughts" and "winged fugitives." Locke framed this point about immediate recording of fleeting ideas as part of the division of labor between memory and judgment or reason. He told Bold that his problem was not "Strength of Mind" but the inability

of Memory to retain a long Train of Reasonings, which the Mind having once beat out, is loth to be at the Pains to go over again; and so your Connection and Train having slip'd the Memory, the Pursuit stops, and the Reasoning is neglected before it comes to the last Conclusion.

Locke concluded with the following affirmation, which stands as a testament to his lifetime of note-taking:

If you have not tryed it, you cannot imagine the Difference there is in studying with and without a Pen in your Hand. Your Ideas, if the Connections of them that you have traced be set down, so that without the Pains of recollecting them in your Memory you can take an easy View of them again, will lead you farther than you can expect. Try, and tell me if it is not so.[252]

Conclusion

In the English draft of the "New Method," Locke said he was offering a "descripsion of my way of makeing collections for the help of my bad memory."[253] Considering the range of opinion we have encountered, it is fair to say that Locke regarded note-taking mainly as a means to avoid undue reliance on memory, or on recollection; his emphasis was on retrieval of a stable record. In *Conduct the Understanding* (1697), he even seemed optimistic, saying that "writing, [which] is but the copying our thoughts."[254] Such a record might then be transferred to other notebooks and compared with relevant information. Even so, Locke's method depended on remembering, or recalling, the Heads (or Titles) when looking for an entry. An index constructed in accordance with the principles set out in the "New Method," unlike a standard alphabetical index, did not display names and topics that could be scanned, and recognized, by anyone. It merely showed the various letter/vowel codes of the Heads that had been used in the notebook and the pages on which these were located. As Locke said, when choosing a marginal Head he first considered "under what title I thinke I shall be apt to looke for it"; and that this should always be "reduced to some one word" in order to "finde it again."[255] His index was personalized, and it was helpful only if one knew the Head under which a particular topic or entry had been placed. Of course, once Locke himself found an entry, such as an excerpt from a book, there can be little doubt that it must have also prompted his recollection of the material from which it was drawn, or the issues that it represented — as seen in his use of journal entries, in tandem with previous notes, as stimuli for later reflection. Indeed, on the basis of traditional advice from humanists and Jesuits, this additional payoff was more than likely because Locke had, on his own account, made judgments about where entries sat in the maps of knowledge he had devised and refined.

The main reception history of Locke's "New Method," at least for English readers, began soon after his death, with the two translations of 1706.[256] The one published by Greenwood began with comment that this way of "making *Common-Places*" carried the authority of "that Great Master of Reason and Method, the late learned Mr. Lock."[257] This edition also included, as a preface, a translation of a section from Jean Le Clerc's *Ars critica* (1697) in which he declared that Locke's manner of keeping notes achieved the right balance between memory and judgment.[258] In comments that echoed some of the advice of figures such as Duport and Holdsworth, Le Clerc advised against copying long passages, instead preferring brief notes that allowed an easy return to the source: hence the need "to mark the Place of the Author from whence you Extract it."[259] He implied, without a detailed justification, that Locke's method was well suited to this aim. However, in another major work addressed to the European Republic of Letters, Vincent Placcius cast some doubt on this.[260] At the start of his *De arte excerpendi* (1689), he listed eight major discussions of the art of scholarly note-taking; of these, Locke's anonymous contribution was the last, and most recent. Placcius included an illustration of the "New Method" index and summarized Locke's explanation of it.[261] While acknowledging that its mode of indexing was new and notable, he argued that its merits were outweighed by its disadvantages. In his view, the method was cumbersome because one might have to check many scattered pages of a notebook to find a particular Head within the many words covered by one of the letter/vowel combinations.[262] Placcius agreed that speedy retrieval was a prime consideration, but asserted that his own elaboration of Harrison's indexes was more effective.[263]

There are also earlier indications of the parameters of this reception among some of Locke's correspondents. The issue of easy retrieval from a personal notebook was mentioned by John Freke, one of Locke's acquaintances in the Netherlands, who reported on Benjamin Furly's reaction:

> I deliverd your methode of common placing to Mr Furly who is very much pleased with it and says his own was of much the same nature. He this morning lent it to the Burgemaster whom he found busie contriving a methode of putting his Notions in an order that he might know where to find them soe that it came in season.[264]

It is not clear what Furly meant by "order," and it is not safe to accept that his method resembled Locke's because this was often misunderstood. Thus when Bold followed up on Locke's advice about keeping a pen at hand, he asked this question:

> one thing more I would also desire to know from you, which is concerning your way of Common placing, viz. Whether the two sides in the common place book must be reserved entirely for what may in time be mett with relating to that Head which shall have the first possession, or whether any other Heads beginning with the same Letter and vowel may be inserted in the Margin?[265]

Here it is evident that Bold was more comfortable with the assumption that the first Head allocated to a double opening takes command of the page, and that only other matter "relating to that Head" should be entered there. Indeed, for some, Locke's method of entering diverse topics on the same page was too radical. Another later admirer, Isaac Watts, welcomed the "New Method" in his *Logick* (1725) but recommended that one leave "a distinct Page for each Subject."[266] Presumably he did not mean preallocating Heads to all pages (as seen in Southwell's commonplace book); nevertheless, the notion that subjects or topics, rather than letter/vowel combinations, should determine the entries on a page is a major dilution of Locke's innovation and, on the issue of economy of space, it reinstated the problem he addressed.[267]

Locke was certainly a master of the information he selected, gathered, recorded, and indexed. However, we need to recognize that the scale and purpose of his activity was individual, not institutional. It did not rival that of his French contemporary, Jean-Baptiste Colbert (1619–83), Louis XIV's influential minister. Colbert did make his own notes, but he also acquired or pillaged vast troves of notes and documents made and accrued by others. It was this rapacious behavior that produced the archive he amassed, policed, and used in the interests of the monarchy and the State. In contrast, everything Locke's notebooks contained depended on his personal notice of the material.[268] The potential for collaboration with others remained, as seen in his exchanges with Boyle and in his ability to retrieve information gathered in France for later communication to the Royal Society. However, these instances relied on Locke's decision about what to notice and what to do with it. In most cases, it seems, he did nothing. He accumulated a mass of material across many subjects, some of which he pursued and developed, others which he left unvisited in his notebooks; he hinted at ways of using hypotheses derived from the observations and experiments he recorded; he explored ways of arranging information in schemata and under Heads, but his indexing relied on personal keywords he expected to remember when he wanted to locate an entry. In principle, Locke's notebooks could be patiently mined by someone else and put to work, but he left them as a repository of unexploited data. In the next chapter, I consider the ways in which various individuals, especially Robert Hooke, thought about ways of ensuring that notes taken by individuals would have a collective value.

[8]

Collective Note-taking and
Robert Hooke's Dynamic Archive

In seeking to highlight key elements of the Scientific Revolution, H. Floris Cohen has identified mathematical description; atomism and corpuscularianism; and empirical observation and experiment, as crucial to the new approach to knowledge. But he then asks: "how did such kernels of 'recognizably modern science' manage to stay in the world once they arrived there?"[1] One answer is that dictionaries, encyclopedias, and other compendia summarized and codified science for a wider audience, embedding it in a public culture.[2] But a prior, and more fundamental, factor in the *consolidation* of the new science was the collection, organization, and storage of notes, letters, and papers. Only six years after the foundation of the Royal Society, Sprat announced in his *History* (1667) that it was already the custodian of good intelligence and information.[3] This, he stated, was central to the intention of its members "to make faithful *Records*, of all the Works of *Nature*, or *Art*, which can come within their reach: that so the present Age, and Posterity" could separate error from truth as part of the quest for new knowledge. This material was "now ready in Bank," making the Society "the general *Banck*, and Free-port of the World."[4] However, the challenge was not just to establish an archive that received material from various sources, but also to direct and regulate the ways in which information could be sought cooperatively and stored for future use. This kind of collective note-taking required protocols, and incentives for individuals to adopt them.

As we have seen, individuals such as Hartlib, Evelyn, Boyle, and Locke were masters of their own notebooks, adapting general precepts from the tradition of note-taking, merging commonplace books and journals, and making indexes to supplement their powers of memory and recollection. Boyle's case indicates that even fragments and loose notes could prompt the recollection of their owners, but this power did not extend to second-hand users, even though some notes taken by Boyle were of use to others, as he hoped. The limits of such personal practices for large-scale and long-term Baconian projects were recognized. This is illustrated by Placcius' critical remarks in 1689 about Locke's "New Method" and his preference for Harrison's "Arca Studiorum" with its movable slips and flexible indexes. When notes were shared, their value was not as prompts to individual recollection but as stable records of information that could be communicated, combined, and analyzed. But such information had to be gathered and entered in some regular fashion that ensured easy retrieval and comprehensibility—by individuals not involved in its initial collection. Sprat optimistically presented this scenario as well underway: "The *Society* has reduc'd its principal observations, into one *common-stock*; and laid them up in publique *Registers*, to be nakedly transmitted to the next Generation of Men; and so from them, to their Successors."[5]

In this chapter, I look at various notions of collective note-taking—either projected or attempted by the English virtuosi. I begin with the efforts of some key members of the Royal Society to construct quasi-formal guidelines for both collating and storing information. A central person here is Henry Oldenburg, one of the first secretaries (with John Wilkins), who tirelessly conducted the correspondence that fueled the *Philosophical Transactions*, starting in March 1665. I also discuss Martin Lister, John Ray, and Francis Willughby who shared and compared notes on the study of plants and animals. The final section of the chapter examines Robert Hooke's significant reflections about the possibilities and difficulties of making, storing, and using Baconian natural histories within an institution, such as the Royal Society.

Baconian Models

It is helpful to start with a distinction between collecting notes and prescribing note-taking. The former activity is clearly seen in all the figures treated in previous chapters. It is vividly illustrated in Hartlib's quest to collate the extant notes of deceased scholars; such troves, embodying lifetimes of effort, were alluring prizes. This interest continued, as exemplified in Oldenburg's eagerness to acquire the notes of living savants. He was keen to hold Daniel

Georg Morhof (1639–91) to his promise "to send over the learned notes of the famous Gude [from texts on agriculture] . . . and a catalogue of his other manuscripts on medicine, natural history, and mathematics."[6]

The second notion involved the direction or supervision of the content and form of notes in the interest of a group project. This possibility predated the Royal Society. It was, of course, present in Bacon's writings and in the activities of those inspired by him, such as Hartlib and his circle, which itself intersected with the network of correspondents around Marin Mersenne in Paris. It is evident in the members of the "Club" initiated in Oxford by John Wilkins and Seth Ward from 1648. Within these informal groups there was a broad agreement on a two-stage process: *first*, collate, condense, and commonplace the information already available in books or oral reports; and *second*, orchestrate the collection of new material.[7]

We can see an attempt at the second stage in *The legacy of husbandry* (1652), written by Arnold Boate but issued under Hartlib's name. This included an "Alphabet of Interrogatories" consisting of a list of items (mainly, but not solely, plants and animals) from "Apricocks" to "Wormes." For each item there were queries of the Where, What, When type. For example, "What several sorts of wormes in Ireland, what harme done by them, and how they are destroyed?"[8] This was an early form of questionnaire; however, the items were not held together even by the general Heads suggested by Bacon and Boyle for natural histories. Nevertheless, Hartlib reflected the collaborative spirit of the project, telling Boyle that it could be advanced "if every ingenious head amongst you would take notice of whatsoever worth the observation occurreth in any place, that so by little and little we might perfectly come to understand the natural history of all the parts in that country."[9]

Some individuals publicized their own attempts at coordinated observation and collection of information. Ralph Austen took Bacon's *Sylva Sylvarum* as a model for empirical inquiries, even though Bacon himself was reportedly worried that it might seem an "Indigested Heap of Particulars."[10] This book comprised ten "centuries," each one containing 100 observations or experiments. As such it was an appealing guidebook for his followers, offering a starting point for additions and corrections to what Bacon had initiated. In 1658 Austen issued a set of comments (dedicated to Boyle) on three of these centuries that touched on "Fruit-trees, Fruits, and Flowers," a subject on which Bacon had "left unto us much upon Record in his Naturall History."[11] He gave Bacon's summary of each selected "Experiment," adding his own "Observation" below it. Austen's comments often qualified or extended what Bacon said on the basis of variations in the trials. Thus in experiment 406 Bacon proposed, in Austen's paraphrase, that "Digging, and loosening the earth about the Roots of trees accelerates germination." Austen agreed, but added that this did not assist "their early budding." Experiment 417 concerned the ability of boughs of trees to take root, but Austen declared that Bacon had taken this "upon trust, according to the generall opinion."[12] Some of his remarks are supportive ("This is a good Experiment" for no. 402 at pp. 2–3), whereas on other occasions he simply stated that the basic proposition is wrong-headed. Thus under Bacon's inference that water, not the earth, is the main source of nourishment for plants, Austen wrote: "Simply Water affords but a feeble, and weake nourishment, crude, and cold." Having reviewed a selection of Bacon's items, Austen reduced these to forty-six in "A Table shewing the Principall things contained in the ensuing Experiments, and Observations," with the suggestion that these could direct further research.[13]

In his *Britannia Baconica* (1661), Joshua Childrey reported that he had also taken Bacon as a model. He asked the reader not to make hasty "condemnation against me for setting down severall idle, empty, and useless things (as he may possibly imagine them to be) till he hath read the sixth Aphorisme of Lord Bacon's Parasceve."[14] This was an appeal to Bacon's insistence that his natural histories should embrace "matters so commonplace" that people might think "it would be pointless to write them down."[15] Childrey later told Oldenburg that he set out to use Bacon's list of 130 Titles in the *Parasceve* in this spirit:

> Some 2 yeares before the happy returne of the King, I bought me as many Paperbookes of about 16 sheets a piece, as my L. Verulam hath Histories at the end of his *Novum Organon*, into which Bookes (being noted with the figure & title given them by my Lord) I entred all Philosophicall matters, that I met with observable in my reading; & intend (God willing) to continue it . . . though indeed I first fell in love with the L. Bacons Philosophy in the yeare 1646, & tried severall Experiments (though such as I now reckon not to be of any moment) in 1647. 1648. 1649. & 50. And besides these I have 2 larger Paperbookes in Folio, one of which I call Chronologia Naturalis, and the other, Geographia naturalis; the former containing the time of all droughts, Comets, Earthquakes, &c. & the other the naturall rarities of Countreyes. These Paperbookes cannot be expected to be yet full, & God knows, whether I shall live to see them filled; But . . . I intend to bequeath them to the Society, whensoever I die.[16]

There is no evidence that this bequest happened, but it is a good example of how one eager contributor thought the Society might benefit from a set of notebooks inspired by Bacon, albeit not produced under the collaborative arrangements on which Baconian natural histories were predicated. As Anthony Wood reported, Childrey died on August 26, 1670, "having never been of the said society."[17]

Proverbs as Collective Information

Another intriguing example of an attempt to use an existing, although untapped, source as the basis for future inquiry is John Ray's *Collection of English Proverbs* (1670). Ray explained his method of proceeding over ten years: "I read over all former printed Catalogues that I could meet with; then I observed all that occurred in familiar discourse, and employed my friends and acquaintances in several parts of England in the like observation and enquiry, who afforded me large contributions." He also added "others such as came to my hands or memory, since the finishing of the precedent Catalogues," arranging everything under three main topics or Heads: Health, Husbandry, and Love. Thus among those pertaining to "Health, Diet and Physick" we find these: "one hours sleep before midnight's worth two hours after"; "Drink wine and have the gout, and drink no wine and have the gout too"; "Wash your hands often, your feet seldom, and your head never."[18] One scholar has remarked, "It is characteristic of Ray's versatile mind that one of his most important books could be entirely unconnected with botany or natural history."[19] But the opposite could be asserted: namely, that Ray's friends sent him new proverbs "amounting to some hundreds," thus accumulating common opinion on various matters which could be expanded and corrected, forming a baseline for further research in natural history.[20] With this in mind, Ray filtered out "All superstitious and groundless Observations of Augury, days, hours, and the like . . . because I wish that they were quite erased out of peoples memories, and should be loath to be any way instrumental in transmitting them to posterity."[21] In the second edition he mentioned that he had omitted proverbs containing lewd words, but not if he judged their content, perhaps about "some excrements of the body," to be important.[22]

Ray did some preliminary analysis of his collection. He annotated some proverbs, happily overruling or qualifying commonplace opinion if empirical observations were available, and documenting these corrections by time and place. Of the forty-nine proverbs about husbandry and weather, eleven include such caveats.[23] For example, "If the grass grows in Janiveer, it grows the worse sor't all year" is followed by Ray's comment, based on recent information: "There is no general rule without some exception: for in the year 1677 the winter was so mild, that the pastures were very green in January, yet was there scarce ever known a more plentiful crop of hay then the summer following." And to the proverb "A green winter makes a fat Church-yard" he added: "This proverb was sufficiently confuted Anno 1667, in which the winter was very mild; and yet no mortality or Epidemical disease ensued the Summer or Autumn following."[24]

Encouraged by the way the first edition had stimulated some readers "to examine their own memories and try what they could call to mind," Ray expected that this collection would be augmented over time.[25] He was aware that there was a problem in defining what counted as a proverb, but included material "sent me for such by learned and intelligent persons" because his interest in the empirical content outweighed any concerns about the literary or sociological aspects of the shared opinions he assembled.[26] Although apparently a textual exercise, Ray's collection of proverbs, together with his empirical addenda, resembles the efforts of Austen and Childrey to use Bacon's *Sylva* and "Parasceve," respectively, as starting points for collaborative gathering of information. The challenge of the Royal Society was to encourage a collective approach that did not rely on the specific set of observations Bacon had supplied.

Institutional Note-taking

Sprat reported that a committee had been set up "to read over whatever Books have been written" on subjects within the purview of the Society in order to "digest" the selections "into *Manuscript volumes*." Early thinking about the collection of information in this new institution was framed in terms of the experience of keeping personal notebooks — also indicated by Sprat's account of how the "Preliminary collection" was made. It was the custom of Fellows, he explained, "to urge what came into their thoughts, or memories concerning them, either from observation of others, of from Books, or from their own Experience, or even from common fame itself."[27] By deciding to direct and secure the information gathered in this traditional manner, the Society envisaged a treasure-house richer and more varied than any individual could assemble.

One of the favored instruments was a list of queries, along the lines of the "interrogatories" in Hartlib's publications. The earliest of these was entered in the Register book on December 5, 1661, at the advice of Lord Brouncker and Boyle. It comprised twenty-two questions, many including specific observational and experimental instructions to be sent to Teneriffe.[28] Boyle's "General Heads for a Natural History" shows a similar approach, although it recommended larger categories with the intention of following up with more specific requests.[29] When Oldenburg mentioned that these "Generall Queries" were about to appear, he received an enthusiastic response from Henri Justel in Paris: "I talk everywhere about the fine plan which you . . . have of making a natural history, in order to stir up the spirit of emulation among our learned men and to prompt them to imitate you."[30] Sprat reported on the Society's approach:

> Their manner of gathering, and dispersing *Queries* is this. First they require some of their particular Fellows, to examine all Treatises, and Descriptions, of the Natural, and Artificial productions of those Countries, in which they would be inform'd. At the same time, they employ others to discourse with the Seamen, Travellers, Tradesman, and Merchants, who are likely to give them the best light. Out of this united Intelligence from Men and Books, they compose a Body of Questions, concerning all the observable things of those places. . . . They have compos'd Queries, and Directions, what things are needful to be observ'd, in order to the making of a Natural History in general: what are to be taken notice of towards a perfect History of the Air, and Atmosphere, and Weather.[31]

These calls for information show how the Heads of the commonplace tradition could become *questions* in natural history and experimental inquiry — often similar to Bacon's specific topics. One problem was that some questions invited tales of wonder, and even neutral ones frequently produced reports leaning toward the curious and marvelous rather than notices of different manifestations of the common, or ordinary, phenomena that Bacon requested.[32]

In a letter of 1669 to the Belgian mathematician René Sluse, Oldenburg described the *Philosophical Transactions* as "these philosophical commonplace books."[33] One obvious interpretation of this analogy is that this journal, comprising letters and articles, could serve one of the functions of a personal notebook, especially when indexed by topics. Although it is significant that he made this comparison, a difference of principle must be stressed. Oldenburg's notion of pooling information ran against the grain of traditional commonplacing: rather than sanctioning the selection of choice material from an established canon, he affirmed the potential value of small, and apparently, insignificant memories, observations, or reports that might otherwise not be noted at all. Hooke supported this idea. In an unpublished lecture (1670s?) that belonged to the series endowed for him by Sir John Cutler, he stressed that the "Philosophicall History" envisaged by his patron could not be attained by "the single indeavours of the most able and Accomplisht person in the world," and certainly not "by my meane abilitys." Consequently, he proposed to use contributions from others, promising that "what ever assistance or Information I shall Receive from any such Person I shall most punctually acknowledg by Entring their Description verbatim together with the time I receivd It and Names of the Persons." In a significant qualification, Hooke added that if such submissions contained "any thing new and considerable and Exprest in few words they shall be printed word for word as they are sent." In this way he hoped to record material that may not have seemed suitable, or "voluminous enough," for print, and thereby to "rescue many excellent things from Oblivion."[34] When Robert Plot took up the editorship of *Transactions* after a hiatus of four years following Oldenburg's death in September 1677, he was able to confirm an already established attitude, saying that this publication should be seen as "a convenient *Register*, for the Bringing in, and Preserving many *Experiments*, which, not enough for a Book, would else be lost."[35]

These remarks from key members of the Society were, in effect, invitations to use the Register Book and the *Transactions* as institutional notebooks. By way of encouragement, Hooke proposed that any contributor should be able "to search the Register" to check whether "any history observation or conjecture" had been properly made, in "the order they are Communicated."[36] For assiduous correspondents, the Society could thus function as a repository of what Beale called "promiscuous" observations and experiments. He sent some of his own to Oldenburg in the first year of the *Transactions*, offering in a letter of September 24, 1666, for example, three loosely linked observations about the effect of a recent drought on oak trees, the properties of brackish water, and the abundance of eels after winter frosts.[37] Similarly, Martin Lister was pleased to use the *Transactions* to register his ongoing observations about insects, often taken directly from what he called "my Private Notes." He felt able to ask Oldenburg to keep these up to date: for example, "to annex a late Observation to the last I sent you."[38] As seen in chapter 5, Beale claimed that he did not keep regular notebooks; yet by providing Oldenburg with copy he was able to use the *Transactions* (and possibly the Register book) as a surrogate notebook to which he could return—an option that would have been aided by the "Tables," or indexes, such as the one he compiled for the numbers published between March 1670 and January 1671.[39]

Oldenburg realized that discrete and particular contributions from various correspondents needed to be guided in some way. On January 27, 1666, he reported to Boyle that Hooke had shown him a paper on "a Method for writing a Natural History."[40] Boyle's "General Heads" of April 1666 was part of this discussion; and in a long letter of June 13, 1666, he offered a sketch of "my Designe about Natural History."[41] Whereas in the former paper Boyle gave "very Comprehensive and greatly Directive" topics of inquiry—that is, with regard to the Heavens, Air, Water, and Earth—he now suggested more specific histories of "Bodys," "Qualitys" (e.g., Cold, Heat), "States of matter" (fluid, firm, animate, and inanimate), and natural processes.[42] He also sought to provide potential observers with more specific intellectual frameworks appropriate to different fields of science. Among the desiderata mentioned were instructions for the use of mathematical instruments and "Chymical" operations; lists of "the Names of the chief Authors" on travel and navigation; and a "Summary" or "short survey" of the main "Hypotheses" concerning natural phenomena, such as the "Peripatetik, the Cartesian and the Epicurean." This last idea was not intended to support one of these hypotheses as the "Basis" of the natural history, but rather to counteract the skewing of observations by large systems or theories. An awareness of these powerful "Theorys," Boyle said, might "admonish a man to observe Divers such Circumstances in an Experiment as otherwise 'tis like he would not heed."[43] Boyle's own commitment to this stance was on show in his New Experiments and Observations touching Cold (1665). He explained that he did not assert "any particular Hypothesis, concerning the Adequate cause of Cold" because only completion of "the Historical part" would allow him to "Survey and Compare the Phaenomena." In inviting others to join him in this quest, Boyle was aware that the presentation of material may be influential: on the one hand, he did not wish "to huddle my Experiments confusedly together"; on the other, he did not want to impose an arrangement that favored a certain theory. He therefore disposed his material under twenty-one "comprehensive Titles," encouraging readers to insert "other Experiments or Observations in any of them," or to invent new titles. Like Bacon in the Sylva Sylvarum, Boyle declined to adopt "a strict method" on the grounds that this might dissuade others from adding their own contributions.[44]

Even if these tips about making observations were heeded, there remained the question of how such information should be set out on paper, in the notes and letters that Oldenburg invited and received. In his "Designe," Boyle raised the problem of an appropriate "way of writing" when "delivering the Particulars, that the Body of the History is made up of."[45] Far more specific and ambitious in this regard was Hooke's "Method for making a history of the weather," a table or "Scheme" designed for recording observations. In citing this as evidence of the Society's early achievements, Sprat said it showed what needed to be done to advance "a perfect History of the Air, and Atmosphere, and Weather." As well as columns for wind direction, temperature, humidity, and barometric pressure, Hooke had one for "Notable Effects," such as prevailing diseases, and one for "Faces of the Sky," for which he supplied some qualitative terms, such as "Let *Hairy* signifie a Sky that hath many small, thin and high Exhalations, which resemble locks of hair, or flakes of Hemp or Flax."[46] The minutes of the meeting of the Royal Society in 1663 ordered that copies be "sent to the several persons, who had been engaged in this work of observing the changes in the weather."[47] If taken up, this "Scheme" had the potential to impose some uniformity on the way such observations were entered in personal notebooks. Locke's weather "Register" begun on June 24, 1666, possibly followed either Hooke's model, or Boyle's suggestions of April 1666 about how and where barometric readings were to be taken and then entered in diaries.[48] However, the difficulty of establishing an accepted arrangement is nicely illustrated in Childrey's letter to Oldenburg of mid-1669 (mentioned above). He enclosed astronomical observations from January 1668 to June 30, 1669, made by a friend, a "young melancholicke Gentleman, a Cadet." Childrey wrote that his friend

> promiseth me to continue it, & I believe he will. He is the fitter (I take it) for observations of this nature, because being not Astrologically affected, or (rather) infected, he is more like to be impartiall.[49]

Yet despite this promising disposition, the young man could

> not to be persuaded or reasoned out of his humour. . . . As I found out by offering to his sight the Synopsis of the weather, printed in M. Sprats history; which he would not dare heare of, when I desired he would make it his exemplar, & offered my further directions.

The resistance of Childrey's friend, probably not a solitary instance, indicates the challenge of standardizing the information requested and received by the Society.

The imperative to amass a collection, sufficient for comparative inferences, required contributors to accept some protocols for their inquiries.
Oldenburg and some of his correspondents were aware that this meant going
beyond the undoubtedly useful collation of extant information — something
Hartlib had stressed. Benjamin Worsley went further when he called for methodical collection of new data over longer periods. Speaking about astronomy and natural astrology in the late 1650s, he explained that

> I mean the making of such constant Observation, and keeping such a
> Diary for the doing of that again, and giving us an History or Diary of the
> Observations of the Weather, and its Changes in all Respects, and then an
> Account of the several Places, Motions, or Aspects, each Day.[50]

Although somewhat slanted toward the "unusual," the queries that Plot drew
up in 1674 as a way of collecting material for his *Natural History of Oxford-
Shire* (1677) also acknowledged that a baseline of information was required.[51]
There was no trouble, Plot remarked, in getting accounts of the "Tempest of
Thunder and Lightning" that hit Oxfordshire on May 10, 1666, but he wondered whether it might be better to have good reports of "the whole weather
of the years past, on every day of the month."[52] Indeed, referring to a manuscript in the Bodleian library, he said that William Merle of Merton College had done this between 1337 and 1344. However, Plot was also able to cite
a contemporary, "the learned and observing Dr. Beale," who in the *Philosophical Transactions* of January 1673 called for "Almanak Makers" to record
the quotidian as well as the exceptional features of the weather.[53]

On this point about comparative data, and many other matters, Beale was
one of Oldenburg's most insistent correspondents. Indeed, Oldenburg tried
to deflect some of these unremitting inquiries to Boyle.[54] Some of Beale's letters betray a free association of ideas, but his attitude illustrates a transition
from Hartlib's focus on the collation of manuscripts, books, and testimony to
the conception of routine procedures for gathering new scientific information. In his early work on *Hertfordshire Orchards* (1651), Beale was aware of
the need to test claims made in books against common empirical knowledge:
he remarked that "Some years ago I read a small treatise of Orchards. . . . In
it I found many assertions which seemed to me so strange . . . so discordant
from our daily practice."[55] Although some of his own observations were denoted "promiscuous" (possibly by Oldenburg), Beale did say that these might
be "not unworthy to be recorded for further application in like cases of time
and place."[56] Indeed, despite his claims about not relying on notebooks, or

at least, perhaps, not on commonplace books, Beale clearly did take notes of empirical matters. In January 1666 Oldenburg relayed this report from him about barometric readings: "I doe not now take a deliberate view of my Notes, but I *wondered* once to see, that *in one day it sunk about ¾ of an inch.*"[57] Thus he was able to respond quickly to Boyle's request for confirmation of a fall in air pressure after a "Change of Weather and Wind."[58]

In the 1670s Beale developed the notion that aggregation of information was the only option, for "by this methode we may learne more in fewe years than at random in all the days of our short & brittle lives." For example, in the contribution Plot cited, Beale wished that "some ingenious Almanak Makers would (instead of the conjectures of Weather to come) give a faythfull & iudicious accompt of the Weather, & other remarkable accidents, & phaenomena, as they fell out on the same day at the Moneth of the yeare foregoing." This, he believed, would supply a basis for comparisons, so that "manifold uses may be made from such informations &c."[59] Beale suggested that other information should be collected on this principle: for example, the "highest & lowest" monthly prices of wheat, rye, barley, peas, beans, and oats sold in London; and the "weekly Bills of Births, Burials, plague, of males, & Females."[60] He set an example for how things might be done, reporting in 1670 that "I have examined my *Ephemeris* from May 28, 1664 till now: In all which time I find it not remark'd for any such degree of Cold, as hath been, since Christmas began." With these records at his disposal he was also able to say that "To compare with former Winters, the lowest mark is upon Decemb. 17. 1665."[61] Beale's call for more coordinated marshaling of information were generally well received, but collective note-gathering, of itself, did not guarantee an adequate basis for comparison.[62]

Sharing Notes among Friends

The activities of Ray, Willughby, and Lister offer an example of what collective note-taking, on a smaller scale, might be. As friends and collaborators from the 1660s, they made and shared notes on a range of topics in the natural history of plants and animals. One result was the publication of important catalogs of plants, birds, fishes, and insects. Although their work depended on an existing corpus of texts, such as those of Gessner, Ulisse Aldrovandi, and Magnus Olaus, Ray stressed the need to check several independent sources, to seek reliable testimony and, whenever possible, to use first-hand experience. In explaining the method he and Willughby adopted in their work on birds, he promised readers that they had not "scraped together" extant information but had "inserted only such particulars as our selves can warrant upon our own knowledge and experience, or whereof we have assurance by the testimony of good Authors, or sufficient Witnesses." They were prepared to exclude descriptions that did not meet these criteria, especially because they were worried about unwarranted multiplication of the number of species.[63] In this way, they played their part in what Brian Ogilvie has aptly called the "science of describing," in which detailed accounts, illustrations, and catalogs were the "final stage of a long condensation of observation, memory, and experience."[64]

By about 1600 there were doubts about relying on memory for the detailed descriptions required to assign specimens to species; however, it is clear that this approach had been the norm. Seventeenth-century naturalists appeared in the repertoire of stories about individuals exhibiting prodigious memories; however, their use of notes was also often mentioned. In a record of conversations with Gregorio Bolivar, a Spanish Franciscan living in South America in the early 1600s, Johannes Faber, secretary of the Academy of the Lincei in Rome, remarked how notes triggered his exceptional memory: "I swear to you, my reader, that what I derived from him up to now from his mouth and notes, he described with good recollection and without the aid of any book. And that he brought them together with the aid of a sort of compendium only."[65] The physician and naturalist Tancred Robinson assessed Ray in similar terms: "I am overjoyed that so vast a Memory, so exact a Judgment, and so universal a Knowledge, will be employ'd in compiling a general History of Plants, an Undertaking fit only for your extraordinary Talents."[66] Yet Ray highlighted the significance of notes. Writing to Robinson in 1685, he said,

> If I had Mr *Willugby's* Notes, I doubt not but I could find out a more exact Description of the *Orphus* than will be met with in Authors; for that Fish I am sure was more than once described by us . . . but all my Notes of high and low *Germany* were unfortunately lost.[67]

Ray knew that lost notes meant lost information.

Ray's circle of friends regarded their own notes as fundamental tools in natural history. At Cambridge, Ray and Willughby were pupils of James Duport who stressed the value of succinct notes in pocket-books. In their travels on the Continent with Philip Skippon (one of Ray's students) they relied on careful noting of details. When Ray published an account of their observations in 1673 it was largely in the form of dated entries, presumably from his journal, and that of Willughby for his account of his journey alone though Spain.[68] They spent the winter of 1663–64 in Padua, where they attended anatomy lectures at the university and witnessed the dissection of a woman's body in the house of the anatomist Pietro Marchetti. Willughby's papers contain a booklet of copious notes and commentary on the dissection, which was performed by Marchetti's son, Antonio.[69] The numbering of these "Ob[servations]" begins on p. 12 of this booklet and reaches number 165 at p. 32 of the thirty-five-page notebook. A similar method is seen in Willughby's comments on his journey through Spain. In his *Collection of Curious Travels and Voyages*, Ray edited the accounts of several travelers, and no doubt approved of the avowal of the German physician Leonhart Rauwolff (1535–96) that his description of the Levant was based on regular notes: "What I saw, learned, and experience'd, during the space of Three years, . . . I consigned all in good order, as it occurred daily, in a Pocket-Journal, to keep as a Memorial of my Life." When asked to publish an account he said that "Whereof I looked my Itinerary over again, and whatever Curiosities I had observed, I did transcribe into a peculiar Diary, which I divided into three parts, according as I travelled into several Countries, and committed it to the Press."[70]

Ray's own convictions about the indispensability of meticulous note-taking may well have been confirmed through his friendship with Lister.[71] Their travels had intersected in France in February 1666 and they met up again later that year in Cambridge and began to exchange letters.[72] Ray appreciated Lister as a fellow traveler, determined "to see with your own eyes, not relying lazily on the dictates of any master but yourself, comparing things with books." He quickly sought Lister's advice on additions to his pocket-book catalogue of plants in Cambridgeshire and encouraged his new friend in his own work on spiders.[73] Their discussion of the contribution of the ancients led to an agreement on the necessity of amassing particulars in all domains of natural history. Lister originally lamented the paucity of detailed accounts in Pliny the Elder's *Natural History*, but after Ray defended its value he relented: "I remember you once took away the Prejudice I had against *Pliny*, and I have ever since look'd upon him as a great Treasure of Learning."[74] Indeed, Lister then criticized Bacon for denigrating Pliny's collection of facts as relatively worthless. This was, he charged, a rash censure that undermined Bacon's own stated principles, "for if a particular nature or phenomenon may be in some particular body more bare and obvious, without doubts the greater number we have of particular histories, the plentifuller and clearer light we may expect from them." He went on to say that "I think it absolutely necessary that an exact and minute distinction of things precede our learning by particular experiments" about the internal structures and underlying principles of natural bodies, because such understanding was "subsequent to natural history." Lister tried to convince Ray to incorporate a methodological statement along these lines in the preface of his forthcoming *Catalogus plantarum Anglicae* (1670). Ray agreed about "the usefulness of being particular and exact in natural history," but added that he did not want to swell the preface to a small book. There is no doubt, however, that Ray believed that precise and detailed observation was fundamental to all natural histories, each of which was "work enough for one man's whole life."[75] Hence pooling of notes was essential: as Lister professed, "I am no Arcana man, and methinks I would have everybody free and communicative . . . the shortness of our lives."[76]

Good notes not only secured information over time; they could also could be shared and communicated with others. In 1684 Robinson claimed that "my own private common-place books do afford some odd, and as I think, useful observations and experiments upon plants."[77] This offer of information from a personal notebook was in keeping with the relationship Ray and Lister developed in their early letters. Thus having missed seeing Lister in Cambridge in July 1667, Ray wrote from Middleton Hall, Willughby's estate in Warwickshire, asking that he "take a little pains this summer about grasses, that so we might compare notes."[78] Toward the end of 1669 they swapped notes on a "general catalogue of English plants." Ray promised to "go over all yours, and give you an account which are to me unknown, and which I have not yet met withal in England."[79] Lister sent dried plants, and also his "tables of spiders," saying that Willughby "may freely command my papers at any time."[80] Through Ray, Willughby suggested that Lister's current "thirty-one species of spiders" might well be doubled in England alone.[81] They were able to draw on their notes to furnish answers or prompt queries: thus on the question of the movement of sap in sycamore trees, Lister reported that at the "beginning of July I cut out an inch, or more, square of the bark, at about my height, in the body of the same tree." About two months later, he added that "In my last year's journal I find that, particularly the 17th of December, there was a very copious bleeding."[82] Ray reciprocated by "looking over my notes in 1668" and providing observations of the rates of bleeding in branches cut from willows and sycamores.[83]

As well as communicating the content of notes by letter, Lister used his frequent correspondence with Oldenburg to lodge some material in the *Philosophical Transactions*, occasionally telling Ray that this is where he could find the information: "I have communicated to Mr. Oldenburgh my notes of the bleeding of the sycamore."[84] Later, when Ray was assembling Willughby's notes and papers into *The Ornithology* (1678), Lister wrote to say that "My Notes are very slender on the Subject of Birds."[85] On other subjects, however, they were able to share productively, not only on plants, but also on insects and fishes, material which, together with that on birds, Ray had inherited from Willughby as his executor.[86] Aware of the range of Ray's interests and the scope of the materials at his disposal, Robinson urged that he aim for "a general history of nature."[87] We can surmise from these letters between Lister and Ray that the exchange of notes, dated and bearing circumstantial details, was a crucial component of this encyclopedic venture.

As I have emphasized throughout this book, notes do more than fix information: in the hands of their makers, they can also stimulate recollection and thought. It is noticeable that Ray and Lister often declined to answer queries without checking their own notes, and, that they engaged in a process of reviewing and supplementing old notes in the course of refining their ideas. Thus in 1669 Lister said he had been "adding this last year's notes to the former, and I have found enough to cause me to make considerable alterations and amendments everywhere, and especially in the table (of spiders) I sent you." And on the subject of "juices" from trees he cautioned Ray that "My notes of this nature being, for the most part, but of one year's standing, I am loth to venture raw conjectures."[88] Other comments confirm that Lister revisited his notes: "I have, this last month, writ over a new copy of my History of Spiders (which is the fourth since I put my notes into any order), and inserted therein all last summer's observations and experiments."[89] Furthermore, it seems that these two friends reflected on the use of notes, because Lister clarified that "I sometimes use my notes and sometimes I trust to my memory."[90]

For Ray and Lister, notes were not necessarily conceived as raw data to be exchanged once made; the additional power of notes was their capacity to enrich the experience and understanding of the person who made them. Lister, in particular, was constantly augmenting, reviewing, and synthesizing his notes. He gave Oldenburg a glimpse of this process when explaining that he wanted to reconsider some earlier comments "if I can get leisure to enlarge the notes I sent you."[91] Questions stimulated a return to old notes and these in turn sparked recollection and further queries. As we have seen, this continual reading of, reflecting on, and recollecting from personal notes was also fundamental to Boyle's style of thinking. The intimate relationship between individuals and their own notes was not fully transferable to another person. Yet the understanding between Ray and Lister indicates that over a decade or so they came to integrate each other's notes quite easily within their own inquiries. Outside a small collaborative relationship such as theirs, however, it is doubtful that notes could generally function in this manner. Thus in 1692 when Aubrey sent Ray his large collection of notes on natural history and related matter, the effect was nothing like the regular and cumulative exchange of material with Lister. As Ray replied: "Your *Adversaria Physica* I have read over once, but the variety & curiosity of the matter & observations is such, that I cannot satisfy my self with a single reading."[92] Having read the adversaria again a couple of months later, Ray urged Aubrey to publish them before it was too late for "the benefit & instruction of others." He also stressed that Aubrey should do this "in your lives time" because otherwise executors might do so without due care.[93] The dialogue between Ray and Lister suggests, however, that Aubrey's notes would have been more effective if exchanged, assimilated, and synthesised over time among a small group of collaborators.

Making a "Philosophical Repository"

Collective note-taking under the auspices of an institution, such as the Royal Society, could not easily replicate such intensive sharing of information and ideas. But the Society could accumulate masses of information. After all, as Sprat recounted, there was "a *Register*, who was to take Notes of all that pass'd; which were afterwards to be reduc'd into their *Journals*, and *Register Books*."[94] Michael Hunter has remarked on a "virtual obsession with record-keeping," and Mordechai Feingold has stressed that the process of acquiring and keeping such material was core business for the new institution.[95] But if members and correspondents sent all sorts of information, as requested, what was to be done with it? How was this "Bank" of information, as Sprat called it, to be selected, organized, and stored for present and future use?

There were early indications that this task was recognized, but not well executed. In July 1665, as members left London to avoid the plague, Oldenburg asked Boyle what to do with "the Bookes and Papers belonging to the Society, that are all in my Custody."[96] This implies that there was no agreed physical space for this paper archive as distinct, say, from the "Repository"—the cabinet or museum of objects and specimens housed in a room in Gresham College.[97] Equally revealing is Oldenburg's letter of November that year telling Sir Robert Moray that it would be easier to find what he wanted to know about "the making of small shott" in Hooke's *Micrographia* than in a manuscript held in the Society. He confessed, "yet indeed I cannot open the boxe and look out that paper, without putting all into confusion."[98] Two years later he protested to Boyle about his single-handed effort to manage a flood of material: "I am sure, no man imagins, what store of papers and writings passe to and from me in a week. . . . I confesse, I extend my patience as far as I can, but I am afraid, I have stretcht it so farr already, that it will break."[99]

The location, arrangement, and administration of papers must have been recognized as a problem because Theodore Haak offered a solution. The minutes of the meeting on November 5, 1662, state that he had "proposed a compendious way of repertory, and was desired to communicate it to the society." Haak did so on November 19, showing "the society a specimen of his repertory."[100] It is likely that he recommended something along the lines of Harrison's "Arca Studiorum," which its penniless inventor offered as a way of indexing Parliamentary papers. Some members of the Society thought that there was an urgent need to have its papers properly arranged and accessible: a submission in 1674 claimed that it was important that the king, as founder, be able to peruse "the Journal, & other Books" and that "a Catalogue or Collection [be] discreetly made & shewd" as a way of demonstrating what progress had been made.[101] A council decision of September 24, 1677, said that "all papers and books concerning the Society be kept in the repository or library."[102] However, at this stage it seems that there were contrasting levels of control over these two domains: in 1674 John Hoskyns gratefully remarked of Nehemiah Grew's catalogue of the repository that it allowed one "to find likenesse and unlikenesse of things upon a suddaine."[103] No one, it seemed, could expect this from other parts of the Society's archive.[104]

The early public statements from, or on behalf of, the Society took a latitudinarian stance toward the collection of information. In his *History* (1667), Sprat explained that "as their purpose was, to heap up a mixt Mass of *Experiments*, without digesting them into any perfect model: so to this end, they confin'd themselves to no order of subjects." Defending the value of such a policy, he argued that "if their *Registers* had been more *Methodical*" such ordering would have been premature and counterproductive.[105] In the following year, Glanvill supported this position, affirming the need to "lay up in Bank for the Ages that come after."[106] In 1670 the publisher of Boyle's *Tracts* appeared to endorse a similar attitude. He stated (presumably with Boyle's agreement) that various "Loos Tracts" were included, and explained that this was justifiable given that the author's main Designe in these as well as his other Physicall Writings, was to provide Materialls for the History of Nature, it would be thought enough that they be substantiall and fit for the Work; in what order or Association soever they should happen to be brought into the Philosophicall Repository.[107]

But earlier, in Boyle's *Cold* (1665), the corresponding advertisement struck a different tone, stressing the author's caution about what he made public in order to ensure "that nothing may slide into the Philosophical store, that may prove prejudicial to the Axioms and Theories hereafter perhaps to be deduc'd from thence."[108] There is a difference, of course, between Boyle's works and the archive of the Society, but the salient issue was the degree to which incoming information should be filtered. Boyle adopted a permissive stance in the second volume of his *Usefulnesse* (1671), advising that most material should be stored, even if its full significance was not yet recognized. Indeed he regarded his own work as one such storehouse:

> For I freely declare, that my designe in this present Tome was not to furnish it as well as I could, but to preserve, as in a repository, several scatterd Experiments and Remarks, which I could best spare from the other Treatises I had design'd, which might otherwise probably be lost.[109]

From the early 1660s, Robert Hooke, Boyle's assistant and (from November 1662) curator of experiments for the Royal Society, thought about the best ways of selecting and storing information. He wanted to work on notes and papers as they were collected, before assigning information to *other* notebooks kept by the Society and managed by delegated members. In his view, the process of note-taking—in this case, culling, sorting, and arranging—should not cease once material had been received. Although intensive sharing and comparing of notes, such as that between Ray and Lister, could not be replicated easily at the institutional level, Hooke was searching for ways of making a dynamic archive. As we now know, he took liberties on behalf of this goal, treating the Society's Journal-Books as an extension of his own notes. When Oldenburg died in 1677, Hooke scoured his secretarial minutes, transcribing some sections, making detailed notes, and removing some of the original draft minutes from the Society.[110] Inevitably, Hooke checked that his own priority on various matters was recorded, noting at one point that "the dog has entred nothing but left a blank." Now known as the "Hooke Folio," this collection of manuscripts had been in Hooke's possession at the time of his death in 1703; it was found in a house in Hampshire in 2006 and purchased by the Society.[111]

Among Hooke's papers, Richard Waller found this account of the Oxford philosophical club:

> At these Meetings, which were about the Year 1655 (before which time I knew little of them) divers Experiments were suggested, discours'd and try'd with various successes, tho' no other account was taken of them but what particular Persons perhaps did for the help of their own Memories; so that many excellent things have been lost.[112]

Hooke may well have pledged to guard against such losses in the future. But some twenty-five years later he still had cause to lament the fact that many discoveries "have been lost" because travelers often delayed recording details of their observations "till they have forgotten what they intended." In suggesting remedies, he returned to the Society's earlier message about the need to show people "how to make their Observations and keep Registers or Accounts of them."[113] One point was clear to Hooke: unaided memory could not be entrusted with the details required by Baconian natural histories.

Hooke on Memory and Retrieval

Throughout his life, Hooke was preoccupied with the weakness and fragility of his own memory. We know that he regularly subjected his body to various chemicals and medicines, and so it is not surprising that he also looked to these as a means of enhancing memory.[114] He kept a diary (from March 10, 1672), partly in the hope that the act of making notes would improve his own memory: an early entry records his purchase of a set of mnemonic verses.[115] As mentioned in chapter 1, Aubrey worked this obsession with memory into a character portrait of his friend as one whose "inventive faculty" and memory were out of balance.[116] Hooke knew that this opposition was epitomized in "the almost Proverbial Saying, that good Wits have ill Memories."[117] But as I have emphasized, this competition between memory and reason was attenuated by the use of notes and other writing. Thus in Hooke's case, the availability of his diary meant that in April 1697 he was able to decide "this Day to write the History of my own Life, wherein I will comprize as many remarkable Passages, as I can now remember or collect out of such Memorials as I have kept in Writing, or are in the Registers of the Royal Society."[118]

In seeking recourse to a combination of memory, personal notes, and institutional records, Hooke epitomizes the themes canvassed in earlier chapters of this book. He searched for ways of supporting memory so that mental effort could be channeled toward reasoning and understanding. This preoccupation was a leitmotif in his two major pieces of writing—the *Micrographia* (1665), and the "General Scheme" (c. 1668) published after his death.[119]

In the preface to *Micrographia*, Hooke contemplated a world in which the "Senses" were guided to "an easier and more exact performance of their Offices." He speculated that by means of artificial aids the senses might "recover some degree of those former perfections" lost by "Mankind" both from "a deriv'd corruption, innate and born with him, and from his breeding and converse with men."[120] However, he was adamant that any augmentation offered by instruments, such as the microscope, must be supplemented by an intellectual reformation that extended to the higher faculties: "So many are the *links*, upon which true Philosophy depends, of which, if any one be *loose*, or *weak*, the whole *chain* is in danger of being dissolv'd; it is to *begin* with the Hands and Eyes, and to *proceed* on through the Memory, to be *continued* by the Reason."[121] Memory was the weakest link in this chain, and its improvement posed special challenges: whereas various instruments, or drugs, might magnify the input of the five senses, the role of memory was to retain information from *all* the senses. Given this diagnosis, memory appeared as the trouble spot in Hooke's quest for a way of "rectifying the operations of the *Sense*, the *Memory*, and *Reason*." As he said,

> The next remedies in this universal cure of the Mind are to be applied to the *Memory*, and they are to consist of such Directions as may inform us, what things are best to be *stor'd up* for our purpose, and which is the best way of so *disposing* them, that they may not only be *kept in safety*, but ready and convenient, to be at any time *produc'd* for use, as occasion shall require.[122]

Hooke was not talking about a means of effecting some lasting improvement of natural memory, but rather about a way of ensuring that the information it absorbed could be remembered or recollected when desired. Continuing this deliberation in the "General Scheme," he contended that the construction of Baconian histories demanded the invention of external aids for memory as part of a program in which none of the mental faculties would be "without their Armour, Engines, and Assistants." He therefore proposed that:

the Senses are helped by Instruments, Experiments, and comparative Collections, the Memory by writing and entering all things, ranged in the best and most Natural Order; so as not only to make them material and sensible, but impossible to be lost, forgot, or omitted, the Ratiocination is helped first, by being left alone and undisturbed to it self, having all the Intention of the Mind bent wholly to its Work, without being any other ways at the same time imployed in the Drudgery and Slavery of the Memory, either in calling particular things to Memory, or ranging them in Order, or remembring such things as belong to another Head, or in transposing, jumbling, ranging, methodizing, and the like; for first all things are set down in their Order, . . .[123]

To explicate this cascade of ideas—typical of his thinking—we should notice that Hooke implied two different ways in which information might be arranged: a pragmatic ("best way") in the *Micrographia*, and a more normative ("Natural Order") in his "General Scheme." I think this indicates a choice he entertained between Bacon's suggestions for how natural histories should be collected, and other options. Like Boyle and Locke, Hooke endorsed Bacon's view that the first stage of inquiry required a "Philosophical History" yielding a "Repository of Materials . . . ranged in a convenient Order."[124] However, it is plausible that in speaking of the "most Natural Order," he was alluding to the potential role of an artificial scheme in which the nomenclature mirrored an underlying classification, as in John Wilkins' *Essay* (1668). Indeed, a Royal Society committee, initiated in May 1668, explored such applications, albeit with serious disagreements over the fundamental basis of a philosophical language.[125] Hooke objected to a plan to use as few as "100 notions" as the radicals in such a language, whatever the "great help to the memory."[126] But in 1681 he made it clear to Leibniz that the attraction of a universal character went beyond its prospects as a "supplement for Latine": not only might it be "usefull for Expressing & Remembring of things and notions but to Direct Regulate assist and even necesitate & compell the mind to find out and comprehend whatsoever is knowable."[127] The taxonomy on which such a scheme was predicated would supply the main Heads under which memory could deposit material as it was collected. It is also possible that Hooke sought to combine Bacon's method for natural histories and Wilkins' taxonomy: that is, to collect information under "particular Heads of Inquiry" which might themselves be subsumed under higher-level categories.[128] In principle, these could be derived from an agreed classification that also served as an external aid to memory.

Hooke's approach avoided the competing demands of memory and reason; it alleviated some of the work of memory by allowing it to assume and employ a framework of Heads. If memorized, this could function as a kind of mental wallpaper; otherwise, an externalized version could act as a prompt. However, beyond the functions of a basic division of subjects, or the ordering of material in the Society's Repository, the Baconian histories that Hooke envisaged involved detailed collection of information that was too particular, too circumstantial, to be usefully handled by an established set of categories. Moreover, the imperative for flexible re-sorting of data — or, as Hooke said, "transposing, jumbling, ranging, methodizing, and the like"—clashed with any prescriptive taxonomy.

The need to devise collective ways of managing information is highlighted by what Hooke said about the physical basis of memory. About fifteen years after the *Micrographia* and the "General Scheme," he devoted one of his Cutlerian lectures to the subject of "Light."[129] Having dealt with the concepts of medium, vibration, ray, radiation, and pulse, Hooke shifted to a seemingly tangential speculation about how the sense of time depended on memory.[130] He offered a hypothesis about how ideas were stored as physical impressions in "the Repository or Organ of Memory" and worked on by "the Soul," acting like a sun shining its light into the corners of this space.[131] For my purposes, the two salient issues he addressed are the load on memory and the manner in which it stored material. This first of these followed from the assumption that memory was a corporeal faculty and that its retention of ideas entailed material space. Hooke acknowledged that the capacity of memory was necessarily constrained by the volume of the brain that housed it.[132] He did not flinch. After deducting time when asleep, during which he assumed no ideas were received, he calculated the number of ideas stored in the memory of an average person over a year. He concluded that the mind added to its "Store" by "about one Million of Ideas" each year. This figure frightened him and he settled on "one hundred for every Day," so that a person would gather almost two million ideas over fifty years. More positively, Hooke looked to his microscopic observations for reassurance that "we shall not need to fear any Impossibility to find out room in the Brain" even for hundreds of millions of ideas. We only had to consider "in how small a bulk of Body there may be as many distinct living Creatures as here are supposed Ideas."[133]

On the second issue, however, it is clear that whatever Hooke thought about the capacity of memory, he doubted its ability to recall ideas and trains of thought when required. Regarding the order in which these ideas were stored, he speculated that memory received ideas in temporal order, linked to one another and so forming a "Chain of Ideas coyled up in the Repository of the Brain."[134] Each idea carried its content and a marker of its position in the chain of ideas, "disposed in some regular Order; which Order I conceive to be principally that according to which they are formed." This chain was thus laid down in chronological order; it did not mirror the true patterns in nature. The challenge, as Hooke saw it, was to reorder ideas held in memory. To a significant extent, this is what the mind did: "So that Thinking is partly Memory, and partly an Operation of the Soul in forming new Ideas" by combining elements in various ways.[135] However, there remained the problem of putting particular ideas at the disposal of reason or judgment when required. In addition to the tendency to forget, the memory, as Hooke put it, "cannot so well propound all it does remember, to be examin'd at once by the Judgment; but prefers some things first in order, before others, and some things with more Vehemence and greater concern."[136] He concluded that there had to be an external storehouse of material able to be analyzed and rearranged by reason. In this way, scientific analysis would transcend the temporal or associative patterns of natural memory.

This analysis demanded the use of external records, for two reasons. First, this was necessary because the mind could not retain the details needed in a proper natural history. Hooke acknowledged that the scale of data far exceeded individual memory. Even the number of Heads themselves, not to mention their contents, would be difficult to memorize: he predicted that for the history of "Air" alone there would be "a multitude of heads."[137] The *second* reason for external records was that they afforded a degree of flexibility not available to natural memory. Even if information might be condensed and abbreviated, Hooke stressed that it needed to be frequently resorted. This followed from Bacon's advice that the preliminary set of Heads should be tentative: "The Method of distributing the Matter of Philosophical History . . . need not be very nice or curious, they being in them laid up only in Heaps as it were, as in a Granary or Store-House; from thence afterwards to be transcribed, fitted, ordered and rang'd, and Tabled."[138] Hooke was adamant that the Heads would need revision as further "Observations and Experiments" superseded earlier material, but he was reluctant to discard anything prematurely; instead, it would be "good when [material] obliterated in one place, to be inscrib'd in another, where at least it may keep its place till some other thing much more significant to the same purpose, may give occasion to displace it."[139] Whereas Descartes, in the *Regulae*, focused on the inability of memory to handle long chains of deduction, Hooke stressed that memory was unable to reclassify and reshuffle bundles of information under various Heads. Any fixed set of Heads, whatever their mnemonic value, would be an impediment. For this reason, notebooks should not act simply as prompts for what might be committed to memory, such as major taxonomic categories. Instead, their function was to provide a way of manipulating information *manually* so that the reason could proceed without recourse to an overtaxed memory.

Notes and Information

Hooke's proposals for managing empirical information relied on methods of note-taking and retrieval advocated by Bacon and practiced, although with some different assumptions, by Hartlib. Indeed, he stands as their most ardent successor, one who understood the need for organized, yet flexible, repositories of written material. Hooke regarded methodical note-taking as central to the business of the Society, and his concern with this is evident from the 1660s and 1670s in various ideas proffered for what he called the "Designe" of the Royal Society. One of these manuscripts recommends "That a certain number of the Society be appointed to read over Antient & modern authors that treat of naturall & Experimentall knowledge. each person making choice of the book he will read over and epitomise as to all things considerable for the Societys Designe."[140] This emphasis on note-taking was a matter of general agreement, but beyond this Hooke's ideas ran counter to the more passive reception of available information—the quasi-official position sanctioned in Sprat's *History*. In *Micrographia*, Hooke had already warned that "the storing up of all, without any regard to evidence or use, will only tend to darkness and confusion," and he declared that "there should be a *scrupulous* choice, and a *strict examination*, of the reality, constancy, and certainty of the Particulars that we admit."[141] In the "General Scheme," Hooke recognized the danger of premature systems, as Boyle did, but he also identified a contrasting tendency, namely, an appetite for innumerable particulars, gathered from different sources, that might stultify inquiry. He admonished those people who were so "confounded with Particulars, that they have only proceeded, groping on after other Particulars, thinking at last they may by chance light upon something that may afford them Information in what they look after."[142]

From the 1670s, Hooke developed a more critical stance about how notes should be made. He suggested that members ("a tenth part of the whole number") of the Royal Society be assigned to examine books for extant information. In this task, they should be selective and discriminating: "approving & confirming what they find true & reall and damning all that they find fals and fictitious, and then for recording them in their proper places & methods."[143] In about 1680–81, in "Philosophicall Scribbles," he urged that every person exercise similar vigilance, being "very cautious of what he takes in to lay up for the future," to "examine what he has already layd up," jettisoning any "false" information.[144] Later, in a paper read before the Society in February 1692, Hooke acknowledged that the judicious selection of material was not easy because so much had been written or reported on the strength of the authority "of some other Person reputedly eminently skilful in this, or that Part of Knowledge." The prospect, as he saw it, was daunting: "With which Kind of Information, how full are the Authors that have treated of some Subjects? and that not one or two, but Hundreds, nay, Thousands, if we consider natural Philosophy and Physick, with the Arts subservient thereunto."[145] Nevertheless, the accumulated weight of past writings would be lightened by vigorous culling. If much of existing knowledge was erroneous, or irrelevant, it became possible to imagine a new start.

How was information, once selected, to be retained and arranged for later use? In our twenty-first-century terms, Hooke recognized a close relationship between inputs and outputs. The manner in which information was received in memory influenced both retention and the ability to recollect by way of prompts. Although endorsing Bacon's criticism of the ancients, Hooke paused to say that

> I do not here altogether reject *Logick*, or the way of Ratiocination already known; as a thing of no use. It has its peculiar Excellencies and Uses . . .
> And affords some Helps to some kinds of Invention . . . as well as to the Memory, by its Method.[146]

The task was to find a way of achieving these advantages for the new empirical sciences. Hooke believed that information recorded on paper could be disposed in a manner that profiled essential elements, thus facilitating committal to memory. He accepted the standard view that visual images made strong sense impressions, so that in his remarks about Moses Pitt's *The English Atlas* (1680) he praised the good proportions of the maps as a "true representation of the Universe and the severall parts therof in picture." He continued: "it may also more easily imprint that Idea the deeper in the memory, which is the Principall use of such a work. There being nothing more conducive to the assistance of the understanding and memory then a plaine simple, cleer and uncompounded Representation of the Object to the sense."[147] In his own work, Hooke extended this principle to other kinds of data. His weather "Scheme" displayed records so that (as Sprat reported) those of "a whole Moneth, may at one view be presented to the Eye," a feature achieved by the sample he provided.[148] Hooke envisaged the use of such an illustration as a step toward "a theory of the weather soe as to be able to predict what will be the subsequent varietys of this Proteus."[149] Similar attention to visual aids is evident in his tree-diagram showing the key parts of "Hydrography," arranged in Ramist fashion, with heavy, eye-catching ink for the key terms (see fig. 8.1).[150] This might be one instance of his proposal that "the Histories belonging to any one Inquiry may be placed so as to appear all at one View."[151] These examples—maps, charts, registers, and tables—relied on the imprinting of a visual image, but Hooke also considered the reduction of other written information, including discursive argument, into a form more easily held in memory.

This preference for *brevitas* was strained by the *copia* of data that Hooke envisaged, and ambiguities ensued. He counseled against the anxieties provoked by the prospect of the many "Volumes" of natural information, declaring that "the whole Mass of Natural History, may be contain'd in much fewer words than the Writings of divers single Authors." This was achievable if "all kinds of Rhetorical Flourishes" were avoided.[152] However, his own suggestions about a complete "Philosophical History" entailed exhaustive collection: there was, he said, "no Body or Operation in the Universe" below notice; and furthermore, "the most precious are here not more considerable, nor perhaps so much as the most trivial and vile." Indeed, he urged that meticulous notice of details of time and place, even with respect to well-known phenomena, should be regarded "as if they were the greatest Rarity." He insisted that these be recorded because of the possible importance of "some of the meanest and smallest Circumstances" which memory could not retain. Yet, at the same time, he suggested that instead of registering all experiments "such ought to be chosen and pick'd which are as it were the Epitomy of the rest."[153] Hooke attempted to resolve this mixed advice about the collection of "Materials" by explaining that

> Care ought to be taken that they are sound and good, and cleans'd and freed from all those things which are superfluous and insignificant to the great Design; for those do nothing else but help to fill the Repository, . . . yet notwithstanding, Brevity is not so much to be studied, as to omit many little Circumstances.[154]

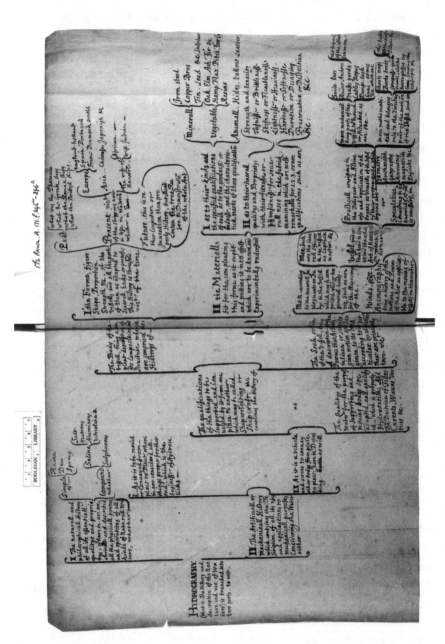

Figure 8.1. Hooke's diagram for Hydrography, c. 1685. MS Rawlinson, A. 171, fols. 245ᵛ–246ʳ. By permission of the Bodleian Libraries, the University of Oxford.

Like Bacon, Hooke wanted the best of both worlds. He therefore sought

> some very good Short-hand or Abbreviation, whereby the whole History
> may be contracted into as little Space as is possible; for this, . . . is of huge
> Use in the Prosecution of Ratiocination and Inquiry, and is a vast Help to
> the Understanding and Memory, as in Geometrical Algebra, the expressing
> of many and very perplex Quantities by a few obvious and plain Sym-
> bols.[155]

This is a reference to the "Philosophical Algebra"—his tantalizing, albeit undelivered, innovation for a general art of reasoning not limited to mathematics.[156]

Philosophical Notes

In his *Philosophy of the Inductive Sciences* (1847) William Whewell remarked
that Hooke's "General Scheme" could be seen as "an attempt to adapt the
Novum Organon to the age which succeeded its publication." He approved
of Hooke's quest for "*a Philosophical Algebra*" for the purpose of "finding unknown causes from known facts, by means of certain regular processes,
in the same manner as Common Algebra." However, he objected that like
Bacon, Hooke was too preoccupied with seeking the "nature" of things instead of "investigating first the measures and the laws of phenomena."[157] This
charge seems unfair when we consider the enormous note-taking and collation of information that Hooke prescribed for the preliminary natural histories of various subjects. It was precisely in this context that he speculated
about the possibility of enhancing Baconian "Natural Inquiry" by seeking a
higher level of abstraction, one that might "not improperly be call'd a Philosophical Algebra, or an Art of directing the Mind in the search after Philosophical Truths."[158] Hooke's first allusion to such an "Algebra" occurs in some
of his early Cutlerian lectures, dating from 1665, and it surfaces in various
lectures and papers that precede and follow the "General Scheme" of 1668.
One constant theme in all of these was his call for cooperative gathering of
data, a task that must involve "the joynt unanimous and Regulated Labour
of a multitude."[159] But the corollary of this was that such information had to
be deliberately arranged as soon as it was assembled. Thus, in a lecture most
likely delivered in 1665 or 1666, Hooke underlined the "care & diligence"
required to compile a "Philosophical History." The information was to be
ranged under Heads "according to the Nature of the several bodys," be they

celestial, mineral, or vegetable. The best way of proceeding was from "the most obvious propertys & qualitys of a body to the most abstruse, & intricate texture constitution & virtues of it."[160] As Whewell later discerned, the goal was to isolate essential properties of matter or, as Hooke said, "to find out the true nature of them"—but only after a large compilation of observations and experiments. The mention of two stages matches Hooke's directive in the "General Scheme" that due attention must be paid to the *first* of the "two main Branches" of this "Method of a Philosophical Algebra," namely, the assembling of "fit Materials to work on."[161]

What was to be done in the second stage? Various possibilities have been canvassed concerning Hooke's unexplicated notion of an algebraic procedure applicable to empirical sciences. There are two leading contenders. The first, suggested by Waller, involved some reduction of data to basic units that could be manipulated by a "Philosophick Algebra."[162] The second interpretation is that it involved the tabulation of alternative generalizations or hypotheses drawn from a collection of information, say from a subset of a natural history, such as hydrography or geology.[163] On the available evidence it is difficult to decide among these possibilities.

We do not know how much Oldenburg understood about Hooke's intentions when he told Boyle that "Mr Hook has also ready (having shewn it to me and others) a Method for writing a Naturall History, which, I think, cutts out work enough for all Naturalists in the World."[164] Oldenburg made the comparison with Boyle's own "General Heads," about to appear in the *Transactions*. The expectation of reducing knowledge to simple units is not usually associated with Boyle, but it is possible that he discussed Hooke's ideas in the period leading up to the composition of the "General Scheme." In a manuscript from about this time, Boyle suggested that "We may also give Symbolical marks to our *Data*, and other Particulars and by adding, subtracting &c. in a way suitable to the nature of this Physical Algebra, we may frame new Propositions, whence will oftentimes result new Truths." This follows his discussion of how to manage "loos Experiments" about "Phaenomena." Boyle thought that if we "look upon these as our Data" we have to decide how to arrange them "into the best Order" so that we can perceive whether the information is "sufficient or no."[165] He may have regarded the use of marks or symbols as a way of labeling various particulars, propositions, or Heads so that they could be sorted and compared. This option may be compatible with the prospect of treating small units algebraically, but the context of how best to deal with multifarious data fits more easily with the tabulation of topics, queries, and hypotheses, which Hooke did in fact employ. In *Micrographia*, he emphasized the promise of arranging material so that "in a moment, as 'twere, thousands of Instances, serving for the *illustration*, *determination*, or *invention*, of almost any inquiry, may be *represented* even to the sight . . . [and asked] How neer the nature of *Axioms* must all these *Propositions* be which are examin'd before so many *Witnesses*?"[166] This is a reminder that the optimism of members of the Hartlib circle about reducing data for ease of memory, recollection, and thinking was not wholly alien to some key figures in the Royal Society.[167] Crucially though, Boyle and Hooke believed that any method of this kind depended on preparatory notes and a special setting out of Baconian histories.

Hooke made suggestions in the "General Scheme" about the collection, arrangement, and storage of information. But even if papers were recorded in the "Register" or published in the *Philosophical Transactions*, he envisaged yet another sifting and collating process aimed at facilitating comparison and inference. In contrast with Boyle's loose notes, Hooke wanted a well designed set of notebooks governed by institutional protocols.[168] This plan required an array of notebooks, movable slips of paper, and other devices.[169] It is fair to say that Hooke's stipulations were not entirely clear, and his verbal descriptions might well have been enhanced by the visual aids he often advocated.[170]

First he wanted "a small piece of very fine Paper" on which to write short notes under "Schedules" containing "the abbreviated and complicated Histories of Observations and Experiments," perhaps along the lines of the Baconian natural histories he listed. Second, since these schedules were to be on single sheets of fine paper, Hooke suggested that they be stuck "with Mouth Glew" in "a large Book" bound in the manner of those used for "keeping Prints, Pictures, Drawings." He called this book "a Repository." This procedure was meant to facilitate the process of sifting and re-sorting—apparently in spite of the use of "Glew." Hooke's idea was that the pieces of paper could be moved around so that information "which was plac'd first may be plac'd middle-most, or last, or transpos'd to another Head, or a little remov'd to suffer another to be interpos'd." It is not clear whether he meant that schedules (that is, lists of topics or inquires), or more particular observations, were to be moved in this way, but his reference to placing something under a different "Head" implies that it was the lower-level data. Third, the "Queries" that an individual "propounds to himself" regarding a certain "History" were to be recorded in "some other small long Book, or else better in a single Sheet of Paper." These queries could be checked against the schedules in the public "Repository," allowing for adjustment of the queries "as further Information shall give occasion." Fourth, "Pictures of things" should be given, partly because a picture is more reliable and efficient than words. Fifth, Hooke anticipated that it might be helpful to add "small Schedules of particular Deductions, or Conjectures" to the larger "Schedules of history." If so, then these should be "exprest in a very few words" so as not to "disturb the Mind," and better still, they should be "written with an Ink of some other Colour, as Red or Green, or the like, for this will much assist the Memory and Ratiocination, as I shall afterwards manifest more at large." The "General Scheme" ends with this reference to a method of using notes. In his closing comment on this work, Waller glossed this method as "the way of ranging Observations and Experiments," saying that, like "the reasoning part of his Philosophick Algebra," he believed it was "never performed."[171] However, it is plausible that whereas Waller distinguished between these two unfulfilled promises, Hooke may have conceived them as parts of the one process.

Despite the absence of further details from Hooke, we can relate this methodical note-taking to his other remarks about how information should be managed. It is clear that this ideal set of institutional notebooks incorporated information and ideas from individuals, in keeping with Hooke's call for contributions to be duly registered. In another account from the same period, he advised that when "making Experiments" it was essential "to register the whole Process of the Proposal, Design, Experiment, Success, or Failure," and that all material and circumstantial details be "fairly written in a bound Book" to be consulted at meetings.[172] Although it is unlikely that this "Book" was the same as the "large Book" or "*Repository*" he described in the "General Scheme," it is plausible that any results and suggestions of "new Difficulties and Queries" were to be absorbed into the series of "Schedules" and notebooks. Yet while Hooke acknowledged the indispensability of cooperation, he implied that the classification and analysis of pooled information was best entrusted to a few individuals, perhaps to a single person. In an undated paper he specified this division of labor: "The next is how these observables are to be recorded & Ranged and Preserved. And thirdly how a Person ought to be qualifyd to be fitt for such an undertaking."[173] In what might have been intended for a Cutlerian lecture, he indicated that he would show "on some other occasion" how "I shall epitomise & range them [the data] afterwards in tables to make them benficiall."[174]

This use of the first-person singular is telling. Hooke wanted to process information as it was gathered; unlike Sprat and others, he was not content to amass a stockpile. We might say that Hooke assumed the crucial architectural role he ascribed to the figure of the "*Understanding*," or Reason, in his account of how it should "order all the inferior services of the lower Faculties," not "as a *Tyrant*" but as a "lawful *Master*." Just as Reason "must *examine, range,* and *dispose* of the bank which is laid up in the Memory," so, too, must one person command the institutional memory of the Society.[175] This emphasis on the need for frequent re-sorting of information — a process that could not be accomplished without external aids to memory — explains the various *transfers* of notes within the set of notebooks. Such reassignment of information was crucial in the application of any higher level analysis, however that was conceived. But it is not clear whether Hooke made provision for retaining the notes submitted by others.

Hippocrates' Complaint

Hooke repeatedly stressed that anyone involved in Baconian inquiries had to accept that nothing of much value could "be expected from the single Endeavours of any one Man." This was because a "Man's Life will be well near half spent . . . before a sufficient Supellex [equipment] can be gathered" by his own efforts.[176] Even a focus on one topic did not remove this problem. In a lecture of 1674 on the motion of the earth, Hooke acknowledged that even though this was "one Subject of millions that may be pitched upon," the job of writing "an exact and compleat History thereof, would require the whole time and attention of a mans life, and some thousands of Inventions and Observations to accomplish it." Thus "no man is able to say he will compleat this or that Inquiry, whatever it be."[177] For Hooke, the solution lay in the collation of information in an institutional archive over time. In *Micrographia*, he offered his "small Labours" as hopefully taking "some place in the large stock of *natural Observations*, which so many hands are busie in providing." But he also indicated that this stock might be augmented by including *past* observations, and he suggested that a collection of medical histories made it feasible that a single physician "has not only a perfect *register* of his own experience, but is grown *old* with the experience of many hundreds of years, and many thousands of men." Hooke said that some of his contemporaries were aware of "this convenience" and had "registered and printed some few *Centuries*"; the problem was that most aimed for "more for *Ostentation* then *publique use*."[178] This criticism was aimed at medical case histories that omitted failures in treatments, but about three years later Hooke envisaged far more general Baconian natural histories centered on topics and queries, not individuals.

What was to be done with such a mass of notes? Here I think Hooke welcomed another of Bacon's pronouncements about dealing with the gulf between short lives and the time required for scientific advances. Bacon declared that "it is the dutie and vertue of all knowledge to abridge the infinitie of individuall experience, as much as the conception of truth will permit, and to remedie the complaint of *vita brevis, ars longa*." This could be accomplished, he affirmed, because the sciences formed an intellectual pyramid: natural history supported natural philosophy and so on up to the most abstract science, metaphysics, which "is charged with least multiplicitie."[179] Bacon contended that this "connexion, or concatenation" of knowledge gave some assurance that "the Notions and conceptions of Sciences" can be reduced to a manageable number. Hence the individual, although confined to one short life, and able to master only a limited portion of knowledge, could appreciate the harmony of its parts. However, Hooke maintained that this progress toward comprehensive understanding would not occur if unprocessed data were simply accumulated. Steps toward generalization had to be taken whenever possible, thus adding stones to the pyramid. In some statements, Hooke acknowledged (like Evelyn, Boyle, and Locke) that the discovery of fundamental properties and processes of nature would need to be entrusted to future generations. In other remarks, however, he anticipated ways of hastening this progress, implying that one person might be able to distill crucial patterns in the data supplied by collaborative effort. We can therefore read his cheeky pun in *Micrographia* either as an acceptance of slow, long-term research, or as disingenuous humility: "I have at length cast in my Mite, into the vast Treasury of *a Philosophical History*."[180]

[9]

Conclusion

In his *Popular Lectures on Scientific Subjects* (1873), the German physiologist Hermann von Helmholtz remarked that his predecessors had established standard ways of collecting, organizing, and retrieving information.

> This organization consists, in the first place, of a mechanical arrangement of materials, such as is to be found in our catalogues, lexicons, registers, indexes, digests, scientific and literary annuals, systems of natural history, and the like. By these appliances, thus much at least is gained, that such knowledge as cannot be carried about in the memory is immediately accessible to anyone who wants it.[1]

Seven years earlier, in 1866, Thomas Huxley pondered one consequence of this expanding archive of scientific knowledge and information: namely, that "if all the books in the world, except the *Philosophical Transactions*, were destroyed, it is safe to say that the foundations of physical science would remain unshaken, and that the vast intellectual progress of the last two centuries would be largely, though incompletely, recorded."[2] This statement can be taken as a confirmation of the foresight of those, especially Henry Oldenburg, who pioneered these *Transactions* (from 1665) during the Scientific Revolution, imagining them as collective notebooks, or as a "bank" of information, growing richer and more effective over time. However, in Oldenburg's time, the creation of an institutional archive depended on individual contributions that were only just beginning to be produced in standard and routine ways suited to collaborative inquiry. This process also required thought about the kind of information that could not "be carried about in the memory" and about the relationships between memory, notes, and records. Indeed, the reflections of the English virtuosi on this issue, closely linked with their own practice, constitute another bequest to Helmholtz's generation.

In the early modern period, the practice of note-taking was closely involved with consideration of the role of memory, and by extension, with the opposition of memory and reason (or understanding). Seventeenth-century thinkers knew this dichotomy as a very old trope, based in part on ancient doctrines of bodily humors and of the soul and its various faculties. As seen in Aubrey's *Brief Lives*, the inevitable tension between memory and reason provided a framework for character portraits. By the mid-eighteenth century, Edward Gibbon thought that this contest had been settled, and he reflected on the demise of memory from its former high status. In his *Essay on the study of literature* (1764), he reported that this was occurring in his own time:

> Our Philosophers have ever since affected to be astonished, that men can pass their whole lives, in acquiring the knowledge of mere words and facts, in burthening the memory without improving the understanding . . . a mind capable of thinking for itself, a lively and brilliant imagination, can never relish a science that depends solely on the memory.[3]

A low estimate of memory, he asserted, was characteristic of the moderns, and this was part of a shift in which "Natural Philosophy and the Mathematics are now in possession of the throne: their sister sciences fall prostrate before them."[4] This linking of memory with historical and literary studies, and reason with the sciences, was also evident in Jean le Rond d'Alembert's "Preliminary Discourse" (1751) to the *Encyclopédie*, which adopted Bacon's classification of knowledge in terms of memory, reason, and imagination. However, when pressed to assign what he called "the art of remembering" to this map of knowledge, d'Alembert placed it with the subjects linked to the faculty of Reason. He explained that this "art" had "two branches, the science of memory itself, and the science of supplements to the memory," one of which was writing, as Bacon had insisted.[5] From this perspective it was possible to avoid a simplistic dichotomy between memory and reason: instead, information could be disposed on paper in ways that relieved memory, prompted recollection, and focused attention and reason. Some early modern figures had already arrived at this position.

Francis Bacon became the mentor on note-taking for the English virtuosi. Of course, these people knew about the Renaissance humanist legacy, especially as epitomized in the method of commonplacing; and, in principle, they had access to the Jesuit additions to this advice. Bacon did not provide detailed guidance on note-taking, but he gave it a rationale suited to the empirical sciences. He contended that this practice could be part of a process of discovery rather than merely a way of collecting copious material from texts. He regarded note-taking as "writing in discontinuous sections," a crucial technique for the collection of "particulars" in large-scale natural history projects. He also admitted that the plethora of data required by these natural histories would strain individual memory, even weakening the power of notes to stimulate recollection. In any case, Bacon recognized that notes only worked in this way for the mind of the person who made them and since his program was emphatically a collaborative one, he directed attention toward retrieval from shared records. His list of "Titles" (or Heads) under which information might be accumulated was influential in the early Royal Society. In his account of "Salomon's House," he also sketched a division of labor that involved various kinds of notes, starting with excerpts from books and concluding with a setting out of data in "titles and tables."[6] In these ways, Bacon's pronouncements supplied a touchstone, a point of reference for those thinking about note-taking in the sciences; however, they did not constitute an easy-to-follow script on the question of how best to combine reliance on memory, notes, and permanent records.

For some of the English virtuosi, the German émigré Samuel Hartlib was a filter and interpreter of the latest ideas, often from Continental Europe, about the collection and management of information. William Petty, John Aubrey, John Beale, John Evelyn, and Robert Boyle had some acquaintance with the vast web of correspondence centered on Hartlib and his own diary, the Ephemerides; others, such as John Locke and Robert Hooke, probably knew only of his reputation as an "intelligencer" in pre-Restoration London. With his correspondents, Hartlib hatched plans to apply conventional commonplacing to larger projects, such as compendia of theological notions, chemical substances and processes, and medical cures. His energetic copying and redaction of information from both print and manuscript delivered some empirical information and advertised a way of dealing with it. He also showed what it might mean to engage in collective note-taking without sacrificing the traditional nexus between memory (via recollection) and notes.

Thus he was keen on ways of reducing information, once assembled, to a scale amenable either to memorization or, more usually, into a format able to trigger recollection. The appeal of Thomas Harrison's "Arca Studiorum" was that it allowed collaborative use, and its indexes promised a way of refining information down to basic, or radical, components—thus supplying keys, or short cuts, to new discoveries. There are signs of this aspiration in Beale's attempts to nudge Boyle into systematizing his work, and his notes, with a view to an imminent completion of the sciences.[7] Among the Hartlib circle there was also another way of hoping that perhaps one lifetime was not too short. This was the idea of deep immersion, from an early age, in a chosen subject. Beale professed to master knowledge in memory through systematic arrangement and intense meditation, even without notes. Petty believed that a lifetime of reading and observing, captured by intensive note-taking, would produce a profound storehouse of experience that might enable one person to crystallize patterns and hence make discoveries. They both regarded Boyle as such a person.

However, the dominant response to Hippocrates' complaint—"*vita brevis, ars longa*"—was a commitment to the accumulation of observations and experiments over successive generations. Most scientific virtuosi acknowledged that the contributions of individuals, made in one short life, were unlikely to produce immediate advances, especially in the subjects pursued by the methods advocated in Baconian natural histories. But this acceptance of the long-term nature of scientific research went hand-in-hand with a desire to maximize the contribution of each generation. For this reason, the well-known capacity of personal notes to support recollection and to record, and hence secure, the stages of the reasoning process, could be harnessed to collective projects. There was recognition that intensive recourse to notes and the sharing of these in small circles—as seen in the work of Martin Lister and John Ray—enabled individuals to assimilate and assess empirical information and to produce intellectual advances. Thus Boyle was concerned that Bacon's eagerness to refine data in tables might overlook the need to retain and review the preliminary notes so crucial for his own recollection and thinking. Boyle, Locke, and Hooke thought that decisions had to be taken about the collection and management of information, and we can discern the choices they made in their own note-taking. Boyle violated some of the standard precepts about note-taking, often lost or mislaid his papers, yet was able to recollect significant details and circumstances of both observations and experiments from the notes he possessed. His relationship with his own "loose" notes showed that these had the power to stimulate recollection of related ideas and episodic memories of the contexts in which they were noted. Boyle was aware that his notes could not work in this way for others, but he maintained that even poorly ordered fragments of information were useful in long-term collaborative inquiries.

Locke was celebrated as a master of information by those who promoted his "New Method" (published in 1686) as a guide to note-taking. In both Chambers' *Cyclopaedia* and the *Encyclopédie*, his model was presented without any mention of the humanist and Jesuit legacy. In breaking with the topical or subject arrangement of material in commonplace books, Locke's method entailed two consequences: it largely rejected the notion of using commonplaces as aids to memory and rhetorical performance, and it opened the way for notes to be entered under labels that made sense only to the individual who made them. Locke claimed that his own notes were a substitute for his poor memory, but it is very likely that some of these, such as his early medical notes, served as triggers for his recollection of additional and related material. It is also plausible that his own maps of knowledge acted as deep intellectual scaffolds for the thousands of detached notes he made on diverse topics. Yet in principle, and as Locke explained in print, his "New Method" was geared to the retrieval of entries that usually contained carefully recorded citations which, in turn, pointed him back to sources, or to cross-referenced entries in other notebooks. Unlike Boyle, Locke does not seem to have used his notes to consolidate episodic memories of material, at least not in a deliberate manner. Retrieval of information was expedited by his innovative way of indexing, provided that he could remember the relevant keyword. His lifetime of note-taking, which included the use of journals as well as commonplace books, therefore did offer a lesson for those engaged in inquiries that demanded collaboration and pooling of information: namely, that reliable aids for searching and retrieving were crucial. Locke himself collected more than he communicated, but he *was* able to retrieve material and pass it on, and he was also aware that some of Boyle's notes and papers had to be recast if they were to be of collective value. His register of the weather and related data was part of an increasing awareness about the need to establish common procedures for gathering information under the aegis of a scientific institution.

Of the individuals I have discussed, it is Hooke who has left the most specific thoughts about the management of both memory and scientific information. We have very few of Hooke's notes, apart from those in his diary, but among his papers preserved in the Royal Society and in some of his publications, he addressed the question of how notes might be best taken and stored for collective, albeit closely monitored, use. Hooke worked within the traditional medical and philosophical model in which the Senses and Memory were charged with the tasks of absorbing and retaining information for use and analysis by Reason (or the Understanding). Perhaps even more acutely than Bacon, Hooke appreciated that numerous and diverse empirical details could not be managed by memory alone. His prescriptions for both individual note-taking and institutional protocols followed from this diagnosis: namely, that information had to be disposed in ways that aided both memory and reason, possibly using some agreed nomenclature and classification as a framework; and that it must be recorded in concise terms, often immediately, and in a form (such as loose slips) that allowed physical sorting and reclassification. Hooke welcomed Boyle's efforts to advertise "General Heads" as guidelines for collective information gathering, but he was worried about the prospect of a continuing stream of unfiltered material. Even so, he struggled to resolve his own dilemma of wanting both copia of circumstantial details and brevity of description.

What I call the empirical sensibility of the English virtuosi comprised a realization that note-taking was crucial for the collection of the kind of information required by Baconian natural histories. This conviction involved a preference for particulars over premature systems, and an acknowledgment that diverse sources of information must be collated and compared. Despite indulging in some opportunistic antibookish rhetoric, the figures I discuss agreed that past writings should be scrutinized for relevant material—ideas, observations, testimonies, and experiments that could serve as starting points and constitute a baseline for comparisons. In accepting this task, they were able to draw on the note-taking tradition deriving from Renaissance humanism. Moreover, the available techniques were by no means limited to the kind of commonplacing taught by Erasmus, Vives, and other pedagogues. Scholars and polymaths, such as Scaliger, Crusius, Gessner, Casaubon, and Selden had developed ways of dealing with the ever-increasing number of books, in part by extracting and condensing nuggets of information that could be appropriated, organized, and indexed in large compendia. However, the English virtuosi diagnosed a somewhat different problem: namely, that there were too many books of the wrong kind, and insufficient information of the right kind. They therefore acknowledged an imperative to gather new information about bodies, fluids, and phenomena not yet adequately observed, recorded, or discussed. They did so by combining features of the commonplace book and the journal, at the same time expanding the types of entries made in notebooks to include lists, queries, weather records, reports of conversations, observations, experiments, and personal thoughts — as well as the usual excerpts from books. They accepted Bacon's call for the accumulation of "particulars": that is, elementary data on topics such as those named in his list of 130 natural histories. This was a different task from the reduction and abbreviation of knowledge already collated to some extent under topics, maxims, sententiae, or doctrines.

The collection and analysis of empirical information demanded continuing collaboration and long-term storage. As secretary of the Royal Society, Oldenburg assured correspondents that the plan was to ensure that this process was ongoing, to "render it everlasting."[8] However, one acute difficulty was that initial inquiries sought "particulars" not yet strongly anchored in agreed categories. For some topics, there was extant information in oral or written form, but in others not even basic propositional content was available. Thus in order to make a start, Ray was willing to seek factual content in collections of proverbs as a basis for further investigation. His efforts suggest that the dismissal of erroneous accounts was easier than the construction of useful generalizations from miscellaneous pieces of information. Hooke analyzed the difficulties in this process, realized that there were various approaches but, in any case, stressed that notes had to serve as clear, stable, and retrievable records for others. This did not mean that personal notes were no longer important, but rather that command of one's own memory, aided by notes, could not be the primary goal. As Evelyn's disclosure made clear, a lifetime of collecting and thinking did not guarantee mastery of even a single subject, in part because new contributions from others were constantly appearing. This led him to contemplate "what inseperable paines it will require to insert the (dayly increasing) particulars into what I have already in some measure prepar'd."[9] A similar perception led Boyle to admit that natural philosophy (unlike even theology) was a subject that could never be properly comprehended by one person.

Indeed, an acceptance of the Baconian program entailed a lifetime commitment to a project that could not reasonably be completed in that lifetime. In response, the people I have discussed in this book confronted the task of establishing adequate information as a basis for preliminary theories; however, their perceptions of likely outcomes differed. Some, such as those in Hartlib's circle, believed that the proper refinement of data would quickly yield stable patterns; Hooke acknowledged the need for collective effort over generations, but expected that the combination of carefully gathered information and mathematical analysis would deliver powerful explanations; others, including Evelyn, Boyle, and Locke doubted that information, however treated, could generate durable theories.[10] And by the twentieth century, their doubt was confirmed, possibly in ways that the early modern virtuosi did not anticipate: the vocation of science came to be seen as one in which the small contributions of an individual were inevitably, sometimes rapidly, superseded. Max Weber construed this as the existential predicament of the modern scientist who knows that what he has "worked through will be out of date in ten, twenty, or fifty years . . . [that each scientific advance] . . . asks to be surpassed and made obsolete."[11]

Both Huxley and Helmholtz were correct in emphasizing the function of institutional archives as ways of retaining information that could not be encompassed by individual memory. However, it is important to recognize that members of the early Royal Society, who set this process in motion in the 1660s, believed that storage and retrieval of information could assist, rather than supersede, the memory and recollection of individuals. In their view, whatever the power of an archive, regular note-taking was part of the intellectual process that enabled occasional breakthroughs in understanding, as well as delivering small additions to collective knowledge over long periods. The practice of making and keeping notes encouraged individuals to record what might be unique experiences and observations, thus safeguarding against the vagaries of memory; it documented the various sources of this information, such as the senses, instruments, testimony, or books; it laid down a rich stock of experiences in individual memory that enhanced judgment about new phenomena; and it created the possibility of collating and communicating a body of information large enough to allow proper comparative judgment. The note-taking of the English virtuosi, and their reflections on concomitant questions about memory and information, shaped an early and crucial part of the modern scientific ethos.

[Acknowledgments]

I have received significant support for this project, both from institutions and individuals. Griffith University provided extensive research time and study leave, and hosted a Professorial Fellowship awarded by the Australian Research Council (ARC). This Fellowship allowed the completion of some publications on the Enlightenment, and gave me the chance to become an early modernist. A further ARC grant supported the final research and writing of this book. I have gained much from my time as a Visiting Fellow in both the USA and UK: at the Dibner Institute for the History of Science and Technology at MIT in the first half of 2003, and at All Souls College, Oxford for two terms in 2004–5. In April–May 2008 I also worked with Karine Chemla and her colleagues at the National Centre for Scientific Research, University of Paris–Diderot. As well as giving me the chance to present my work and attend seminars, these fellowships put me close to libraries and collections essential to my research: the Houghton Library, Harvard; the Burndy Library within the Dibner Institute; the Beinecke Library, Yale University; the Bodleian Library, Oxford; the Library of the Royal Society of London; and the British Library, London. I am extremely grateful to all these places and the people who represent them.

Closer to home, in July 2008 the ARC Network for Early European Research funded a symposium on "Notebooks and Note-Takers in Early-Modern Europe." This was hosted by the Griffith Centre for Cultural Research at the State Library of Queensland, Brisbane. I learnt much from the speakers and participants, and I thank all those involved. I also acknowledge the staff at the two libraries I rely on: Griffith University Library, and especially its interlibrary loan staff (in particular, Elspeth Wilson) who rarely failed to find their quarry; and the University of Queensland Library, which I have been able to use through an honorary affiliation with the Centre for the History of European Discourses. I thank Peter Cryle and Peter Harrison, two successive directors of the Centre, for this privilege. I acknowledge the collegiality of the history of science (liberally construed) community within Australia and, in particular, Randall Albury, Luciano Boschiero, Alan Chalmers, John Forge, Ofer Gal, John Gascoigne, Stephen Gaukroger, Peter Harrison, Rod Home, David Miller, David Oldroyd, John Schuster, Vanessa Smith, John Sutton, and Charles Wolfe. I also thank Louise Goebel and Robin Trotter for their conscientious research assistance.

My debts to the work of many scholars are registered in the notes. Here I wish to thank, especially, the following people for their assistance, conversation, and encouragement during this project, and for the example of their own scholarship: Peter Anstey, Ann Blair, Alberto Cevolini, Michael Hunter, Jamie Kassler, Rhodri Lewis, Ian Maclean, Noel Malcolm, Paul Nelles, Victor Nuovo, Richard Serjeantson, Jacob Soll, Lyn Tribble, and Angus Vine. In addition, I thank John R. Milton for reading a near-final draft of chapter 7, and for giving me the benefit of his unrivaled knowledge of John Locke's papers. Finally, I am very pleased to acknowledge the expert advice of Ann Blair and Anthony Grafton, who kindly disclosed their role as readers for the Press. Their helpful and insightful comments (and corrections) were invaluable in the final stages of preparation.

For help in locating and reproducing images, I thank the following staff and the libraries at which they work: Kathryn James at the Beinecke Rare Book and Manuscript Library, Yale University; Patricia Buckingham at the Bodleian Library; Jacky Hodges at Special Collections, Sheffield University; Felicity Henderson and Joanna Hopkins at the Library of The Royal Society; Julie Price at Rare Books and Special Collections, the University of Sydney Library; and the staff of the Manuscript Room of the British Library. From the tentative inception of this book several years ago, Karen Merikangas Darling, my commissioning editor at Chicago, has been an unfailing source of assurance and guidance, for which I am most grateful. I also thank Therese Boyd for her thoughtful copyediting and Alan Walker for his finely grained index.

Chapter 5 draws upon material previously published in "Memory and Empirical Information: Samuel Hartlib, John Beale and Robert Boyle," in

The Body as Object, ed. Charles T. Wolfe and Ofer Gal (Dordrecht: Springer, 2010), 185–210. This has been used with kind permission from Springer Science+Business Media B.V. Chapter 6 draws upon "Loose Notes and Capacious Memory: Robert Boyle's Note-taking and Its Rationale," previously published in *Intellectual History Review* 20 (2010): 335–54.

I owe the largest thanks to my family: to my daughters, Gillian and Claire; and to my wife, Mary Louise, for reading, thinking, and editing; for sharing doubts and discoveries; and for much more.

[Notes]

Preface

1. Grafton, *The Footnote*.

2. Aczel, *Descartes' Secret Notebook*, 10–15; Brian Selznic, *The Invention of Hugo Cabret* (New York: Scholastic Press, 2007).

3. Galileo to Kepler, August 19, 1610, cited and trans. from Johannes Kepler, *Gesammelte Werke* (Munich: C. H. Beck, 1937–), 16:329, in Grafton, *Defenders of the Text*, 2.

4. Bacon, *Parasceve*, 457.

5. Hereafter I do not italicize this word unless required by a quotation; I explain its English connotations in chap. 1.

6. See, for example, Grafton, *Defenders of the Text*; Maclean, *Logic, Signs and Nature*; Siraisi, *History, Medicine, and Traditions*.

7. Locke to Thomas Herbert (Earl of Pembroke), November 28, 1684, *Locke Correspondence*, vol. 2, no. 797. Admittedly, the context here was Locke's wish to say that he had not been involved in political cabals, preferring the company of "bookish not busy men."

8. Locke, *Essay*, II.xiii.27.

9. Locke, *Conduct of the Understanding*, in *Works*, 3: 236.

10. For the classic statement, see C. P Snow, *The Two Cultures* (1959).

11. Eisenstein, *Printing Press*. See the critique of this position in Johns, *Nature of the Book*.

12. Love, *Culture and Commerce of Texts*; Yale, "Marginalia, Commonplaces, and Correspondence."

13. For a pertinent story about how the "paperless office" has not eventuated because of the ways in which print and digital media support each other, see Brown and Duguid, *The Social Life of Information*, 176–83.

14. Galton, *Art of Travel*, 26.

265

15. Grafton, "Martin Crusius Reads His Homer."

16. Blair, *Too Much to Know*; including chap. 2 on note-taking.

17. For Luhmann's preoccupation and its antecedents, see Cevolini, *De arte excerpendi*, 417–28; Krajewski, *Paper Machines*, chap. 4. For a related story, see Yeo, "Before Memex."

18. For a twentieth-century plea for the relevance of older techniques to new social sciences, see Webb, "Art of Note-Taking."

19. I explain the distinction between natural history and natural philosophy in chap. 1. Apart from Hooke, the people I discuss were not closely involved in the physico-mathematical disciplines of astronomy, mechanics, and optics although they kept up with major discoveries in these fields.

20. For curiosity as a passion, see Krzysztof, *Collectors and Curiosities*; Swann, *Curiosities and Texts*; Benedict, *Curiosity*; Kenny, *Uses of Curiosity*.

21. Bacon, *Novum Organum*, Book II, aphorism 29, p. 299.

22. Boyle, *Christian Virtuoso, First Part* (1690), 304. Notwithstanding Boyle's prescription, this could often resemble what Lorraine Daston has called "militant empiricism," or the collection of matters of fact "without *a priori* prejudices." Daston, "Baconian Facts," 349.

23. On matters of fact as brief notices, see Daston, "Perché i fatti sono brevi?" 752–53, 757–59.

24. In the seventeenth century, *datum/data* still referred to the "given," or starting, point of an argument or a proof. However, the term "particulars" roughly approximated to our contemporary sense of "data" as material that may constitute "information" after additional processing. See Floridi, *Information*, 23. I discuss the early modern usage of "information" in chap. 3.

25. See Cook, *Matters of Exchange*; Schiebinger and Swan, *Colonial Botany*; Smith and Findlen, *Merchants and Marvels*.

26. For Newton's ruthless determination in this respect, see Schaffer, "Newton on the Beach."

27. "The British Association for the Advancement of Everything in General and Nothing in Particular," *Punch* 3 (1842): 6–7. This drew on Charles Dickens, "Full Report of the first Meeting of the Mudfog Association for the Advancement of Everything," *Bentley's Miscellany* 2 (1837): 397–413.

Chapter One

1. Pepys, *Diary*, September 2 and 8, 1666, 7:271, 282.

2. See Sherman, *Telling Time*, 35, on daily entries.

3. MS Locke d. 9. Locke began the Register on a double opening near the back of this notebook and then moved toward the front. He did not number these pages. All my references are to the date of the entry. Locke kept this Register most regularly in 1666–67; he sporadically kept one in London between 1669 and 1675. He lived in France between November 1675 and May 1679. He resumed the register for Oxford on March 14, 1681, albeit with many gaps; the last entry is June 30, 1683, not long before he went into exile in Holland. See chap. 7 for further discussion.

4. Ibid. "Weather" is in the second-to-last column on the right before the one for the readings of humidity from a hygroscope. By mid-September (not shown in fig. 1.1) he

had moved this last column to the left of "Wind," thereby grouping the three measurements from instruments, and leaving the two qualitative entries on the right.

5. See chap. 7 for Locke's ways of indicating time of day.

6. MS Locke d. 9.

7. Ibid., pp. 40, 228–29, under "Aer."

8. Boyle, *History of the Air*, 104–32. Locke altered his note on the Fire to read: "This unusual Colour of the Air, which without a Cloud appearing, made the Sun-beams of a strange red dim Light, was very remarkable. We had then heard nothing of the Fire of *London*: But it appeared afterward to be the Smoak of *London* then burning, which driven this way by an Easterly Wind, caused this odd Phenomenon" (at 106); also in *Boyle Works*, ed. Hunter and Davis, 12:72.

9. Shapin, *Social History of Truth*; Shapiro, *Culture of Fact*; Ogilvie, *Science of Describing*.

10. For suggestions and examples, see Daston, "Taking Note(s)"; and Blair and Yeo, "Note-taking in Early Modern Europe."

11. For example, Clement Draper (1542–1620), a London merchant, chemical and medical devotee, took piles of notes from books and about his own experiments while in debtor's prison in Southwark. See Harkness, *The Jewel House*, 195–205.

12. Ibid., xv–xviii.

13. Hartlib, *A Further Discoverie of The Office of Publick Addresse for Accommodations* in HP47/10/40B. The other terms he defined in this passage were "art," "prudence," and "wisdom." Bacon continued to use the term "science" in the traditional way but he hoped, in the case of the study of nature, to overcome the fact that "commonly Sciences receive small or no augmentation" once reduced to principles (*Advancement of Learning*, 29).

14. Locke, *Essay*, IV.xii.10. See also IV.iii.26 where it is clear that this declaration covered "*experimental* Philosophy." More generally, see Anstey, *Locke and Natural Philosophy*, 19–30. Locke's opponent, the Aristotelian John Sergeant, was able to agree with him, and then twist the knife by saying that the experimental method was "utterly Incompetent or Unable to beget Science." See [Sergeant], *Method to Science*, sig.d4,2^{r-v}.

15. Boyle, *Usefulnesse* (1663), 301–2, 306, 313; BP 38, fols. 80^{r-v}, reprinted in Boyle, *Excellencies of Robert Boyle*, 25–26; Boyle, *Memoirs . . . Humane Blood*, 5.

16. Dryden, *Of Dramatic Poesy*, 37.

17. Peacham, *The Compleat Gentleman*, 104–5; Blount, *Glossographia*; Ray, *Observations Topological, Moral, & Physiological*, 395.

18. This Society was formed toward the end of 1660 as a club from earlier groups in Oxford and London. The charter nominating "The Royal Society" was granted on July 15, 1662, and revised on April 22, 1663. John Evelyn applied the adjective "Royal," before it was awarded, in his dedication to Lord Clarendon in Naudé, *Instructions Concerning Erecting of a Library*, sig.A3^{r}.

19. Pepys, *Diary*, April 28, 1662, 3:72.

20. Aubrey, *Brief Lives*, 2:141. Aubrey's manuscript was written c. 1680–81.

21. Hartlib to John Worthington, December 17, 1660, in Worthington, *Diary*, 1:246, 249; Hartlib to Worthington, January 1, 1660/61, 257. See Purver, *Royal Society*, chap. 4, on the differences between the plans of Comenius, to which Hartlib alluded, and those of the Royal Society.

22. Boyle to Locke, June 2, 1666, *Locke Correspondence*, 1:279; *Correspondence of Robert Boyle* (hereafter *Boyle Correspondence*), 3:165. See chap. 7 for details.

23. Boyle, "Experiments and Notes about the Mechanical Origine or Production of Electricity" (1675), in *Boyle Works*, ed. Hunter and Davis, 8:519; and BP 27, p. 45. See also John Wallis to Boyle, August 17, 1669 (with reference to Lady Mary Hastings), *Boyle Correspondence*, 4:145. See Evelyn, *Memoires*, 56, for "Virtuoso Ladys."

24. Boyle, *Robert Boyle by Himself*, xlii, cited in Hunter, *Between God and Science*, 236.

25. Shapiro and Frank, *English Scientific Virtuosi*; [Hill], *Familiar Letters*, 83, for February 5, 1685/6; Hunter, *Aubrey and the Realm of Learning*; Poole, *John Aubrey and the Advancement of Learning*. Hoskyns was elected FRS in December 1661.

26. Middleton, *Lorenzo Magalotti*, 134–40. Magalotti was a member of the Accademia del Cimento founded by Leopoldo de' Medici in 1657, but dissolved in 1667. On Evelyn, see Hanson, *The English Virtuoso*, 4–5, 59–69.

27. Houghton, "The English Virtuoso," pt. 1, 70–72; pt. 2, 201.

28. Sprat, *History*, 417. Sprat began this work in 1663 and its completion in 1664 was supervised by John Wilkins as one of the first secretaries of the Society. On its role as an official apologia, see Hunter, *Establishing the New Science*, 50–55.

29. Shadwell, *The Virtuoso*, II.i.304, V.iii.78–79. See Lloyd, "Shadwell and the Virtuosi"; Anstey, "Literary Responses to Robert Boyle's Natural Philosophy," 159.

30. Hooke, *Diary*, 235, June 2, 1676; also 239. This reaction belies Houghton's opinion that the "real scientists, men like Boyle and Hooke, Ray and Newton, were then as always beyond the reach of ridicule" (Houghton, "The English Virtuoso," pt. 2, 211). Locke recorded buying *The Virtuoso*, along with some other plays, for his friend James Tyrrell. See MS Locke f. 5, Journal, May 1681, p. 54.

31. Petty, "Of Ridicule," in *Petty Papers*, 2:198. On Evelyn's reaction in a letter to his friend, Margaret Blagge, see Hunter, *Science and Society*, 177–78; also Evelyn, "To the Reader," in *Sylva*, 3rd edn, sig.a^(r–v).

32. Wotton, *Reflections upon Ancient and Modern Learning*, 419.

33. Chambers, *Cyclopaedia*, vol. 2.

34. Yeo, *Encyclopaedic Visions*, 38–40, 163–69.

35. Newton, "An Hypothesis explaining the Properties of Light discoursed of in my several Papers," read December 9, 1675, at the Royal Society; in Birch, *History*, 3:249. On the outbreak of this tension after Newton's death, see Cohen, *The Newtonian Revolution*, 61–78.

36. For comments on this, see Stimson, *Scientists and Amateurs*, 36; Hunter, *Science and Society*, 64, 68; and Hunter, *Establishing the New Science*, 28.

37. For the "macaroni" in relation to the virtuosi, see Gascoigne, *Joseph Banks*, 59–62. Banks was president until 1820.

38. Leibniz, *New Essays on Human Understanding*, 527. This work was completed in 1704 but not published until 1765 — as *Nouveaux Essais* in a collection of his works.

39. Bacon, *De augmentis*, 293–99. For overviews, see Yeo, "Classifying the Sciences," 241–66; Harrison, "Natural Philosophy," 117–48.

40. Bacon, *Parasceve*, 475–85.

41. See Pomata and Siraisi, *Historia*; Sorell, Rogers, and Kraye, *Scientia in Early Modern Philosophy*.

42. Bacon, *Descriptio globi intellectualis*, 105; also *A Description of the Intellectual*

Globe, 507–8; and Robert L. Ellis' commentary, in *Works of Francis Bacon*, ed. Spedding, Ellis, and Heath, 3:715–26. See Pomata, "*Praxis Historialis*," 111, for the shift in Renaissance Aristotelianism "from *historia* as knowledge without causes to *historia* as knowledge preparatory to the investigation of causes."

43. Bacon, *Instauratio Magna, Part II*, 41; also *Advancement of Learning*, 104–5.

44. Bacon, "Natural and Experimental History," in *Instauratio Magna, Part III*, 12–13 and 5; also *Works of Francis Bacon*, ed. Spedding, Ellis, and Heath, 5:134 and 127.

45. For the "methods of bookishness," see Blair, *Theater of Nature*, 49–81.

46. Bacon, *Advancement of Learning*, 58; Descartes, *Discourse on Method*, 3; also Ian Maclean's introduction, xxiv, lv; John Evelyn to William Wotton, March 29, 1696, BL, Add. MS 4229, fol. 59^r, in Boyle, *Boyle by Himself*, 88.

47. Aubrey, *Brief Lives*, 1:43 and 2:144. Aubrey appears to check this point in a summary of his "life" of Petty. See Aubrey to William Petty, July 17, 1675, in Fitzmaurice, *Life of Sir William Petty*, 168.

48. Aubrey, *Brief Lives*, 1:349.

49. Aubrey, *Aubrey on Education*, 86. This is a transcription of the manuscript in Bodl. MS Aubrey 10. See also Lewis, "Robert Hooke at 371."

50. Petty to Southwell, April 3, 1677, *Petty-Southwell Correspondence*, 23; and BL, MS Sloane 2903, fol. 8^r.

51. *Petty Papers*, 2:90.

52. Sprat, *History*, 97, 336–38. On the tension between solitude and sociability for bookish scholars, see Algazi, "Food for Thought."

53. This position was spelled out in an anonymous memorandum (c. 1680): "And till there be a sufficient collection made of Experiments, Histories, & Observations, there are no debates to be held at the weekly meetings of the Society concerning any Hypotheses or principles of philosophy, nor any discourses made for explicating any phenomena" (BL, Add. MS 4441, fol. 1^v). For a similar draft of a printed news-sheet, see BL, Sloane 1039, fols. 112–13, as discussed in Hunter, *Establishing the New Science*, 56 and 242, n. 211.

54. Sprat, *History*, 119, 120.

55. See Beale to Oldenburg, June 1, 1667, in *Correspondence of Henry Oldenburg* (hereafter *Oldenburg Correspondence*), 3:427; the editors (at p. 429) identify the offending comments in South, *Sermons preached upon Several Occasions*, 1:373–75. This may have been delivered before Sprat's book appeared in October 1667.

56. Glanvill to Oldenburg, July 19, 1669, in *Oldenburg Correspondence*, 6:137. See also Wallis to Oldenburg, July 16, 1669, ibid., 129–30. Both refer to South's address in Oxford.

57. For example, Stubbe, *A Censure upon Certain Passages Contained in the History of the Royal Society*. Oldenburg joked about the hyperbole of these attacks; Oldenburg to Evelyn, July 8, 1670, in *Oldenburg Correspondence*, vol. 7, no. 1482, 58. See Hunter, *Science and Society*, 140, for Stubbe's claim that he was attacking only the lesser virtuosi, "these *Prattle-boxes*."

58. Feisenberger, "The Libraries of Newton, Hooke and Boyle"; Harrison, *Library of Isaac Newton*; Rostenberg, *Library of Robert Hooke*; Harrison and Laslett, *Library of John Locke*; Malcolm, "The Library of Henry Oldenburg"; Avramov, Hunter, and Yoshimoto, *Boyle's Books*.

59. Pepys, *Diary*, 7:283, 297. Nathaniel Hardy gave this sermon. The vulnerability of

such treasures was reinforced by a later work: Clavell, *A Catalogue of Books printed in England since the dreadful Fire of London in 1666.*

60. Descartes, *Discourse on Method*, 13, 15–16.

61. Naudé, *Instructions Concerning Erecting of a Library*, sig.A4ᵛ. This is Evelyn's translation of *Advis pour dresser une bibliothèque* of 1627.

62. Edward Bernard to John Collins (1626–83), April 3, 1671, in Rigaud, *Correspondence of Scientific Men*, 1:158–59. Collins earned his living as a clerk and accountant in various government offices; he was mathematically adept and had extensive correspondence with other mathematicians. Bernard refers to him as "the very Mersennus and intelligence of this age" (p. 159).

63. See Lewis, ed., *Petty on the Order of Nature*, 26, 82, for his theological reading.

64. RS MM 4.72, fol. 1ᵛ. This is signed "A. B" and dated "Oct 19. 1674." See transcription in Hunter, *Establishing the New Science*, 230. Hooke made the same point; see chap. 8.

65. Sprat, *History*, 95–97. On these combinations, see Johns, "Reading and Experiment in the Early Royal Society."

66. Bacon, *Advancement of Learning*, 53.

67. EEBO gives 1566 as the earliest use of "note boke."

68. *OED* gives the earliest usage as 1617; but the first *title* (in EEBO) containing this term was published in 1661.

69. Takarangi and Loftus, "Dear Diary, Is Plastic Better than Paper?"

70. See, for example, Nelles, "Seeing and Writing"; and Sankey, "Writing the Voyage of Scientific Exploration."

71. *OED* dates this from 1561.

72. Ong, "Typographic Rhapsody," 429–32; Lechner, *Renaissance Concepts of the Commonplaces*, including the preface by her teacher, Walter J. Ong; Moss, *Printed Commonplace-Books*, 2–13.

73. Comenius, *Orbis sensualium pictus*, 201.

74. There were late medieval predecessors, such as the *florilegia*—manuscript compilations of choice quotations from famous authors, indexed for retrieval by author and/or topic. See Moss, *Printed Commonplace-Books*, 43; Blair, *Too Much to Know*, 124–26.

75. Blair, *Too Much to Know*, 35

76. Lechner, *Renaissance Concepts of the Commonplaces*, 103.

77. On Agricola, see Ong, *Ramus*, 92–130.

78. Erasmus, *De Ratione Studii*, 672.

79. Erasmus, *De Copia*, especially 605–9, 627, 635–48.

80. Moss, *Printed Commonplace-Books*, viii.

81. Vickers, introduction to *Bacon: Major Works*, ed. Vickers, xlii. See also Décultot, "Introduction. L'art de l'extrait," 7–28.

82. Cave, *The Cornucopian Text*. For examples and excellent commentary, see Havens, *Commonplace Books*.

83. Marbeck, *Booke of Notes and Common places*. Marbeck wanted to use this technique to assemble material for anti-Papist polemics (sig.Aiiiᵛ). Ong, "Typographic Rhapsody," 432.

84. See BL, Add. MS 36354. For a printed version, see *A Common-Place Book of John Milton*; also Mohl, *Milton and His Commonplace Book*, 11–26.

85. Hanford, "The Chronology of Milton's Private Studies," 289. See Poole, "The Genres of Milton's Commonplace Book," 374–77, for the point that this was not a collection of excerpts for memorization. Many of the notes were entered by amanuenses.

86. Newton, "Questiones quaedam Philosophicae," CUL, Add. MS 3996, fols. 87^r–135^r. See McGuire and Tamny, *Certain Philosophical Questions*, 6–7, for the list of Heads. See also Hall, "Newton's Note-Book."

87. See Yeo, "Ephraim Chambers's *Cyclopaedia*," 175, for a printed "Common Place Book" of 1855 professing to contain a million facts.

88. Lewis and Short, *Latin Dictionary*, 48: "adversarious," hence "adversaria in mercantile language, a book at hand." By the seventeenth century, in English usage, this term indicated that the notes stood in some relation to other texts, with no necessary implication of adversarial contest.

89. Chambers, *Cyclopaedia*, vol. 1. See also "Common-Places" (vol. 1) where "Adversaria" is given as the equivalent.

90. Johnson, *Dictionary of the English Language*, vol. 1.

91. They also largely discontinued the use of the term "lemmata" to refer to short bibliographical notes—in contrast with "adversaria" as topical notes. See Placcius, *De arte excerpendi*, 55–56, 61; Moss, *Printed Commonplace-Books*, 233–34. However, Locke (see chap. 7) did retain "lemmata" as a label for some of his notebooks.

92. *Aubrey on Education*, 35 and n. 5, 161.

93. Hobart and Schiffman, *Information Ages*, 96–106. For the continuing appeal of commonplace collections in the 1700s, see Allan, *Commonplace Books and Reading*.

94. Fuller, *Holy State*, 176.

95. Moss, *Printed Commonplace-Books*, 276.

96. Pomata, "Observatio ovvero Historia."

97. Bacon, *Advancement of Learning*, 70. Typically, he also cited the ancients, remarking that "the Journall of *Alexanders* house expressed every small particularitie." See also Bacon, *De augmentis*, 310, for the observation (from Plutarch) that matter "of greater or less concern, was promiscuously entered in the Journals as it passed."

98. Bacon, "Of Travaile," in *Essayes*, 56. The first edition of the *Essayes* appeared in 1597, the extended and amended one in 1625.

99. Stagl, *A History of Curiosity*. On Jesuit diaries, see Nelles, "Seeing and Writing."

100. Pepys, *Diary*, 5:25, 6:295; Hooke, *Diary*, 339.

101. The section on bookkeeping, *De computis et scripturis*, was soon extracted and published in various translations. I use the version in Geijsbeek, *Ancient Double-Entry Bookkeeping*.

102. For the admonition to keep "a Journall or Diary" of "all Gods gracious dealings with us," and mention of merchants and their account books, see Beadle, *Journal or Diary of a Thankful Christian*, 10, 48.

103. Aho, "Rhetoric and the Invention of Double Entry Bookkeeping"; Poovey, *A History of the Modern Fact*, 33–38; Aho, *Confession and Bookkeeping*; Quattrone, "Accounting for God."

104. Geijsbeek, *Ancient Double-Entry Bookkeeping*, 39. There could be more than one

of each type; the first of each type was marked with "the holy cross," then others were labeled alphabetically. See also p. 89, based on Pacioli.

105. Ibid., 41. In 1586, in his elaboration of this *accounting* method, Don Angelo Pietra stipulated that only one bookkeeper could sign the journal and ledger. Geijsbeek, *Ancient Double-Entry Bookkeeping*, 89. For the relation between humanist and book-keeping methods, see Soll, "From Note-taking to Data Banks."

106. BL, Add. MS 27278. For a rich account, see Vine, "Commercial Commonplacing," 197–218.

107. Spedding, "Private Memoranda," in *Works of Francis Bacon*, ed. Spedding, Ellis, and Heath, 11:22; and Bacon, "Comentarius solutus," 62.

108. Bacon, "Comentarius solutus," 60, 61, 62. The "Comentarius" was no. 24 of his notebooks.

109. One example is Adams, *Writing Tablets with a Kalender for xxiii yeres*, in Hough-ton Library, Harvard. See also Chartier, Mowery, Stallybrass, and Wolfe, "Hamlet's Tables and the Technologies of Writing."

110. Woudhuysen, "Writing-tables and Table-books," 3–4.

111. Hooke, BL, MS Sloane 1039, fol. 117ʳ.

112. MS Locke f. 2, p. 291 (October 7, 1677).

113. MS Locke f. 2, p. 292 (October 8, 1677). Locke cites "Mr. Brisban" (John Brisbane) of the British embassy in Paris as his informant. See Lough, *Locke's Travels in France* 176–77, n. 4.

114. RS MS 41, fol. 52ᵛ.

115. Pepys, *Diary*, May 10, 1667, 8:208.

116. See Chartier, *Inscription and Erasure*, for the association of erasure with either the fear of losing fragile texts or the anxiety about the proliferation of texts in print. This does not capture the focus on transfer of information in early modern scientific circles.

117. Richter, *Notebooks of Leonardo da Vinci*, 220.

118. The term soon became *de rigeur* for portable books, such as almanacs. In 1699 John Playford added "pocket" to the subtitle of his successful *Vade Mecum*, first pub-lished in 1679: *Vade Mecum: or, the Necessary Pocket Companion*.

119. Petty, *Advice*, sig.A2,2ʳ; Aubrey, *Brief Lives*, 2:146; Southwell to Petty, September 15, 1677, *Petty-Southwell Correspondence*, 34.

120. On March 7, and April 4, 1667, the Royal Society discussed, respectively, "several lists of particulars," and that "*Elenchus* of experiments" mentioned by the "Ve-netian philosopher, Franceso Travagino." Birch, *History*, 2:153, 164.

121. Keller, "Accounting for Invention."

122. Referring to his numbered lists of "observations" on medical and political topics, Petty said there was value "in every one of these 100 Observations." See Petty, *Petty Papers*, 2:226.

123. Jonson, *Timber, or, Discoveries*, 521; Jonson's manuscript was published in the 1640–41 edition of his works. Bacon, *Sylva Sylvarum*; the title page has 1626 but an added engraved title page is dated 1627.

124. Bacon's chaplain and literary executor, William Rawley, stated that this "Fable" had "so neare Affinity (in one Part of it) with the Preceding Naturall History." See Bacon, *New Atlantis. A Work Unfinished* (1627), bound with *Sylva Sylvarum*, sig.a2ᵛ.

125. Fothergill, *Private Chronicles*, 14–20.

126. Cited in Dekker, *Egodocuments in History*, 7. Others include personal letters and autobiographies. Presser was expelled from his post during the Nazi occupation, went into hiding, and subsequently was among the first readers of Anne Frank's diary.

127. Grafton, "Martin Crusius Reads His Homer"; Pattison, "Diary of Casaubon"; Grafton, *Worlds Made by Words*, 216–30.

128. See North, *Notes of Me*, for a transcription of BL, Add. MS 32506. This was completed in the 1690s but not published until the late nineteenth century. See also North, *General Preface*, 78; Kassler, *Roger North*, 77–80, 89–93.

129. Watts, *Logic*, 72.

130. Levy, "How Information Spread among the Gentry."

131. See RS MS 190, fols. 146ʳ–166ᵛ for Boyle's "weather observations" for December 1684 to January 1686 within a notebook (albeit not strictly a commonplace book) of experiments and other material.

132. Bacon, *Advancement of Learning*, 113, 118. See also see Blair, "Humanist Methods in Natural Philosophy," 549–50; Beal, "Notions in Garrison," 138–39; Moss, *Printed Commonplace-Books*, 269–72.

133. Bacon, *Advancement of Learning*, 118. He also commented that: "Some have certain Common Places, and Theames, wherein they are good, and want Variety" (Bacon, "Of Discourse," in *Essayes*, 103).

134. Bacon, *De augmentis*, 435–37. See also chap. 3.

135. Moss, *Printed Commonplace-Books*, 117.

136. Bacon, *Description of the Intellectual Globe*, 508; Blair, *Theater of Nature*, 69–77.

137. Bacon, *Parasceve*, 457

138. "An Account of all the Lord Bacon's Works," in Bacon, *Baconiana*, 48.

139. Bacon, *Parasceve*, aphorism 3, p. 457.

140. Bacon, *Novum Organum*, Book I, aphorism 98, p. 155, and aphorism 82, p. 129.

141. Bacon, *Advancement of Learning*, 126 and 66. One target here was Petrus Ramus.

142. Rossi, *Francis Bacon*, 213.

143. See also Moss, *Printed Commonplace-Books*, 271.

144. For notes as part of what Bacon called "*experientia literata*," see chaps. 3 and 6.

145. Carruthers, *Book of Memory*, 8. It seems likely that Aquinas did not use notes from texts but rather composed either from memory or by direct access to the books he needed. See Blair, "The Rise of Note-taking in Early Modern Europe," 308.

146. Clanchy, *From Memory to Written Record*, 172–77.

147. Cited in ibid., 179.

148. Blair, *Too Much to Know*, esp. chap. 3.

149. H. Wanley [unsigned] to unknown recipient, September 16, 1710, in Heyworth, *Letters of Wanley*, 259. See also Garrett, "Legacy of the Baroque"; Garberson, "Libraries, Memory, and the Space of Knowledge."

150. Scaliger, *Scaligerana ou bons mots*; Selden, *Table-Talk*; Wolf, *Casauboniana*. Strictly speaking, the latter work is a collection of excerpts rather than table talk recorded by others. Casaubon did not teach after 1600 and this reduced the supply of student memories and records (I thank Anthony Grafton for this point). More generally, see Grafton, "The World of Polyhistors." Chambers' *Cyclopaedia* (1728) defined "Ana" as a "Latin Termination" used in the title of books that are "Collections of the memorable

Sayings of Persons of Learning, and Wit; much the same with what we otherwise call *Table Talk*."

151. Heinsius, "Funeral Oration," 94, 96.

152. Bayle, *Dictionnaire Historique et Critique*; see "Bodin, Jean," 1:583. In the entry on Bodin (at note E), Bayle quotes from Jacques Auguste de Thou, who was appointed Royal librarian by Henry IV. Cited in Blair, "Note Taking as Transmission," 97.

153. Bacon, *Advancement of Learning*, 3. Bacon's own table talk was admired by the earl of Dorset who relied on a secretary "to remember and to putt-down in writing my lord's sayings at table." Aubrey, *Brief Lives*, 1:67.

154. Carruthers, *Craft of Thought*, 8, 19.

155. Hippocrates, *Hippocratic Writings*, 81–82, 84–86, 262; see Huarte, *Examination of Mens Wits*, 60–62, 64, on performance of memory, imagination, and judgment as affected by different tempers of the brain.

156. Huarte, *Examination of Mens Wits*, 63.

157. Aubrey, *Brief Lives*, 1:411.

158. See the expositions in Gaukroger, *Descartes' System of Natural Philosophy*, 204–6; and Clarke, *Descartes's Theory of the Mind*, 91–99. For the significance of localized versus distributed, and representational versus dispositional, memory, see Sutton, *Philosophy and Memory Traces*.

159. Anon., *Polite Gentleman*, 64–65, 66–67, 68. See also Lund, "Wit, Judgment, and the Misprisions of Similitude."

160. Anon., *Polite Gentleman*, 69–70. Locke also made an unfavorable contrast between "wit," assisted by a quick memory, and "Reason and Judgment." *Essay*, II.xi.2.

161. Sorabji, *Aristotle on Memory*, 52–60. Bloch, *Aristotle on Memory and Recollection*, chap. 3, esp. 72–77.

162. Huarte, *Examination of Mens Wits*, 63.

163. For a nice account of this point, see Lewis, "A Kind of Sagacity," 158–59.

164. See Harvey, *Inward Wits*, 10, 12. However, Andreas Vesalius (1514–1564) and Thomas Willis (1621–75) shifted the likely location from the cavities of the brain to the tissues of the cerebral cortex. See Thomas Willis, *Cerebri Anatome* (London: Jo. Martyn and Ja. Allestry, 1664); and Dewhurst, *Thomas Willis's Oxford Lectures*, 134–50.

165. Isaac Newton, "Questiones quaedam philosophicae," CUL, Add. MS 3996, fol. 108^r, in McGuire and Tamny, *Certain Philosophical Questions*, 392. For Corvinus, the Roman author, see also Montaigne, *Complete Essays*, II.17, p.739.

166. Fuller, *Holy State*, 176.

167. Sorabji, *Aristotle on Memory*, 451b54–55. He also stated, more severely, that "recollection is neither the recovery nor the acquisition of memory" (451a53), but most early modern authors defined recollection as the finding of lost or blocked memories. See also Carruthers, *Book of Memory*, 19–21.

168. Hobbes, *Leviathan*, 2:38–42. Compare Locke, *Essay*, II.xix.1, on "Recollection" as involving "pain and endeavour."

169. Reynoldes, *Treatise of the Passions*, 13, 17. In later editions his name is spelled Reynolds.

170. Walker, *Of Education*, 136–37.

171. Reynoldes, *Treatise of the Passions*, 17.

172. Anon., *Ad Herennium*. The title indicates that it is addressed to Gaius Heren-

nius. In his *Institutio Oratoria* (end of the first century CE), Quintilian's support of the technique was muted. Yates, *Art of Memory*, 37–41.

173. Anon., *Ad Herennium*, Book III, 207–25, 233. For an early modern adoption of this point, albeit one that could not resist giving suggestions, see Platte, *The Jewell House*, 82–83.

174. On visual encoding as the key feature, see Schacter, *Searching for Memory*, 46–48. For Shereshevski, the mid-twentieth century Russian mnemonist, see Luria, *The Mind of a Mnemonist*, 23–29.

175. Sutton, "Porous Memory," 136–37.

176. See Coleman, *Ancient and Medieval Memories*, chap. 4, for the catalogues of memory feats in Pliny the Elder's *Natural History*.

177. Hunter, "Lengthy Verbatim Recall," 207.

178. Small, *Wax Tablets of the Mind*, 202. But Carruthers, *Book of Memory*, 87–89, says that accuracy in recollection of texts was highly valued in medieval culture.

179. Quintilian, *Institutio Oratoria*, XI.ii.27, as discussed in Carruthers, *Book of Memory*, 74–75; Coleman, *Ancient and Medieval Memories*, chap. 3.

180. Quintilian, *Institutio Oratoria*, XI.ii.29–30, 32.

181. Agrippa, *Uncertainty and Vanity of the Arts and Sciences*, 24–25.

182. Spence, *Memory Palace of Matteo Ricci*, 4.

183. Fuller, *Holy State*, 174.

184. HP27/8/12B, undated. These are notes in Hartlib's hand on "Preaching, from Perkins."

185. William Petty to Robert Southwell, August 16, 1687, *Petty-Southwell Correspondence*, 284.

186. For the persistence of mnemonic skills in early modern society, see Thomas, "Literacy in Early Modern England."

187. Willis, *The Art of Memory*, sig.A3ᵛ; and 1–12 for his explanation of topical memory. This is a partial translation of his Latin work, *Mnemonica* (1618).

188. Bacon, *Novum Organum*, Book II, aphorism 26, p. 287.

189. Aubrey, *Brief Lives*, 1:70, 83; also 331, 334–35, 351.

190. *Aubrey on Education*, 57.

191. BL, MS Sloane 2903, fol.6ᵛ.

192. These points are exemplified in chaps. 2 and 6.

193. The last term ("prospective") concerns remembering to remember. See Kliegel, McDaniel, and Einstein, *Prospective Memory*. More generally, see Danziger, *Marking the Mind*; Foster, *Memory*; Bernecker, *Memory*, 13–22.

194. Casaubon, *Generall Learning*, 168. Meric was the son of Isaac Casaubon, mentioned earlier.

195. Donald, *A Mind So Rare*, 16. See also Hurley, *Consciousness in Action*; Menary, *The Extended Mind*.

196. Hutchins, *Cognition in the Wild*; Tribble, "Distributing Cognition in the Globe"; Sutton, "Distributed Cognition: Domains and Dimensions."

197. "Exogram" is the counterpart of "engram"—a coined by Richard Semon and then used by Karl Lashley to denote "a single entry in the biological memory system." See Semon, *The Mneme*; Lashley, "In Search of the Engram," 478–505; Donald, *Origins of the Modern Mind*, 314–16.

198. Tribble and Keene, *Cognitive Ecologies*.

199. Wilkins, *Essay*, 289; see 22–288 for his forty genera in tables; also 289–96 for "explication of the foregoing Tables."

200. Ibid., 21; also Epistle, sig.a1ᵛ. Thus in the phonetic, spoken version the word "Zana" combines "Za" (the genus fish); "n" indicating "squamous river fish," the ninth "Difference"; and "a," pinpointing the second species, the salmon "of a reddish flesh." Wilkins, *Essay*, 142.

201. Andrew Paschall to Aubrey, June 11 and July 8, 1678, Bodl. MS Aubrey 13, fols. 31–32, cited in Turner, "Andrew Paschall's Tables," 349–50.

202. Aubrey to Ray, July 9, 1678, in Ray, *Philosophical Letters*, 144–45; and Slaughter, *Universal Languages*, 177.

203. *Aubrey on Education*, 56; Bodl. MS Aubrey 10, fol. 94ʳ.

204. Thomas Pigot to Aubrey, December 28, 1676, Bodl. MS Aubrey 13, fol. 106, cited in Lewis, *Language, Mind, and Nature*, 216–17; and chap. 5 on the *Essay*.

205. Clark and Chalmers, "The Extended Mind."

206. In contrast, "free recall" involves remembering items without a specific prompt and in no prescribed sequence. See Murdock, "The Serial Position Effect of Free Recall."

Chapter Two

1. Plato's *Phaedrus*, 274B–275.

2. Small, *Wax Tablets of the Mind*, 129.

3. Freud used the recently invented "Wunderblock" (or magic notepad) to illustrate his views on the relations between "conscious, preconscious and perceptual-conscious systems." Freud, "A Note upon the 'Mystic Writing-Pad,'" 227–32.

4. Saint Augustine, *Confessions*, Book 10; also Coleman, *Ancient and Medieval Memories*, chap. 6; Carruthers, *Craft of Thought*, 31–32.

5. Petrarch, *The Secret*, 98–100. See also Quillen, *Rereading the Renaissance*, 182–216.

6. Montaigne, "On Liars," in *Complete Essays*, I.9, pp. 32–33; Montaigne, "Of Presumption," in ibid., II.17, p. 738; Montaigne, "On Experience," in ibid., III.13, p. 1240.

7. Descartes to Frans Burman, April 16, 1648, in *Philosophical Writings*, 3:334.

8. Bolgar, *Classical Heritage*, 271–75. What it meant to memorize was not necessarily equivalent for medieval and humanist readers. See Carruthers, *Book of Memory*, chap. 5; and Hunter, "Lengthy Verbatim Recall."

9. Heinsius, "Funeral Oration," 84. These notes were more likely Scaliger's published comments and annotation on texts; of course, he did keep other notes, some of which were lost. I owe this point to Anthony Grafton.

10. Wilkins, *Essay*, 196.

11. Gassendi, *Mirrour of True Nobility*, 188 (in Book VI).

12. Ibid., 191. See also Miller, *Peiresc's Europe*.

13. Aubrey, *Brief Lives*: "prodigious memorie" (1:416–17); "great memorie" (2:37, 277); "prodigious memorie" (2:174); 1:293, 2:288.

14. Ibid., 1:106. The word "chambers" may refer to rooms in the college rather than niches in the wall of a building, such as the chapel. Compare Aubrey's comment about Bacon's house in Goramberry in which the glass windows of the portico were "all painted; and every pane with several figures of beast, bird, or flower: perhaps his lordship might use them as topiques for locall memory" (1:82).

15. For the shopping illustration, see Willis, *The Art of Memory*, 58–64; see also Platte, *The Jewell House*, 84–85.

16. Aubrey, *Brief Lives*, 1:257; 2:67, 223.

17. Fuller, *Holy State*, 174, 175–76. On Fuller, see Lewis, *Language, Mind, and Nature*, 68–69.

18. Bacon, *Advancement of Learning*, 118; and Bacon, *Advancement and Proficience of Learning*, 254. See Bacon, *De augmentis*, 435, for the more colorful expression: "inviting the memory to take holiday."

19. Pope, *The Life of . . . Seth* [Ward], 192, 194–95, on his final loss of memory.

20. Brinsley, *Ludus literarius*, 51. He followed this with a caution against excessive demands.

21. Foxe, *Acts and Monuments*, 677, cited in Collinson, "The Coherence of the Text," 89. For other instances of reciting Scripture "without book," see *The Unabridged Acts and Monuments Online* or *TAMO* (HRI Online Publications, Sheffield, 2011). Available at: http://www.johnfoxe.org.

22. In the Middle Ages, rote memory had been seen as a lower form of memory. See Carruthers, *Craft of Thought*, 30; also Small, *Wax Tablets of the Mind*, 133 for the point that the word "verbatim" did not appear until the fifteenth century.

23. These sermons were themselves a product of rhetorical training: tutors and Fellows, usually destined for the clergy, practiced by performing "commonplaces." In Clarke, *Lives of Thirty-Two Divines*, 115, these are defined as "a Colledg-exercise in Divinity, not different from a Sermon, but in length"; cited in Fletcher, *Intellectual Development of Milton*, 2:57.

24. Hanford, "Chronology of Milton's Private Studies," 254 n. 5.

25. [Duport], "Dr. Duport's Rules," CUL, Add. 6986, [pp. 6–7]. This copy does not have folio or page numbers, so I have assigned page numbers. I thank Christopher Preston for digital copies and for his advice. See another copy, "Rules to be observed by young scholars in the University" (1660), Trinity College, Cambridge, MS: 0.10A.33, p. 9.

26. See Newton's "Fitzwilliam notebook," Fitzwilliam Museum, Cambridge, MS 1-1936, pp. 3^{r-v}, transcribed in Westfall, "Short-Writing." I take no. 11 to refer to lack of due care in memorizing college sermons. Westfall rightly suggests (p. 16) that "committing" can also mean "to put in writing," but he does not mention memorization.

27. Casaubon, *Generall Learning*, 126, 160.

28. Walton, *Life of Dr Sanderson*, sig.k2^v.

29. Sacchini, *De ratione libros* (1615), 62. This work went through several editions and translations; see Cevolini, *De arte excerpendi*, 145–62; Blair, *Too Much to Know*, 70–85.

30. Sacchini, *De ratione libros*, 84. See the report on Pliny's incessant note-taking in Pliny the Younger, *Letters of the Younger Pliny*, 89.

31. Sacchini, *De ratione libros*, 71–72. See Sherman, *Used Books*, on those who did in fact mark pages.

32. Newton was a serial dog-earer: his copy of *A Catalogue of Chymicall Books* (London: W. Cooper, 1675), held in the Babson Collection at the Huntington Library, San Marino, California, exhibits at least nine instances of this habit.

33. Sacchini, *De ratione libros*, 75–76, 81, 93–106.

34. On the *Aurifodina*, see Moss, *Printed Commonplace-Books*, 232–37;

35. Drexel, *Aurifodina*, 3, 69, 75. See also Blair, *Too Much to Know*, 77–79.

36. Blair, *Too Much to Know*, 79.

37. *Aubrey on Education*, 36, 56–57. See Bodl. MS Aubrey 10, fol. 94ʳ for his marginal reference to the *Aurifodina*.

38. Carruthers, *Craft of Thought*, 8, 29–31.

39. Seneca, "On Gathering Ideas," 277–81; Moss, *Printed Commonplace-Books*, 105; Havens *Commonplace Books*, 13–15.

40. Erasmus, *De copia*, 607, 620, 635, 636.

41. Seneca, Letters 27.6–78, cited in Small, *Wax Tablets of the Mind*, 129. This was not unusual: some Roman politicians and lawyers owned educated slaves, referred to as *graeculi*, or little Greeks, who memorized relevant facts and prompted their masters. See Schönpflug and Esser, "Memory and Its *Graeculi*."

42. Quintilian, *Institutio Oratoria*, XI.iii.142. I thank Lyn Tribble for this reference.

43. Carruthers, *Book of Memory*, 83, 172, 242.

44. Cave, *The Cornucopian Text*, chap. 1.

45. See Carruthers, *Book of Memory*, 19–20, 61, 89–90.

46. Sacchini, *De ratione libros*, 99–102.

47. Nelles, "Note-taking Techniques," 85. On advanced note-taking, see Soll, "Amelot de la Houssaye."

48. Foxe, "Titulorum omnium cui hoc libro continentur," in *Locorum communium tituli*.

49. Foxe, "De usu et ratione tabulae," in *Locorum communium tituli*, 23–27. This introductory matter comprises thirteen unnumbered pages with an alphabetical list of topics, A–Z. The Bodleian copy (shelfmark Bod 4°B16(6)Th) does not include the blank pages. See Moss, *Printed Commonplace-books*, 192–94; Sherman, *Used Books*, 130.

50. In his influential textbook, *Logicae artis compendium*, 26–27, Robert Sanderson gave these as: substantia, quantitas, qualitas, relatio, actio, passio, ubi, quando, situs, habitus.

51. See the suggestions in Walker, *Of Education*, 139–69.

52. Foxe, *Pandectae*, trans. in Rechtien, "Foxe's Comprehensive Collection," 84; Moss, *Printed Commonplace-books*, 194–95.

53. Foxe, *Pandectae*. This four-page index begins after fol. 604; it is not paginated. Similarly, Conrad Gessner added an alphabetical index of *loci* to the subject classification given in his *Pandectarum sive partitionum universalium* (1548). See Nelles, "Reading and Memory in the Universal Library"; Krajewski, *Paper Machines*, 9–13.

54. Foxe, *Pandectae*, fol. A2, cited in Rechtien, "Foxe's Comprehensive Collection," 85.

55. Moss, *Printed Commonplace-Books*, 195. For Locke's confirmation of this, see *Essay*, III.x.14.

56. Foxe, *Pandectae*, 361.

57. Ibid., fol. A2ᵛ; cited in Rechtien, "Foxe's Comprehensive Collection," 85–86.

58. Bacon, *De augmentis*, 422; also Bacon, *Advancement and Proficience of Learning*, 111.

59. Quintilian, *Institutio Oratoria*, V.x.20–22, discussed in Carruthers, *Book of Memory*, 62.

60. Bacon, *Novum Organum*, Book II, aphorism 26, p. 287.

61. Phillips, *Studii legalis ratio*, 153–55. Roger North recognized the value of commonplacing in legal study, and also the choice between memorizing and retrieving. See Kassler, *Roger North*, 63–65.

62. Cicero, *De Oratore*, 1:288–89; cited in Skinner, *Reason and Rhetoric*, 117.

63. Moss, *Printed Commonplace-Books*, chaps. 6–8.

64. Farnaby, *Index Rhetoricus*, appendix. For the adoption of such a list in a classroom, see Hoole, *A New Discovery of the Old Art of Teaching Schoole*, 133.

65. On social memory, see Halbwachs, *On Collective Memory*; Wilson, "Collective Memory."

66. Costello, *Scholastic Curriculum*, 14–31, 52–58. See also Mack, *Elizabethan Rhetoric*, 65–73.

67. [Duport], "Dr. Duport's Rules," CUL, Add. 6986, [p. 17]. For his reputation, see Monk, "Memoir of Dr. James Duport."

68. Waterland, *Advice to a Young Student*, 9. For topics and questions, see Johnson, *Quaestiones Philosophicae*, 15–16.

69. Bacon, *Advancement and Proficience of Learning*, 73–74; also in *Works of Francis Bacon*, ed. Spedding, Ellis, and Heath, 4:288. This passage closely follows the corresponding one in *Advancement of Learning*, 59.

70. Bacon, *Advancement and Proficience of Learning*, 74. See also Bacon, *De augmentis*, 288–89.

71. Bacon, *Advancement of Learning*, 60.

72. Bacon, "A Letter, and Discourse, to Sir Henry Savill," 229.

73. Holdsworth, "Directions" is a transcript (edited by H. F. Fletcher) of Emmanuel College, Cambridge, MS 1.2.27(1). The "Directions" was composed between 1615 and 1637.

74. Holdsworth, "Directions," 639 (no. 25), 641 (no. 30). The numbers refer to the sections given in the transcription.

75. Ibid., 640 (no. 26), 641 (no. 30), 639 (no. 25). "Conning" referred to the process of getting texts or arguments by heart. Holdsworth implies that it involved inadequate understanding.

76. Ibid., 634 (no. 16), 639 (no. 25), 650 (no. 49).

77. Holdsworth's attitude matches that of his contemporary, Jeremias Drexel, discussed above. See also Blair, "Note Taking as Transmission," 98.

78. Holdsworth, "Directions," 650–52 (nos. 50–53).

79. Ibid., 651 (nos. 50–51).

80. Ibid. (nos. 51–52).

81. [Barlow], *A "Library for Younger Schollers,"* 50.

82. [Duport], "Dr. Duport's Rules," CUL Add. 6986, [pp. 20 and 22]. He also counseled against copying out large passages: "Transcribe not whole sayings or stories at length, for that is teadious & endless; but make short references to book & page."

83. Walker, *Of Education*, 138–39.

84. [Duport], "Dr. Duport's Rules," CUL Add. 6986, [p. 25].

85. Morhof, *Polyhistor*; Placcius, *De arte excerpendi*. See Zedelmaier, "De ratione excerpendi."

86. See Todd, *Christian Humanism*, 54, 74–77, and 82–92, on university commonplace books showing reading from humanists more so than from neo-scholastics. See also Mack, *Elizabethan Rhetoric*, 44–45, 104–14; and more generally, Kearney, *Scholars and Gentlemen*, 163–67.

87. Mack, *Elizabethan Rhetoric*, 104 n. 6; see *Renaissance Commonplace Books*. The force of this generalization may depend on whether the notes were linked to some kind

of pedagogy or made by mature scholars who often flouted conventions; see Sherman, *John Dee*, 70–71.

88. Drexel, *Aurifodina*, 83, sig.[A8]ʳ, cited in Blair, *Too Much to Know*, 85.

89. Meric Casaubon to Oliver Withers, February 24, 1645, in Casaubon, *Generall Learning*, 194–95. See also Wheare, *Method and Order of Reading*, 324: "I should be too long if I should here attempt to describe the Form of Common Place-books, or describe their Methods."

90. Holdsworth, "Directions," 651 (no. 52). See also "Some Short Directions for a Student in the university, . . . by John Merryweather BD . . . Cambridge 1651/2," in Bodl. MS Rawlinson D. 200, fol. 41ʳ. See Trentman, "The Authorship of *Directions for a Student in the Universitie*," 180–81, for the argument that this is not by Merryweather, but is a copy based on Holdsworth's updated version of his earlier manuscript.

91. North, *Life of Dr John North*, 107, 134–37, cited in Kassler, *Roger North*, 54.

92. James Spedding suggested that this letter was written by Bacon on behalf of Robert Devereux, the second Earl of Essex. See "[Advice to Sir F. Greville on his Studies]," *Works of Francis Bacon*, ed. Spedding, Ellis, and Heath, 9:21. Snow, "Francis Bacon's Advice to Greville," gives strong reasons for concluding that Bacon was the author. I cite from Snow's transcription of an earlier manuscript (PRO. SP. 14/59, fols. 4ʳ–5ᵛ), as modernized in Vickers and titled "Advice to Fulke Greville on his studies," *Bacon: Major Works*, ed. Vickers, 102–6.

93. Bacon, "Advice to Fulke Greville," 102–4. Bacon was not averse to collecting certain material "by the labor of a servant in part"; see Bacon, "Comentarius Solutus," 62. He said that "some *Bookes* also may be read by Deputy, and Extracts made of them by Others; But that would be, onely in the lesse important Arguments, and the Meaner Sort of *Bookes*." Bacon, "Of Studies," in *Essayes*, 153.

94. Bacon, "Advice to Fulke Greville," 105.

95. Placcius reported the dismay of those who hoped for more from the notes of the legal scholar, Hermann Conring. See Placcius, *De arte excerpendi*, 185; Blair, *Too Much to Know*, 112.

96. Wood, *Athenae Oxonienses*, vol. 3, col. 936; cited in Richard Serjeantson's introduction to Casaubon, *Generall Learning*, 56.

97. See Wolf, *Casauboniana*; Pattison, *Isaac Casaubon*, 427, 482.

98. Pattison, *Isaac Casaubon*, 428. For reproductions of some of his annotations, see Blair, *Theater of Nature*, 196–97. Casaubon gave the name "adversaria" both to his marginalia in printed books and to the loose sheets on which he scribbled notes about them. However, Pattison's understanding of the full scope of Casaubon's note-taking was limited. For recent accounts, see Sherman, *John Dee*, 71–73; Grafton, *Worlds Made by Words*, 225–28.

99. Bacon, "[Advice to the Earl of Rutland]." Spedding ascribed the first two letters to Bacon, but expressed a doubt about the third (*Works of Francis Bacon*, ed. Spedding, Ellis, and Heath, 9:4–6); Vickers, "Authenticity of Bacon's Earliest Writings," 255, reviews the evidence for crediting all three to Bacon. The first letter was published in Essex and Sidney, *Profitable Instructions* (1633), 27–73, and was said to be by the "late Earle of Essex."

100. Bacon, "[Advice to the Earl of Rutland]," 13, 18, 19, 20.

101. Bacon, "Of Studies," in *Essayes*, 153.

102. Aubrey, *Brief Lives*, 2:154, 174.

103. MS Locke f. 8, p. 207 (October 31, 1684); see also Dewhurst, *John Locke*, 261. Cranston, *John Locke*, 237-38, assumes this remark fits with Locke's favorable impression of Dutch medical training; but Cook, *Trials of an Ordinary Doctor*, 63-64, interprets it as Locke's disappointment at the lack of "competitive oral argument" in these "intellectually lightweight" performances. I doubt this, given Locke's loathing of Oxford disputations and his avoidance of the Paris medical school on his visit in 1677.

104. Locke, *Some Thoughts*, 233 (sec. 176).

105. Locke, *Mr Locke's Reply to the Right Reverend the Lord Bishop of Worcester's Answer to his Second Letter* (1697), in *Works*, 4:212; see also *A Second Vindication of the Reasonableness of Christianity* (1697), in *Works of John Locke*, 7:197.

106. For Pierre Bayle's point that many quotations were inaccurately transcribed and thus not useful, see Moss, *Printed Commonplace-Books*, 280 n. 45, citing the entry "Langius" in his *Dictionnaire*.

107. Evelyn, *Memoires*, 43-44.

108. Ibid., 44. This is an allusion to Bacon's contrast between ants and bees. See Moss, *Printed Commonplace-books*, 280 n. 44.

109. Levine, *Between Ancients and Moderns*, chaps. 1-2.

110. Gassendi, *Mirrour of True Nobility*, sig.A4^r.

111. Evelyn to Pepys, April 28, 1682, in De la Bédoyère, *Particular Friends*, 129, cited in Darley, *Evelyn*, xiv.

112. See 1 Thessalonians 5:21 in the King James version, which reads "Prove all things; hold fast that which is good."

113. Bodl. MS Aubrey 4, fol. 120^v, crossed through; cited Hunter, *Science and the Shape of Orthodoxy*, 68.

114. Evelyn, *Diary*, 2:1, 10. The "Kalendarium" (BL, Add. MSS 78323-25) covers his life from 1620 to 1697. In about 1700 Evelyn made another version of this diary, "De Vita Propria," treating the years from 1620 to 1644.

115. For such details, see de Beer's introduction to the *Diary of Evelyn*, 1:45-46, 69-74. This diary was not intended for publication, although half of it was printed in 1818.

116. Evelyn to "Platona," December 20, 1649, BL, Add. MS 78298, fol. 39^v. For Evelyn as "the great minute-taker of the Restoration," see Howarth, *Lord Arundel and his Circle*, 126.

117. On the sermons, see de Beer's Introduction to the *Diary of Evelyn*, 1:82-83; "Vade Mecum" [1649-51], BL, Add. MS 78327; and "A collection in two parts, the first medical and culinary recipes, the second of skills and crafts" (BL, Add. MS 78340). See also Hunter, *Science and the Shape of Orthodoxy*, 72.

118. Evelyn to Sir Richard Browne, June 14, 1649, in BL, Add. MS 78221, fol. 45^r. See also Hunter, "The Library of John Evelyn," 82-102.

119. Evelyn, *Memoires*, 68.

120. See BL, Add. MS 78330, fols. 1-71^v for theology; fols. 72^r-149^v for sciences; fols. 150^r-171^v for liberal arts; fols. 172^r-180^v for law and politics. Each of these is divided into chapters, given as Cap. I, Cap. II, and so on. See also BL, Add. MS 15950, fol. 125, for "Cap. IIII, Metaphysica," an elaborate breakdown of that subject, which may be a stray leaf from one of the large commonplace books, as Hunter suggests in *Science and the Shape of Orthodoxy*, 74.

121. In Evelyn's "Vade Mecum" (BL, Add. MS 78327) the sequence, based on Alsted, begins with grammar, rhetoric, logic, natural philosophy, metaphysics, and ethics.

122. Bodl. MS Bodley 878, fol. 5^r [late 1640s] cited in Beal, "Notions in Garrison," 139–40. This was part of letter to Christopher, 1st Baron Hatton (1605–70).

123. BL, Add. MS 78330, fol. 72^r.

124. See, for example, BL, Add. MS 78330, fols. 1^v–2^r, 6^v–7^r, 108^r. For Pepys on Hoare, see Pepys, *Diary*, May 10, 1660, 1:132; Hunter, "The Library of John Evelyn," 84; Darley, *Evelyn*, 320 n. 58. On Evelyn's "note-hand," as opposed to his "letter-hand" and "book-hand," see de Beer, Introduction to *Evelyn, Diary*, 1:48–51.

125. BL, Add. MS 78333, "Adversaria Historical, Physical, Mathematical," fol. 1^r. This entry has a cluster of marginal Heads — "Ingenium, Aer, Scientia, Ars, Innovatio, Inventio." It is also significant that the stray leaf with subheadings for "Metaphysica" opens with an excerpt from "DesCartes Methode"; see BL, MS 15950, fol. 125^r, in n. 120 above.

126. Hunter, *Science and the Shape of Orthodoxy*, 74, remarks that his book collection was "abnormally cosmopolitan."

127. Casaubon, *Generall Learning*, 168.

128. Evelyn, *Memoires*, 37; also 40–43, 48–51.

129. This was published in 1640 as *Of the Advancement and Proficience of Learning*.

130. BL, Add. MS 78330, fol. 72^v.

131. More generally, Levine nominates Bacon as the author who showed Evelyn how to reconcile the "humanist *paedia* with the new science." Levine, *Between Ancients and Moderns*, 23.

132. Bacon, *Advancement and Proficience of Learning*, 37. Evelyn's transcription is almost verbatim, but omits italics and alters punctuation. See BL, Add. MS 78330, fol. 72^v.

133. BL, Add. MS 78327, fol. 1^v.

134. See BL, Add. MS 78337, and BL, Add. MS 78340.

135. BL, Add. MS 78333, fols. 17^r–20. This section is toward the end of the notebook after many blank pages. See Evelyn, *Memoires*, 38, for the value of a household book of medical "Receits."

136. See "Adversaria Historical, Physical, Mathematical," BL, Add. MS 78333, fol. 1^r.

137. Evelyn to George Evelyn, October 15, 1658, BL, Add. MS 78298, fol. 88$^{r–v}$.

138. Evelyn had recorded this passage under "Adversaria" in one of the commonplace books: "Quid est quod negligenter scribamus adversaria . . . Cicero pro Rosaio Comoedo." (BL, Add. MS 78330, fol. 82^r). See also Cicero, *Pro Roscio comoedo*, II, 7, in *Orations of Cicero*, 1:88–90: "Therefore no one ever produced memoranda at a trial; men do produce accounts and read entries in books."

139. Evelyn, *Memoires*, 52–53, 63.

140. He referred to this in BL, Add. MS 78330, fol. 1^r, in the top right corner in pencil: "note that the black lead figures of the pages agree onely with The Universall Table of all three Tomes."

141. BL, Add. MS 78331, fol. 4^v for the example of "Anima." For a similar habit, see Paschall to Aubrey, March 2, 1679/80, Bodl. MS Aubrey 13, fols. 39–40, in which there are twenty-two marginal words (some repeated) against the first twenty-seven lines of his letter.

142. BL, Add. MS 78331, fol. 1^v ("Advertisement"). See also BL, Add. MS 15950, fol. 125^r, where the margin has thirty-nine Heads words.

143. BL, Add. MS 78331, fols. 1–92^r ("Index Locorum Communium"), 94^r–103^v; 104^v–107^r; 108^r–113^r.

144. Blair, *Too Much to Know*, 6.

145. Sharpe, *Reading Revolutions*, 69–89; for Harvey, see Grafton and Jardine, "Studied for Action"; on Hartlib, see chap. 4.

146. He was the eldest son of an Italian doctor, Cesare Adelmare who, as one of the royal physicians, was nicknamed "Caesar" by Queen Mary and Elizabeth I. His son, Julius, then became "Julius Caesar." See *ODNB*; Sherman, *Used Books*, 127–28.

147. BL, Add. MS 6038 is numbered to fol. 616. The last of Foxe's printed Heads are on fol. 611; his index is at fols. 612–13, with numerous Heads added in the margin by Caesar, who also inserted two additional pages containing an extra 146 Heads (fols. 614–15) and a backing page (fol. 616).

148. Sherman, *Used Books*, 132–37.

149. While Hoare was helping with the large commonplace book, Evelyn began compiling a small "Vade Mecum" containing anecdotes and maxims under seven headings, such as "Sententiae" and "Auditiones quotidinanae." This is in the Houghton Library, Harvard University, MS Eng. 992.7. The first page is dated 1650 and carries the final two words of Evelyn's motto ("meliora retinete").

150. See BL Add, MS 78333, fol. 4^r for notes added while transcribing from a manuscript copy of Aubrey's *The Naturall Historie of Wiltshire*, as discussed in Yale, "Marginalia, Commonplaces, and Correspondence," 199.

151. For papers related to "Elysium Britannicum," see BL, Add. MSS 78342–44. This work was never published in its entirety, but part of it appeared as *Britannicum: Kalendarium hortense* (1664) and *Acetaria* (1699).

152. BL, Add. MS 15950, f. 80^r.

153. Naudé, *Instructions Concerning Erecting of a Library*, 35–36. Naudé advised that smaller, precious manuscripts be kept more securely (81–82).

154. BL, MS Sloane 2891–2900. The nine notebooks are bound in seven volumes; each bound volume is approximately 20×18 cm.

155. [Hill], *Familiar Letters*, vi. The exception is [Abraham Hill], "Some Account of the Life of Dr. Isaac Barrow." (Signed A. H.). This is the preface to *The Works of the learned Isaac Barrow*, ed. John Tillotson (London: M. Flesher for B. Aylmer, 1683–86), vol. 1, sig.a^r–d^v.

156. [Hill], *Familiar Letters*, x. When Tillotson became archbishop of Canterbury in 1691 he asked Hill to take the office of "comptroller."

157. Birch, *History*, 1:396 (March 16, 1663/4), 4:65. See Purchas, *Purchas His Pilgrimage*.

158. Hill was treasurer, 1663–65 and 1679–99, and secretary, 1673–75. See Maddison, "Abraham Hill," 177, for Hill as a source of information; Aubrey, *Brief Lives*, 1:101 on consulting Hill about Henry Billingsly; 1:135 about Samuel Butler; and 2:69 for details about John Milton. See also Wood, *Life*, 3:295.

159. Beinecke Library, MS Osborn b112. This manuscript is 19×16 cm. There are no dated entries. The inside front cover bears a date that may be 1653, but this does not reliably indicate when the notebook was started. Havens, *Commonplace Books*, 69, suggests that it was used between c. 1653 and 1675. Southwell left several journals: see, for example, BL, Egerton MSS 1632 and 1633.

160. BL, MS Sloane 2899, fols. 26ᵛ–27ʳ (or in Hill's numbering, fols. 818ᵛ–819ʳ) has various references to memory squeezed into two pages. In the index (MS 2900, fol. 74ᵛ) Hill made subdivisions under "Memory" for temperance, citation, Inquisition, understanding, Augustus, and Petrarch.

161. BL, MS Sloane 2899, fol. 26ᵛ (or 818ᵛ).

162. BL, MS Sloane 2899, fol. 27ʳ. The likely passage is Erasmus, *De Ratione Studii* (1512), 671. See also fol. 819ʳ: "Lock Essay Abridg 34 Lips polit 51." Hill cited the 1661 partial translation of Willis' *Memonica* (1618).

163. See the extended essay in the notebook of Richard Cromleholme Bury (Beinecke Library, MS Osborn b182), begun October 19, 1681, in Havens, "*Of Common Places, or Memorial Books,*" 1–2. This is marked "per W. H," which I conjecture might indicate that content is drawn from both Obadiah Walker and Richard Holdsworth. See also "Memory" in an anonymous commonplace book, c. 1690, in Havens, *Commonplace Books*, 70.

164. Beinecke Library, MS Osborn b112, p. 169: "Walkers Education a good book."

165. Ibid., 292. The suggestion was from "Mr Godolphin," the diplomat, Sir William Godolphin (1635–96).

166. Bacon, "Intellectual Powers," 225–31. The edition of 1671 is noted on the inside front cover of Southwell's commonplace book, and under "Inventio" on p. 290.

167. Beinecke Library, MS Osborn b112, p. 362, cites Watts' edition of *Advancement and Proficience of Learning*, 74. This entry continues on p. 363.

168. "Memorandum recording the Advice of Lord Chiefe Baron Hale," Beinecke Library, MS Osborn Files 14242. At this time, Southwell was a clerk to the Privy Council. Hale was made chief justice in 1671.

169. Beinecke Library, MS Osborn Files 14242, fol. 1ᵛ. Hale's contrast between youthful commonplace books and the more judicious selections suited to a profession, is also found in Holdsworth, "Directions," 651–52 (no. 52).

170. For example, Southwell thought that a book should be read several times before censuring it or accepting the negative opinion of others. Beinecke Library, MS Osborn b112, p. 575 (in added entries).

171. Remarkably, though, there was very little extrapolation of regular dating of entries from the journal format.

172. Naudé, *Instructions Concerning Erecting of a Library*, 77 and 82.

173. See Yeo, "Loose Notes and Capacious Memory," and chap. 6 on Boyle's notes.

Chapter Three

1. Boswell, *Life of Samuel Johnson*, 456, for April 18, 1775. Johnson offered this dictum while perusing the spines of books in Richard Owen's library in Cambridge. See also De Maria, *Johnson*.

2. Johnson, *The Idler*, no. 74 (September 15, 1759); this was subsequently published as an essay called "Memory Rarely Deficient."

3. Gibbon, *Essay*, 77–86; this originally appeared as *Essai sur l'etude de la Literature* (Londres: T. Becket and P. A. Hondt, dans le Strand, 1761).

4. This term was not used until 1859 (and 1870 in French). These works included thesauri, concordances to the Bible, and various Latin lexicons of technical terms in the professional disciplines of law, medicine, and theology. In addition, from the sixteenth

century there were general works of reference, albeit also aimed at readers of Latin. See Blair, *Too Much to Know*.

5. Wood, *Life*, entry for July 24, 1692, 3:396.

6. For Latin predecessors, see Blair, *Too Much to Know*; for English ones known to Johnson, see Yeo, *Encyclopaedic Visions*.

7. Gibbon, *Memoirs*, 147. The substance of this work was published in 1796, but composed, at intervals, from about 1788.

8. Bas van Fraassen writes about *The Empirical Stance*, but he is mainly concerned with epistemology rather than with how to gather information; he deals with philosophers (such as David Hume), not physicians and scientists.

9. Johnson, *Dictionary of the English Language*, vol. 1. In the fourth edition of 1773, the entry remained unchanged.

10. Nunberg, "Farewell to the Information Age," 111.

11. For warnings about this usage, see Brown and Duguid, *Social Life of Information*, chap. 1.

12. Instead, data refers to "a lack of uniformity" or, as Gregory Bateson (1904–80) put it, "the elementary unit of information—is a difference that makes a difference" (cited in Floridi, *Information*, 23); see also p. 1 for Claude Shannon (1916–2001) on different meanings. See also Hobart and Schiffman, *Information Ages*, 201–34.

13. Burke, *Social History of Knowledge*.

14. Aubrey, *Brief Lives*, 2:25 (life of Ralph Kettell). See Petty to Southwell, February 1679: "You know I doe not deale in News," in Petty, *Petty-Southwell Correspondence*, 67.

15. Boyle, *History of the Air*, 46; also in *Boyle Works*, ed. Hunter and Davis, 12:35.

16. In English, this plural form died out by about 1700, whereas it still exists in some other European languages: thus *les informations*; *gli informazioni*.

17. Bacon, *Advancement of Learning*, 165.

18. Beale to Boyle, November 9, 1663, *Boyle Correspondence*, 2:171.

19. See de Beer, introduction to Evelyn's *Diary*, 1:83–84.

20. Milton, "Date and Significance of Two of Locke's Early Manuscripts," 75–76.

21. Brook Bridges to Locke, January 18, 1704, *Locke Correspondence*, 8:169. For the Bank of England, established in 1694 and supported by Locke, see Pinkus, *1688: The First Modern Revolution*, 371. See also Dooley and Baron, *Politics of Information*, on political "news" as needing constant evaluation.

22. "Professor De Carro's Letter to Dr. Pearson," *Medical and Physical Journal* (January–June 1801): 159, cited in Bennett, "Note-taking and Data-Sharing," 426.

23. But even the sense of the term we use in everyday speech was not common until the mid-twentieth century. Before 1937, the *OED* records no sense of the word "information" as separate from the person who conveyed it.

24. Bacon, "The Praise of Knowledge" (c. 1590–92), in *Works of Francis Bacon*, ed. Spedding, Ellis, and Heath, 8:125.

25. Beale to Oldenburg, December 21, 1662, in *Oldenburg Correspondence*, 1:481.

26. Beale to Boyle, November 9, 1663, *Boyle Correspondence*, 2:174. For Boyle's possible reference to Beale as a "particular friend, (a great virtuoso of the Royall Society)," see RS MS 187, fol. 32^v, cited in Hunter, *Robert Boyle: Scrupulosity and Science*, 230 n. 27; reprinted in *Boyle Works*, ed. Hunter and Davis, 11:lxvii.

27. Beale to Boyle, November 21, 1663, *Boyle Correspondence*, 2:207–8. Wotton was provost of Eton between 1624 and 1639, during which time Beale was a student.

28. Beale to Boyle, July 30, 1666, ibid., 3:195.

29. For analysis of this concept, see Rosenberg, "Early Modern Information Overload."

30. Seneca, "On Tranquillity of Mind," in *Dialogues and Letters*, 45.

31. Blair, *Too Much to Know*, 55–61; also Hobart and Schiffman, *Information Ages*, 87–96.

32. Leibniz, "Precepts for Advancing the Sciences and Arts" (as a later editor entitled it), in *Leibniz: Selections*, 29–30.

33. Bayle, *Dictionary Historical and Critical*, 1:4. See the remarks by John Collins in Rigaud, *Correspondence of Scientific Men*, 1:119–25; also the editors' introduction to *Oldenburg Correspondence*, 7:xxv.

34. Glanvill, *Plus Ultra*, 91.

35. Bacon, *Advancement and Proficience of Learning*, 75. See also Bacon, *De augmentis*, 290. In some fields, such as history, Bacon said the ancients were "to be preferred before the best of our moderns" (Bacon, "Advice to Fulke Greville," 105).

36. [Locke], "De Arte Medica (1669)," in Dewhurst, *Dr. Thomas Sydenham*, 82. See Anstey and Burrows, "John Locke, Thomas Sydenham."

37. See Oldenburg to Johannes Hevelius, January 31, 1667/8, in *Oldenburg Correspondence*, 4:136–38.

38. Burke, *Social History of Knowledge*, 119, 139.

39. See de Vivo, *Information and Communication in Venice*.

40. John Graunt used these in his *Natural and Political Observations* (1662). Aubrey, *Brief Lives*, 1:272, 274, believed that Graunt "had his hint" from "that ingeniose great virtuoso, Sir William Petty."

41. Anon., "Observazioni Meterologiche fatte in Firenze (December 1654–March 1670)," 1–223.

42. Fischer, *Francis Bacon*, xi.

43. Passmore, *Hundred Years of Philosophy*, 19.

44. Yeo, *Defining Science*, esp. 177–80.

45. See Waldow, "Empiricism and Its Roots."

46. Bacon, *Novum Organum*, Book I, aphorism 95, pp. 152–53.

47. Bacon, *Advancement of Learning*, 10.

48. Locke to Denis Grenville, March 13/23, 1678, *Locke Correspondence*, vol. 1, no. 374.

49. Browne, *Pseudodoxia Epidemica* (1646).

50. Casaubon, *Generall Learning*, 187.

51. Boyle, "A Proemial Essay," in *Certain Physiological Essays*, 12–14. Boyle said that this was written in 1657.

52. Power, *Experimental Philosophy*, 184, 192. See also Anon., *A Brief Vindication of the Royal Society*, 3, on systems versus the judicious care advised by Bacon.

53. For related perspectives, see Wolfe and Gal, *The Body as Object*.

54. Williams, *Keywords*; Wierzbicka, *Experience, Evidence*, 6–11, 50–52.

55. Dear, "Totius in Verba," 152; also Dear, *Discipline and Experience*, 3–6, 11–15; and his "Meanings of Experience," 112–15.

56. Shapin, *Social History of Truth*, chap. 5; Daston, "On Scientific Observation."

57. Boyle, *Sceptical Chymist*, 342; also his *Usefulnesse* (1663), 383; and his *Memoirs* . . . *Humane Blood*, 66.

58. Locke, *Essay*, I.iv.25.

59. Harvey, *Anatomical Exercitations*, sig.a4, 2ʳ; ¶4, 2ᵛ; also sig.¶ᵛ. See also Shapin, *Social History of Truth*, 201.

60. Hooke, "Discourse of Earthquakes," 338. This collection was published after Hooke's death by Richard Waller. See also Poole, *World Makers*, 104–7.

61. Ray, preface to *Ornithology*, sig.A2, 2ʳ, 2ᵛ.

62. Ray to Aubrey, May 1678, MS Aubrey 13, fol. 170ʳ, taken from the copy in BL, MS Egerton 2231, fols. 113–46. John Tradescant, the Elder (1570–1638), and his son, the Younger (1608–62), were botanists and gardeners.

63. Christopher Merrett, "Observations Concerning the Uniting of Barks of Trees Cut, to the Tree It Self," *Philosophical Transactions* 2 (1666): 453–54; also his second communication on p. 455 for observations and experiments on the fruiting of cherry trees.

64. Hooke, *Diary*, 169.

65. RS CP.xx.50, fol. 97ʳ, cited in Hunter and Wood, "Towards Solomon's House," 74; and Hunter, *Establishing the New Science*, 217.

66. Hooke, "General Scheme," 63.

67. Harvey, *Inward Wits*, 43–46. For Aristotle's summary of the senses receiving "impressions," see *De Anima*, in *Aristotle: the Complete Works*, 1:424a18–424a23; and *De memoria et reminiscentia*, in Sorabji, *Aristotle on Memory*, 50–52.

68. Bacon, *Descriptio globi intellectualis*, 97.

69. The powers of *imaginatio*, *cogitatio* (comparing and combining), and *memoria* were also located in the cerebral ventricles (Harvey, *Inward Wits*, 10–12, 17).

70. Bacon, "A Description of the Intellectual Globe," 504; and Bacon, *De augmentis*, 292–93.

71. Bacon, "A Description of the Intellectual Globe." 503. For a similar account, see Burton, *The Anatomy of Melancholy*, 1:152: "*Memory*, lays up all the species which the senses have brought in, and records them as a good *Register*, that they may be forthcoming when they are called for by *Phantasie* and *Reason*." Bacon's formulation omits some of the complexities in the passage from senses via imagination and memory to cognition, as found in earlier writings. See Smith, "Picturing the Mind."

72. Boyle, *Christian Virtuoso, The First Part* (1690–91), 326.

73. Boyle, *Origine of Formes* (1666), 298.

74. Boyle, *New Experiments and Observations Touching Cold* (1665), 216; also 233.

75. Boyle, *Usefulnesse* (1663), 204.

76. Boyle, *The Christian Virtuoso, The Second Part*, 438. This second part was composed c. 1691–92 but published posthumously. See also *Christian Virtuoso, First Part*, 294, where Boyle argues that atheism can result from "want of due Information."

77. Boyle, *Christian Virtuoso, First Part*, 307.

78. Ibid., 306. This passage shows that Boyle had to distinguish his position from that of "empiricks," who also professed opposition to general systems. See Galen, *Three Treatises*, 23–45.

79. Boyle, *Christian Virtuoso, First Part*, 307, 308.

80. Ibid., 304.

81. Boyle, "Cogitationes Physica" (1670–80s), BP 8, fol. 210ʳ.

82. Boyle, *Three Tracts* (1670), 338; and "Advertisement," 319, for the point that he had done his best to check "the Informations of Others" and to communicate with "credible Persons."

83. Hobbes, *Leviathan*, 2:124: "The Register of *Knowledge of Fact* is called *History*" whereas the "Registers of Science, . . . contain the *Demonstrations* of Consequences of one Affirmation, to another."

84. Boyle, *An Examen of Mr. T. Hobbes in his Dialogus Physicus* (1662), in *Boyle Works*, ed. Hunter and Davis, 3:174 (translation, 185). Here Boyle quotes the words of a speaker in a dialogue from Hobbes' own *Examinatio et emendatio mathematicae hodierne* (1660). The *locus classicus* for this contest is Shapin and Schaffer, *Leviathan and the Air-Pump*.

85. Hooke, "General Scheme," 8–9.

86. Oldroyd, "Some 'Philosophicall Scribbles,'" 17–19.

87. Fuller, *Holy State*, 176, warned against trusting solely in memory because its contents could be erased by "a violent disease" that might "rob and strip" one of everything.

88. Hooke, preface to *Micrographia*, sig.a1^v; emphasis in original.

89. Hooke, "General Scheme," 22–26, for the kind of topics and particulars expected in Baconian natural histories.

90. Bacon, *Novum Organum*, Book I, aphorism 102, p. 159.

91. Bacon to Baranzano, Junii ultimo 1622, *Works of Francis Bacon*, ed. Spedding, Ellis, and Heath, 14:374–77.

92. Pliny the Elder, *Naturalis Historia*, 1:13. See also Murphy, *Pliny the Elder's Natural History*; Doody, *Pliny's Encyclopedia*.

93. Bacon, *Advancement of Learning*, 84.

94. See also Bacon to Baranzano, Junii ultimo 1622, *Works of Francis Bacon*, ed. Spedding, Ellis, and Heath, 14:377; and Sanderson, *Logicae Artis Compendium*, 22–23, on reasoning from universal to particulars.

95. Bacon, *Novum Organum*, Book I, aphorisms 19, 22, 24, 25, pp. 71–75.

96. Ibid., Book II, aphorism 28, p. 297.

97. See Daston and Park, *Wonders and the Order of Nature*, 120–22, for the concept of the "preternatural"; and da Costa, *The Singular and the Making of Knowledge*, 20–24, 119, for the notion of "the singular."

98. Bacon, *Novum Organum*, Book I, aphorism 119, p. 179.

99. Ibid., Book II, aphorism 28, p. 297. In 1612 Bacon emphasized that this "natural history is not about single objects." See Bacon, *Descriptio globi intellectualis*, 101.

100. Bacon, *Novum Organum*, Book I, aphorism 112, p. 171, and aphorism 24, p. 73.

101. Bacon, *Parasceve*, Title no. 10, 475.

102. Bacon, *Novum Organum*, Book I, aphorism 119, pp. 178–79; also *Parasceve*, aphorism 6, p. 465.

103. For Bacon's first mention of division by faculties, Bacon, *Advancement of Learning*, 62. See also Kusukawa, "Bacon's Classification of Knowledge," 47–74.

104. Bacon, *Advancement of Learning*, 118–19.

105. Bacon, *De augmentis*, 435–48.

106. Ibid., 435.

107. For his first use of the term "*Experientia literata*," see Bacon, *Advancement of Learning*, 111 (where the other stage is "*Interpretatio Naturae*"). See also chap. 6.

108. Bacon, *Novum Organum*, Book I, aphorisms 112 and 113, p. 171, aphorism 118, p. 177.

109. Bacon, *Parasceve*, aphorism 8, pp. 467–69.

110. See Daston and Park, *Wonders and the Order of Nature*, 223–24, for Bacon's awareness about the danger of open-ended collection of particulars.

111. Bacon, "Advice to Fulke Greville," 103. Elsewhere he acknowledged that "the inquiries are so mixed up with one another that some of the things inquired fall under different titles." ("Abecedarium Naturae," in *Works of Francis Bacon*, ed. Spedding, Ellis, and Heath, 5:211)

112. Bacon, *Parasceve*, aphorism 10, pp. 471–73.

113. Ibid., aphorism 3, p. 457.

114. Ibid., aphorism 6, p. 465; see also *Novum Organum*, Book I, aphorism 120, p. 179, for citation of Pliny's defense of the inclusion of vile and ugly things.

115. Bacon, *Parasceve*, aphorism 5, p. 463, aphorism 6, p. 465. Bacon cross-referenced this discussion to *Novum Organum*, Book I, aphorisms 99, 118, 119, 120, pp. 157, 177–81.

116. Bacon, *Novum Organum*, Book I, aphorism 82, p. 129, aphorism 125, p. 189. In speaking of the "ancients," Bacon meant the Greeks.

117. Ibid., aphorism 110, p. 169.

118. When the Society discussed an earlier draft, Oldenburg was concerned "whether there be enough said of particulars" (Oldenburg to Boyle, November 24, 1664, *Boyle Correspondence*, 2:416).

119. [Oldenburg], "An Account of some Books. The *History . . .* by Tho. Sprat," *Philosophical Transactions* 2 (1667): 502.

120. Ibid., 503.

121. Sprat, *History*, 318–19.

122. Anstey, "Experimental versus Speculative Natural Philosophy," 215–42, contrasts "experimental" and "speculative" philosophy. I think the other kinds of empirical data (which Oldenburg listed in 1667) were aligned with the "experimental" side of this contrast.

123. Boyle, *Christian Virtuoso, First Part*, 292, 307.

124. Oldenburg, "A Preface to the Third Year of These Tracts," *Philosophical Transactions* 2 (1666–67): 414.

125. Shadwell, *The Virtuoso*, III.iii.102–10.

126. See *Philosophical Transactions* 1 (1665–66): 360–62; also "Inquiries for Suratte and other parts of the East-Indies," *Philosophical Transactions* 2 (1667): 415–19; Sprat, *History*, 156–57. See also Anstey, "Literary Responses," 150, 159–60.

127. Oldenburg to Benedict Spinoza, August 4, 1663, in *Oldenburg Correspondence*, 2:103, cited in Henry, "Origins of Modern Science," 106. The same stricture applied, of course, to any comments that might embroil the Society in religious controversy.

128. Oldenburg to Curtius, November 6, 1668, in *Oldenburg Correspondence*, 5:133; also 2:111.

129. Oldenburg, "A Preface to the Third Year of these Tracts," 410 (see n. 124 above).

130. BP 9, fol. 19ʳ [1670s]. Hunter, *Boyle Papers*, 47, identifies this as being in the hand of Boyle's chief amanuensis, Robert (called Robin) Bacon.

131. Boyle, "Observations about the Barometer," 181–84; also in *Boyle Works*, ed. Hunter and Davis, 5:504–7.

132. Zilsel, "Genesis of the Concept of Scientific Progress."

133. Glanvill, *Plus Ultra*, 90–91.

134. Sprat, *History*, 194–95; see also 60–62.

135. Evelyn, "To the Reader," *Sylva*, 3rd ed., sig.A2ᵛ.

136. See Evelyn to Beale, July 27, 1670, BL, Add. MS 78298, fol. 182ᵛ, letter no. cccxxix; cited in Levine, *Between Ancients and Moderns*, 223 n. 45.

137. Evelyn to Beale, August 27, 1668, BL, Add. MS 78298, fol. 173ʳ.

138. Evelyn to Beale, July 11, 1679, BL, Add. MS 78299, fol. 2ᵛ. Evelyn was referring to his unfinished "Elysium Britannicum" which he discussed with Thomas Browne in 1655; he announced his intention to start it in 1659.

139. Hippocrates, *Aphorismes of Hippocrates*, sig.Bʳ. For a modern translation, see *Hippocratic Writings*, 206: "Life is short, science is long; opportunity is elusive, experiment is dangerous, judgement is difficult."

140. Seneca, "On the Shortness of Life," in *Dialogues and Letters*, 59.

141. This is nicely treated in Weinrich, *On Borrowed Time*.

142. Galen, *Three Treatises*, 11. See also *Galen: On the Therapeutic Method*; and Mattern, *Galen and the Rhetoric of Healing*.

143. Galen, *Three Treatises*, 34–35, 101, 103–4. See also Pomata, "*Praxis Historialis*," 112.

144. Siraisi, *History, Medicine, and Traditions*, 63–64, 187–89.

145. Galen, *Three Treatises*, xxx and 33.

146. Mattern, *Galen*, 11. On Galen's usage, see Flemming, "Commentary," 323–54, 324–27.

147. Bacon, *Novum Organum*, Book I, aphorism 92, p. 149. Of course, Bacon also sought an increase in the individual's life span as a direct response to Hippocrates' first aphorism: at the end of *New Atlantis* (1627), "The Prolongation of Life" is the first in a list of desiderata; see "Magnalia Naturae" in Bacon, *Sylva Sylvarum*, sig.g3ʳ.

148. Bacon, *Advancement of Learning*, 85, 99–100. See editor's "Commentary," 294–95. Compare the less accepting tone in Bacon's unpublished "Masculine Birth of Time," in which Hippocrates stands for excessive empiricism and Galen for undisciplined theory. See Spedding, preface to "Temporis Partus Masculus," *Works of Francis Bacon*, ed. Spedding, Ellis, and Heath, 3:523–26. See also Cantor, *Reinventing Hippocrates*.

149. Schenck, *Observationes medicae . . .* (Frankfurt, 1609), cited in Pomata, "Sharing Cases," 220. This large work appeared in installments between 1584 and 1597.

150. Bacon, *Advancement of Learning*, 99.

151. Bacon, *De augmentis*, 299.

152. Bacon, "The Great Instauration," in *The Instauratio Magna, Part II*, 25.

153. Bacon, *Novum Organum*, Book I, aphorism 49, p. 87.

154. Descartes balanced the two options, deciding to "devote my whole life" to "an indispensable branch of knowledge" and hoping to achieve something despite "the brevity of life or the lack of *empirical information.*" Descartes, *Discourse on Method*, 52.

155. Boyle, *Usefulnesse* (1663), 414; and 544 for translation of Latin quotations taken from an unidentified work by Paracelsus.

156. Boyle, *Christian Virtuoso, First Part*, 308.

157. Hooke, preface to *Micrographia*, sig.d1ʳ.

158. In offering a similar reflection on the accumulation of knowledge, Blaise Pascal was more optimistic, saying that the individual man "derives advantage, not only from his own experience, but also from that of his predecessors; since he always retains in

his memory the knowledge which he himself has once acquired." Pascal, "Preface to the Treatise on Vacuum" [1650s?], in Pascal, *Thoughts, Letters, Minor Works*, 449.

159. Ong, *Ramus*, 225–27; Hotson, *Commonplace Learning*, 43–51.

160. Leibniz, "Advancing the Sciences and Arts," in *Leibniz: Selections*, 31–32.

161. Leibniz, "Towards a Universal Characteristic (1677)," in *Leibniz: Selections*, 22; and "An Introduction to a Secret Encyclopaedia," 5–9; Jones, *The Good Life in the Scientific Revolution*, chap. 6.

162. Leibniz, "Horizon of Human Doctrine (1690)," in *Leibniz: Selections*, 75.

163. Bacon, *Advancement of Learning*, 85. The Oxford mathematician Seth Ward betrayed a similar interest in the prospect of uncovering "simple notions" which could constitute the radicals in a philosophical, and hence universal, language that reflected "definitions of things" in the world. See Ward, *Vindicae Academiarum*, 21; Lewis, *Language, Mind, and Nature*, 76–78.

Chapter Four

1. [Plattes], *Description of Macaria*, 1–2; Webster, "Authorship and Significance of Macaria."

2. Bacon, *Novum Organum*, Book I, aphorism 98, pp. 155–57.

3. Hartlib, *A Brief Discourse*, 45; Purver, *Royal Society*, chap. 4; Webster, *Great Instauration*, 67–77.

4. Hartlib, *A Further Discoverie*, 4. See also HP63/7/8A–9B; HP63/7/10A–12B; and Cagnolati, *Il Circolo di Hartlib*, 83–92.

5. See Solomon, *Public Welfare, Science and Propaganda*.

6. Hartlib, *A Further Discoverie*, 29; another copy at HP14/2/3/1A–19B.

7. Ibid., 26. Tight control of information was a feature of some contemporary utopias; see William Poole, *Francis Lodwick: A Country Not Named*, 85–87, 104 for transcription of relevant passages of BL, MS Sloane 913, fols. 1–33.

8. In Shadwell's *The Virtuoso*, the character of "Sir Samuel Hearty" bears some resemblance to Hartlib (who was never knighted), continually swapping recipes and gadgets. (Shadwell, *The Virtuoso*, III.iv.146–57).

9. Hartlib, "The Epistle Dedicatorie," in [Weston], *A Discours of Husbandrie*, sig.A3,2^r.

10. This was awarded April 26, 1649, although securing its actual payment in the 1650s proved time-consuming and frustrating. See Greengrass, "Hartlib," *ODNB*; also Turnbull, *Hartlib, Dury and Comenius*, 29.

11. "To the right honourable the Commons of England assembled in Parliament, the humble petition of Samuel Hartlib senior" (BL, Add. MS 6269, fols. 30^v–31^r is a scribal copy). For a related account of his life and activities, see Hartlib to Worthington, August 3, 1660, in HP26/1/1A–2B; also in Turnbull, *Hartlib, Dury and Comenius*, 110–11.

12. Trevor-Roper, "*Three Foreigners*," 240 and 258; also 289–90.

13. Purver, *Royal Society*, 196–234. See Young, *Comenius in England*, 59–63, on his dedication of *Via Lucis* (1668) to the Royal Society. Trevor-Roper admitted that Hartlib and Comenius thought the Royal Society was at least a partial realization of their plans for the organization of knowledge, including natural knowledge.

14. Hall, "Samuel Hartlib," 141.

15. See Webster, *Great Instauration*; Greengrass et al., *Samuel Hartlib*. These plans

encompassed the Atlantic colonies; see Irving, *Natural Science and the Origins of the British Empire*, 47–68.

16. Beale, *Herefordshire Orchards*, facing sig.A2ʳ.

17. Greengrass et al., *Samuel Hartlib*, 9–12.

18. Robert Sanderson, *XXXVI Sermons* (1689), 205, cited in Feingold, "Science as a Calling," 81.

19. HP47/10/17B and 24A, undated; Clucas, "Hartlib's *Ephemerides*," 34.

20. Dury, *Reformed Librarie-Keeper*, 10–11, 31. See Greengrass, "Samuel Hartlib and the Commonwealth of Learning," 318.

21. HP30/4/60B–61A, Ephemerides 1640, parts 3 and 4.

22. Although English historians such as Trevor-Roper called these men Puritans, the intellectual disposition (including the millennial aspect) of Hartlib and Comenius was formed on the Continent, especially by Johann Heinrich Alsted (1588–1638) at Herborn. See Hotson, *Johann Heinrich Alsted*. Bacon, *Novum Organum*, Book II, aphorism 93, p. 151 also cited the passage from Daniel.

23. Harrison, *Fall of Man*, 188–89, 196–97.

24. [Hartlib], *Reformed Common-Wealth of Bees*, 44.

25. Beale to Hartlib, November 4, 1661, HP67/22/12A; also HP71/6/3A, dated 1661.

26. Beale to Boyle, June 23, 1682, *Boyle Correspondence*, 5:301.

27. Pell, *An Idea of Mathematicks*, in Dury, *Reformed Librarie-Keeper*, 41.

28. HP29/2/49A, Ephemerides 1634, part 5; HP30/4/4B, Ephemerides 1639, part 1. See also chap. 1 for his skepticism about artificial memory techniques.

29. Greengrass, "Archive Refractions," 35–47.

30. Worthington, *Diary*, 1:55, for Hartlib's first letter to Worthington, who was appointed the Master of Jesus College, Cambridge, in 1650.

31. Hartlib to Worthington, February 6, 1662, Worthington, *Diary*, 2:106–7; also November 2, 1661, 68, for a reference to the loss of "all my best papers" in the custody of another person.

32. Worthington to Dr. Ingelo, June 10, 1667, ibid., 2:230.

33. Turnbull, *Samuel Hartlib*, 72; Turnbull, *Hartlib, Dury and Comenius*, v; Greengrass et al., *Samuel Hartlib*, 4–9.

34. See http://www.shef.ac.uk/library/special/hartlib, on the original deposit of about 25,000 folios. Within this archive there are 4,250 letters; see "Early Modern Letters Online" (http://emlo.bodleian.ox.ac.uk).

35. Dircks, *Memoir of Hartlib*, 9. See Turnbull, *Hartlib, Dury and Comenius*, 88–109, for a list of sixty-five items that includes those of other authors published by, or dedicated to, Hartlib.

36. Dircks, *Memoir of Hartlib*, 15.

37. Greengrass, "An 'Intelligencer's Workshop,'" 51–52.

38. HP29/2/1A, Ephemerides 1634, part 1.

39. Greengrass, "Hartlib and Scribal Communication," 56.

40. HP30/4/2B, Ephemerides 1639, part 1.

41. There are at least eight Heads in HP29/2/2B, Ephemerides 1635, part 1.

42. HP28/2/72A, Ephemerides 1653; HP29/5/9B and HP29/5/17A in 1655.

43. Greengrass et al., *Samuel Hartlib*, 2.

44. See John Beadle's work of 1656 mentioned in chap. 1.

45. HP35/5/154B, a scribal copy of Comenius' "Linguarum methodus novissima," undated. See Greengrass, "Archive Refractions," 44; and Greengrass, "An 'Intelligencer's Workshop,'" 60–61, for a fragment of Comenius' diary that sketches a philosophical program.

46. HP29/2/10B, Ephemerides 1634, part 1: "Ephemerides debent continere Experimentales Observationes. 2. Incidentes Notiones"; HP31/22/10A, Ephemerides 1648, part 1.

47. Greengrass, "An 'Intelligencer's Workshop,'" 51.

48. HP29/2/3A, Ephemerides 1634, part 1.

49. See, for example, HP28/1/46A, Ephemerides 1650, part 1.

50. HP31/22/2B, Ephemerides 1648, part 1 under "Oeconomica. Table-books"; HP28/2/53B, Ephemerides 1653, part 1.

51. HP29/2/28A, Ephemerides 1634, part 3, under "Didactica Scribendi." See also HP30/4/91A, Ephemerides 1643; and HP71/1/5A on recipes for glossy paper to make ink more indelible.

52. HP29/2/29A, Ephemerides 1634, part 3; HP28/2/3B–4B, Ephemerides 1651, part 1.

53. Petty, Advice, sig.A2^r. For a later edition, see Petty, Advice, in Malham, Harleian Miscellany, 6:1–13.

54. Love, Culture and Commerce of Texts; Greengrass, "Hartlib and Scribal Communication."

55. Love, Culture and Commerce of Texts, 183–84. For this attitude, see the contract signed on March 3/13, 1642 by Hartlib, Comenius, and Dury, in Turnbull, Hartlib, Dury and Comenius, 458–60.

56. Swanson, "Undiscovered Public Knowledge," 116; Davies, "The Creation of New Knowledge." Love's own work brought together "two fields which to date have only been studied as isolated phenomena" (Love, Culture and Commerce of Texts, 9). But he did not apply this notion to Hartlib's archive.

57. HP30/4/47B, Ephemerides 1640, part 2, under "Invention." See also HP/29/3/56A; HP29/3/58A; HP29/3/62A. Hartlib's concern is confirmed by the loss of the Iznik tile-making process in the Ottoman empire during the 1600s.

58. HP29/2/30A, Ephemerides 1634, part 3; HP/30/4/5A, Ephemerides 1639, part 1, under "MS Verulamii"; cited in Clucas, "Hartlib's Ephemerides," 50–51.

59. HP29/2/53B, Ephemerides 1634, part 5; HP30/4/64B, Ephemerides 1640, part 4.

60. HP29/2/23A, Ephemerides 1634, part 3. See also a copy of Benjamin Worsley's letter to an unknown person requesting him to send "whatsoever is remarkeable in your Adversaria to mee As I shall willingly accept of any favour of that nature." HP42/1/27A, undated.

61. HP29/2/24B, Ephemerides 1634, part 3: "Sam. Ward, of Ipswich his Brother Nath. Ward having left his Commonplaces with him out of which they may bee had."

62. HP29/2/44B–45A, Ephemerides 1634, part 5. See Hartlib's copy of "Mr Richersons Notes on the Logick Tables," HP24/7/1A–15A; and Clucas, "In Search of The True Logick," 57.

63. HP31/22/8B, Ephemerides, 1648, part 1; under "Agency of Learning; Adversaria."

64. Aubrey, Brief Lives, 1:268–69, 303. For Aubrey's collecting of manuscripts, see Yale, "With Slips and Scraps."

65. HP29/6/23A, Ephemerides, 1657, part 2; HP29/6/20B for the raisins. There are at least eighteen separate entries in 1657 about "the stone." On April 7, 1660, Hartlib told

Evelyn that he was suffering from "the Stone and Ulcer in the bladder." BL, Add. MS 15948, fol. 98^r.

66. Hartlib to Worthington, November 2, 1661, Worthington, *Diary*, 1:68. Hartlib died on Monday, March 10, 1662, in Axe Yard, Westminster and was buried at St Martin-in-the-Fields.

67. Turnbull, *Samuel Hartlib*, 7–71.

68. HP29/3/56A, Ephemerides 1635, part 5; HP29/3/63A, Ephemerides 1635, part 5; HP31/22/2A–2B, Ephemerides 1648, part 1. See Roos, *Salt of the Earth*, 39–40, on Hartlib taking notes from Johann Glauber and discussing these with Robert Child.

69. HP28/1/47A, Ephemerides 1650, part 1, under "Pensa dispensatorium Oeconomicum."

70. Worthington to Hartlib, October 26, 1661, Worthington, *Diary*, 2:65.

71. Clucas, "The True Logick," 68–73; Greengrass, "Samuel Hartlib and the Commonwealth of Learning," 310–12.

72. Hotson, *Johann Heinrich Alsted*, 33–38, 85–90, and his *Commonplace Learning*, chaps. 5–6. For this intellectual tradition, see Schmidt-Biggemann, *Topica Universalis*.

73. HP30/4/60B, Ephemerides 1640, part 3. With Comenius and Dury, he sought to condense theological doctrines as a way of establishing a consensus among Protestant sects.

74. HP30/4/77B, Ephemerides 1641.

75. HP28/1/4A, Ephemerides 1649, part 1.

76. HP29/2/6A, Ephemerides 1634, part 1; and HP31/22/32B, Ephemerides 1648, part 2. John (the elder) Tradescant established the "museum," which was continued by his son. See Tradescant, *Musaeum Tradescantianum*.

77. See HP30/4/63B, Ephemerides 1640, part 4; also HP28/2/56A, Ephemerides 1653, part 2.

78. HP31/22/21A, Ephemerides 1648, part 3. The reference is to *Purchas, his Pilgrimage* (1631), a collection of travel accounts by the English cleric Samuel Purchas (1575?–1626).

79. See HP29/2/2A, Ephemerides 1634, part 1. For an instance in which he could not find what he wanted, see Hartlib to Worthington, September 24, 1661, Worthington, *Diary*, vol. 2, part 1, 40: "but not finding it in my notes, I shall repeat it again."

80. HP29/3/13B, Ephemerides 1635, part 2, under "Didactica studii Theologici Eruditition [*sic*]." See chap. 2 for the method described in Holdsworth's "Directions." There is a reference to Holdsworth in Hartlib's account book of 1640 (HP23/12/1B), and other entries on his views (HP29/2/5A; 29/3/28B and 39A).

81. HP29/3/13B, Ephemerides 1635, part 2

82. HP30/4/28B, Ephemerides 1639, part 4, under "Drexeli Aurifodini." Hartlib's source was the Strasbourg professor Matthias Bernegger (1582–1640), who claimed that Drexel did not candidly explain Lipsius' method.

83. HP30/4/73A, Ephemerides 1641. Also HP29/3/29A, Ephemerides 1635, part 3.

84. HP28/1/4A–4B, Ephemerides 1649, part 1: "Mr. Dury to Mr. Stresso."

85. The original French publication is T. Renaudot, *Recueil général des questions traictées es Conférences du Bureau d'Adresse*, 5 vols (Paris: G. Loyson, 1655–56). This was issued in English as *A General Collection of Discourses of the Virtuosi of France, render'd into English by G. Havers, Gent.* (London: T. Dring and J. Starkey, 1664).

86. HP30/4/43B, Ephemerides 1640, part 1.

87. HP30/4/73A, Ephemerides 1641; under "Hubner; Conferences."

88. HP47/10/42B, undated (edited by Hartlib). See also Rayward, "Some Schemes," 164–66.

89. Hartlib's son-in-law, the chemist Frederick Clodius, was one source of information; see Principe, *Aspiring Adept*, 166.

90. HP29/2/10B, Ephemerides 1634, part 1; HP30/4/12B, Ephemerides 1639, part 2, under "Eruditio." See also under "Plattes."

91. HP30/4/76A Ephemerides 1641. See Clucas, "Correspondence of a Seventeenth-Century 'Chymicall Gentleman.'"

92. Cheney Culpeper to Hartlib, November 20, 1644, HP13/55A, printed in Braddick and Greengrass, *Culpeper Letters*, 203–4. The quotation preserves Culpeper's spelling.

93. Plattes, "A Caveat for Alchymists," 54, 85. Plattes' dedication is dated 1643.

94. Seth Ward to Sir Justinian Isham, February 27, 1651/2, printed in Robinson, "An unpublished letter of Dr Seth Ward," 69.

95. HP28/2/5B, Ephemerides 1651, part 1.

96. William Rand to Hartlib, February 14, 1652, HP62/17/1B.

97. HP30/4/36B, Ephemerides 1639, part 4.

98. HP30/4/46A, Ephemerides 1640, part 2.

99. HP30/4/46A–B, HP30/4/47A and HP30/4/46A, Ephemerides 1640, part 2; see Harrison's stress on "private and public use [usus iam privatus aut publicus]"). BL, Add. MS 41846, fol. 203^r.

100. Malcolm, "Thomas Harrison," 196. Young, *Comenius in England*, 66 n. 2, incorrectly suggested that the author was John Harrison who was "in the service of Queen Elizabeth of Bavaria." Harrison's proposal is described in the BL catalogue as "design for an index cabinet with instructions for use, by a friend of Samuel Hartlib." See BL, Add. MS 41846, fols. 194^r-204^r. The Latin title is "Arca Studiorum sive Repositorium" (fol. 194^v).

101. Index cards, as we know them, did not appear until the late 1700s and not in a sustained way until the early twentieth century. See Zedelmaier, "Buch, Exzerpt," 38–53; and Krajewski, *Paper Machines*, 34–45.

102. Hübner to Johannes Gronovius, March 19/29, 1640, BL, MS Sloane 639, fol. 85^r, cited in Malcolm, "Thomas Harrison," 198–99. On Haak, see Stimson, "Hartlib, Haak, and Oldenburg."

103. Placcius, *De arte excerpendi*, 124–59.

104. For such a project, directed by Matthias Flacius Illyricus, professor of Hebrew at Wittenberg from 1544, see Grafton, "Where Was Salomon's House?"

105. Placcius, *De arte excerpendi*, 72; also 138, 140, 152, 153, 155, for illustrations. See Meinel, "Enzyklopädie der Welt"; Blair, *Too Much to Know*, 94, 99. Jungius (1587–1657) was a professor of mathematics, medicine, and natural philosophy at several German universities; his final post was at Hamburg.

106. HP30/4/46A and HP30/4/47A, Ephemerides 1640, part 2.

107. HP30/4/47A, Ephemerides 1640, part 2.

108. HP30/4/70B, Ephemerides 1641, under "Ars excerpendi." On this technique used by early modern compilers, see Blair, *Too Much to Know*, 213–25.

109. HP30/4/55A, Ephemerides 1640, part 3. See Malcolm, "Thomas Harrison," 230 n. 103.

110. HP29/5/40B, Ephemerides 1655, part 3; HP30/4/46A, Ephemerides 1640, part 2.

Hartlib presumably meant the bishop of Exeter, John Gauden (1599?–1662?), although he usually writes "Gawden," as in this undated list of recipients of one of Comenius' books: HP23/13/1A–2B.

111. Movable notes remained attractive to individual scholars. See the suggestion of the Oxford book collector Richard Rawlinson (1681–1725) for a box of slips or a "rota literaria." Rawlinson inserted a handwritten leaf between pp. 162 and 163 in vol. 2 of Morhof's *Polyhistor*; see Bodleian Library Press Mark 8 Rawl. 599, cited in Malcolm, "Thomas Harrison," 231 n. 122.

112. HP30/4/46A and HP30/4/55A, Ephemerides 1640, parts 2 and 3.

113. The fashioning of these is described in BL, Add. MS 41846, fols. 200^r–201^v; see also Malcolm, "Thomas Harrison," 206–7.

114. HP29/5/40B, Ephemerides 1655, part 3; HP30/4/46A–B, Ephemerides 1640, part 2.

115. BL, Add. MS 41846, fols. 201^v and 200^v respectively; fol. 200^v for "3000 ad minimum"; and "vacuis 300."

116. These comments occur in a letter from London of October 8/18, 1641, to his friends in Leszno, Poland. See Young, *Comenius in England*, 66. Comenius was told that Harrison was "out of London" and they may not have met. For other documents concerning Comenius' visit, see pp. 52–63.

117. Ibid., 67. Comenius refers to "a learned man, N. Harisson" (66).

118. See Čapková, "Comenius and His Ideals."

119. Comenius, *Patterne of Universall Knowledge*, 4 and 8. He stayed in London until June 1642. See also Greengrass, "Jan Comenius," *ODNB*.

120. HP30/4/47B, Ephemerides 1640.

121. Malcolm, "Thomas Harrison," 208–9.

122. HP30/4/46A, Ephemerides 1640, part 2.

123. See the Heads in HP30/4/46B–47A, Ephemerides 1640, part 2.

124. HP30/4/46B, Ephemerides 1640, under "Harrison": "Hase a several way in his Indexes of verborum, 2. sententiarum, et Rerum." Of course, complex indexes were not unknown: Hartlib and Hübner discussed the massive indexation of classical texts by Matthias Bernegger and Johann Freinsheim in Strasbourg. See Malcolm, "Thomas Harrison," 215–16, citing Hübner's letter of 1640.

125. HP30/4/46B; under "Ars Excerpendi."

126. HP30/4/48B, Ephemerides 1640, part 2; HP30/4/47A, Ephemerides 1640, part 2; HP30/4/62A, Ephemerides 1640, part 4.

127. For the numerals assigned to each slip of paper, see BL, Add. MS 41846, fol. 198^r cited in Malcolm, "Thomas Harrison," 208. See also BL, MS Sloane 1466, fol. 165^r, for Harrison's intention to number the quotations.

128. Young, *Comenius in England*, 66.

129. HP30/4/48B, Ephemerides 1640, part 2; HP30/4/55A, Ephemerides 1640, part 3.

130. HP47/20/4A, [Hartlib], notes on London University Public Library, undated; also HP47/20/1A–8B. Harrison's influence might have come to fruition if Hartlib had been appointed at the Bodleian. See Gerard Langbaine (Provost of Queen's) to John Selden, March 16, 1651/52, Bodl. MS Smith 21, fol. 27; Webster, ed., *Samuel Hartlib*, 57.

131. HP30/4/47A, Ephemerides 1640, part 2.

132. BL, MS Sloane 1466, fol. 165^r, in a manuscript headed "1648.12.dec. Harrisoniana."

133. For the application of this approach to information of various kinds, and its

function in Hartlib's "Office of Publike Addresse," see Hartlib, *A Brief Discourse*, 34–46. For collective commonplacing, see Yeo, "Between Memory and Paperbooks," 11–17.

134. Comenius, *Patterne of Universall Knowledge*, 4.

135. "Harrisoniana," November 8, 1647; BL, MS Sloane 1466, fol. 84^v. This is in Hartlib's hand; other copies show his corrections of Harrison's drafts. See Malcolm, "Thomas Harrison," 203.

136. BL, MS Sloane 1466, fol. 164^r.

137. BL, Add. MS 4384, fol. 64^r.

138. Malcolm, "Thomas Harrison," 204.

139. "Harrisoniana 1646. May.8." The catalogue description is "A Protest of Thomas Harrison against the utter loss of his valuable work for the promotion of learning." BL, MS Sloane 1466, fol. 82^v. The partner term, "inventio," referred to the uses to which excerpts were put. See Malcolm, "Thomas Harrison," 203, 206, for Harrison's assertion that his method had not been used before; and BL, Add. MS 41846, fols. 195^r and 197^v.

140. HP30/4/48B and HP30/4/47, Ephemerides 1640, part 2; HP30/4/55A Ephemerides 1640, part 3.

141. BL, MS Sloane 1466, fol. 84^r. For Hartlib's recognition of this application, see HP28/1/20 Ephemerides, 1649, part 2.

142. Culpeper to Hartlib, February 17, 1645/46, letter no. 102 in Braddick and Greengrass, *Culpeper Letters*, 265.

143. HP30/4/21A, Ephemerides 1639, part 3.

144. See HP30/4/22A, Ephemerides 1639, part 3, on Acontius; HP30/4/24B, Ephemerides 1639, part 3, criticizing Comenius; and Hartlib to Boyle, May 8 or 9, 1654, praising Jungius, *Boyle Correspondence*, 1:172–73. See also Clucas, "In Search of *The True Logick*," 58–63.

145. Hartlib to Dury, September 13, 1630, HP7/12/1B–3A. See also Webster, ed., *Samuel Hartlib*, 76–77.

146. HP29/3/13A, Ephemerides 1635, part 2.

147. For Bacon on aphorisms, see Clucas, "A Knowledge Broken."

148. HP30/4/42A, Ephemerides 1640, part 1.

149. HP30/4/53B, Ephemerides 1640, part 3. Hartlib's stance matches the spirit of Bacon's plea for "a greater abundance of experiments" and "of new particulars" as a foundation for further inquiry. Bacon, *Novum Organum*, Book II, aphorisms 100 and 103, pp. 159–61.

150. See HP30/4/3A, Ephemerides 1639, part 1: "Cartes, Bisterfeld Comenius begin their philosophating a priori. . . . Jungius goes more warily and does a posteriori not caring so much to teach as first to find out the truth that may not bee gain-sayed." Jungius also used slips of paper for carefully defined bits of information. See Cevolini, *De arte excerpendi*, 82–83, and Clucas, "*Scientia* and *Inductio Scientifica*."

151. HP30/4/42A–B, Ephemerides 1640, part 1.

152. HP30/4/39A, Ephemerides 1640, part 1. See also HP30/4/25B, Ephemerides 1639, part 3: "Two great Faults have beene committed in our philosophy the one is that it hase not beene made truly universal. 2. that it hase not begun a particularibus."

153. HP30/4/43A, Ephemerides 1640, part 1; HP30/4/64B, Ephemerides 1640, part 4; also 30/4/13A, 1639, part 2.

154. HP30/4/49B, Ephemerides 1640, part 2; this echoes Bacon, *Novum Organum*,

Book II, aphorism 19, p. 71. Not long before this entry Hartlib observed under "Historia" that "The general Notions are but few and drye. But those which occurre in Historys though not altogether so authentical yet carry far more profit along with them" (HP30/4/40B, Ephemerides 1640, part 1).

155. HP30/4/43B, Ephemerides 1640, part 1.

156. HP30/4/47B and 48B, Ephemerides 1640, part 2.

157. HP 30/4/70B, Ephemerides 1641.

158. Haak told Aubrey that Pell "communicated to his friends his excellent *Idea Matheseos* in half a sheet of paper." See Bodl. MS Aubrey 6, fol. 53, included in Andrew Clark's edition of *Brief Lives*, 2:130.

159. See P. J. Wallis, "Early Mathematical Manifesto—John Pell's *Idea of Mathematics*," 140–41. I use Wallis' transcription (at 141–45) of the English broadsheet of Pell's *Idea*, a copy of which is bound with the British Library's copy of Pell's *Tabula Numerorum Quadratorum* (London: Thomas Ratcliffe, 1672). This insertion has the shelfmark 528.n.20 (5'). The *Idea* is printed on both sides of a sheet measuring approximately 33×17.5 cm. There was also a Latin version, now lost, entitled as both "Ideae Mathematicae" and "Idea Matheseos."

160. For the distribution of Pell's *Idea* via intermediaries, see Malcolm, "Life of John Pell," 65–76; also 69 for HP14/1/6A as a "mock-up title page," dated 1634.

161. [Pell], "An Idea of Mathematicks, long since written by Dr. John Pell," 127–34; for letters from Mersenne, Pell and Descartes, 135–45.

162. Pell, *Idea of Mathematicks*, in Dury, *Reformed Librarie-Keeper*, 33–46. Some of Hartlib's main correspondents apparently did not see Pell's *Idea* until later. Aubrey sent Boyle his transcription of a copy, apologizing for not sending it sooner. Aubrey to Boyle, March 15, 1666, *Boyle Correspondence*, 3:111. By this date Boyle would have been able to read the printed version in Dury's book. Hartlib's letters to Boyle include reports about Pell, but without mention of his *Idea*.

163. Bacon, *Advancement of Learning*, 125; in *Works of Francis Bacon*, ed. Spedding, Ellis, and Heath, 3:406.

164. Pell, *Idea of Mathematicks*, in Wallis, "Early Mathematical Manifesto," 142–44. A similar claim that this might be the task of "one man" occurs in Pell's letter to Thomas Goad, August 7, 1638, cited in Malcolm, "Life of John Pell," 65–66.

165. Pell, *Idea of Mathematicks*, in Wallis, "Early Mathematical Manifesto," 144. The notion of "Pocket-learning" had a negative connotation in John Selden's *Historie of Tithes*. See Feingold, "English Ramism," 163.

166. Pell, *Idea of Mathematicks*, in Wallis, "Early Mathematical Manifesto," 144, 145.

167. BL, Add. MS 4415, fol. 27^r, cited in Stedall, "Mathematics of John Pell," 263. He also referred to the basic units as "Common Notions" which "come of necessity from our Nature or Maker" (fol. 27^r).

168. BL, Add. MS 4420, fol. 23, cited in Stedall, "Mathematics of John Pell," 265. See also Lewis, *Language, Mind, and Nature*, 24–29, for the relevance to a philosophical language.

169. [Pell], *Tabula numerorum*. The English subtitle is "A Table of ten thousand Square Numbers."

170. HP29/3/36A, Ephemerides 1635, part 3. See Clucas, "In Search of *The True Logick*," 65, 72.

171. HP29/5/36B, Ephemerides 1655, part 3.

172. HP31/22/3B, Ephemerides 1648, part 1: "Petty is in writing a Memoria Musicalis to remember exactly and readily the Tablatures or the Notes without the helpe of a Booke"; also HP31/22/4A on "the Musical Memorie."

173. However, according to Aubrey, Pell believed that the mathematics had to be understood before the tables could function in the ways he described. See *Aubrey on Education*, 101.

174. Mersenne to Haak, November 1, 1639, printed (with Pell's reply) in Wallis, "Mathematical Manifesto," 145–47; quotation from Mersenne at 145–46. Always the "intelligence" broker, Hartlib noted that Mersenne was "Also a great Promoter of truly Mathematical wits with whom Mr Pell also should keepe Correspondency." HP30/4/5A, Ephemerides 1639, part 1.

175. See the original Latin version in Hooke, *Philosophical Collections*, 144–45; also a partial translation of this letter in Wallis, "Mathematical Manifesto," 147. Malcolm, "Life of John Pell," 72, explains that Hogelande forwarded this letter to Haak.

176. Descartes to Cornelis van Hogelande, February 8, 1640, in Descartes, *Philosophical Writings*, 3:144.

177. Descartes' criticism was probably not widely known until Hooke published it in 1682. John Collins told John Beale that the *Idea* contained some "improbable presumptions" which were "severely censured by Des Cartes, a man exceedingly deceived, as might be instanced, with the like conceits of himself. Mr. Haak, of the Royal Society, hath the censure, which, being epistles of Des Cartes to his friends, I wish they had passed the press amongst the rest; but Mr. Haak, being an admirer of Dr. Pell, will not impart them." Collins to Beale, August 20, 1672, in Rigaud, *Correspondence of Scientific Men*, 1:197. Collins helped Pell in various ways; see Malcolm, "Life of John Pell," 203, 213–23.

178. HP30/4/45B, Ephemerides 1640, part 2; HP30/4/53B, Ephemerides 1640, part 3.

179. Hartlib received this in 1648 as a printed pamphlet. See HP15/1/6A–B for a manuscript copy of the title page in Hartlib's hand.

180. Petty, *Advice*, 2, 3, 19, 20–21.

181. Such techniques fell under what Bacon called "literate" experience. See Bacon, *Novum Organum*, Book II, aphorism 103, p. 161; and my chaps. 3 and 6.

182. Petty, *Advice*, 9. The *OED* gives "nosocomium" as hospital. Petty had recently returned from medical study in France and the Netherlands. See Webster, "English Medical Reformers of the Puritan Revolution."

183. Petty, *Advice*, 12–13, 16, 21. He later filled out some of these desiderata in plans for medical education. See *Petty Papers*, 2:167–70.

184. Petty, *Advice*, 13, 19–24.

185. HP31/22/9A, Ephemerides 1648, part 1.

186. Ibid., 8, 10; and 7 for Pell.

187. Ibid., 8, 21.

Chapter Five

1. Thomas Povey to Robert Boyle, May 8, 1661, *Boyle Correspondence*, 1:454. Povey was a government administrator, connoisseur, and a Fellow of the Royal Society.

2. Petty to Boyle, April 15, 1653, *Boyle Correspondence*, 1:142. Among his favorite maxims, Petty included "Read to exercise your judgment not your memory." BL, MS Sloane

2903, fol. 8^r. For the link between melancholy and excessive study, see Shapin, *Social History of Truth*, 153–56.

3. Petty to Boyle, April 15, 1653, *Boyle Correspondence*, 1:142–43. John Harwood cites these remarks as praise of what he calls Boyle's "*method*" of thinking; but Petty thought Boyle's procedure was faulty (Harwood, "Introduction," in Boyle, *Early Essays*, xli).

4. This is point 15 in biographical notes dictated by Boyle and copied by Burnet. See "Burnet Memorandum," in Boyle, *Boyle by Himself*, 27, from a transcription of BL, Add. MS 4229, fols. 60–63; see also Hunter, *Between God and Science*, 6–7, 48.

5. Beale copied out this point from Pell: "That men may consider what meanes may be used to fortify the imagination, to prompt the Memory, or regulate our Reason, and what effects may be produced by the uniting of these meanes, and the constant exercise of them." BL, Add. MS 4384, fol. 64^r, titled "June xi.1663 About 4 p.m. Covent-Garden." See also Lewis, "*The Best Mnemonicall Expedient*," 125–26.

6. Maddison lists twenty-seven letters from Beale to Boyle between these dates (Maddison, "A Tentative Index to Boyle's Correspondence"); in *Boyle Correspondence* there are thirty letters, two of which are no longer extant. In assisting Birch with his edition of Boyle's works, Henry Miles noticed that Beale's letters were "more in number, Mr. Oldenburg's excepted, than any one correspondent" (BL, Add MS 4229, fol. 88^r).

7. Boyle, *Certain Physiological Essays*, 13–14. Less surprisingly, the other authors he mentioned were Descartes and Gassendi. See Anstey, *Philosophy of Boyle*, 4–7, for the caveat that Boyle was not opposed to systematization as such.

8. Boyle, *History of the Air*, 150, for Title XIX, "Of the Heat and Coldness of the Air"; also in *Boyle Works*, ed. Hunter and Davis, 12:100. The material for this work was gathered from the 1660s onwards (see chap. 7).

9. In this instance, Boyle was not referring to grand systems, but to a doctrine about the nature of air and some of its effects in climatic regions.

10. Boyle to Hartlib, [early 1647], *Boyle Correspondence*, 1:51; Boyle also replied on March 19, 1647 (1:52) to a letter (now lost) from Hartlib. For a notice of Boyle in Hartlib's diary of early 1648, see HP31/22/1A, Ephemerides 1648, part 1. For Boyle's early links with Hartlib's circle, including Beale, Petty, Benjamin Worsely, William Brereton, and John Worthington, see Maddison, *Life of Boyle*, 61–63, 68, 71, and 95; and Hunter, *Between God and Science*, 65–66. See also O'Brien "Samuel Hartlib's Influence"; and Shapin, *Social History of Truth*, 137, 144, 175.

11. Boyle, "An Invitation to free and generous Communication"; also in *Boyle Works*, ed. Hunter and Davis, 1:1–12. This piece was written c. 1647–48. See Rowbottom, "Earliest Published Writing of Boyle."

12. Boyle to John Mallett, March 2[3], 1652, *Boyle Correspondence*, 1:133. For the context of Hartlib's interests, see Bennett and Mandelbrote, *The Garden*, 33–42 and 157–68.

13. See Stubbs, "John Beale," 477–85 and 464, for an estimate of Beale's correspondence as about 400 letters. We now know that there are 152 letters from Beale in Hartlib's extant correspondence. See "Early Modern Letters Online" (http://emlo.bodleian .ox.ac.uk).

14. Beale to Hartlib, undated, HP25/6/1A and 4B; some parts of this letter appear (crossed through) in a fragment from Beale dated March 18, 1656, HP52/4A.

15. Hartlib to Boyle, April 27, 1658, *Boyle Correspondence*, 1:268. Maddison, *Life of Boyle*, 14, says Beale "was a near contemporary" of Boyle at Eton, "though a trifle

earlier." This is understated, since Beale enrolled at Eton in 1622 and Boyle did not do so until October 1635. Beale was twenty years older than Boyle.

16. Hartlib to Boyle, September 8, 1657, *Boyle Correspondence*, 1:233.

17. Hunter, Harwood, and Principe indicate that only about half of Boyle's writings concern natural philosophy or natural history. See Hunter, "How Boyle Became a Scientist," and Principe, "Newly Discovered Boyle Documents," 63–64.

18. For this period (1644–52), see Maddison, *Life of Boyle*, 57–88; Hunter, *Between God and Science*, 59–69, 292–93.

19. Hunter, *Boyle by Himself*, xvi.

20. See Principe, "Virtuous Romance," 379; and his "Style and Thought of Boyle."

21. For Boyle's sense of his "Ethics" (that is, the "Aretology" begun in 1645) as part of his studies, see Boyle to Isaac Marcombes, October 22, 1646, *Boyle Correspondence*, 1:37. Marcombes was Boyle's tutor during his time in Geneva between 1639 and mid-1644.

22. See "Another Advertisement" in *Seraphic Love* (1659), *Boyle Works*, ed. Hunter and Davis, 1:60, for the explanation by a friend that Boyle had not dropped his scientific work, but would continue to publish "those Experimentall Essay's and other Physiologicall Writings, which he is known to have, lying by him." For the shift from moral topics to science, see Hunter, "How Boyle Became a Scientist," 24–34.

23. Boyle, "A Discourse touching Occasional Meditations," in *Occasional Reflections*, 22.

24. Boyle, "An account of Philaretus," in Maddison, *Life of Boyle*, 15; also in Boyle, *Works*, 1:xii–xxvi; and in Hunter, *Boyle by Himself*, 1–22. Boyle's amanuensis, Robin Bacon, said he had a good memory; see BL, Add. MS 4229, fol. 66, cited in Boyle, *Early Essays*, 194 n. 21. But compare Boyle to Hartlib, April 8, 1647, in *Boyle Correspondence*, 1:56: "the treacherousness of my memory."

25. Maddison, *Life of Boyle*, 11, citing a letter of [December] 1635.

26. Boyle, "Biographical notes dictated by Boyle . . . to Robin Bacon," in Hunter, *Boyle by Himself*, 24; transcription of BL, Add. MS 4229, fol. 66^r.

27. See BP 8, fol. 208^r [probably late seventeenth century]. Of his sister, Katherine, it was said that "She hath a memory that will hear a sermon and goe home and penn itt after dinner verbatim." See Verney and Verney, *Memoirs of the Verney Family*, 1:203–4.

28. Boyle, "The Dayly Reflection," in *Early Essays*, 233, transcribed from BP 7, fols. 269–87.

29. Boyle, "The Doctrine of Thinking," in *Early Essays*, 194. The title "Doctrine of Thinking" is supplied by the editor, Harwood, for RS MS 197, fols. 4–43.

30. Ibid., 200–201.

31. Boyle, "Dayly Reflection," 201 n. 1 and 222.

32. See "Sir Peter Pett's notes on Boyle," in Hunter, *Boyle by Himself*, 66, for a transcription of BL, Add. MS 4229, fol. 37^{r-v}. For Boyle's concerns about conscience, see Hunter, *Robert Boyle: Scrupulosity and Science*, chap. 4

33. For these concerns, see Corneanu, *Regimens of the Mind*, esp. 118–19. More could be said about the role of note-taking as a mental discipline.

34. Boyle, "Dayly Reflection," 207, 208.

35. Boyle, *Occasional Reflections*, 32, 33.

36. Boyle to Hartlib, May 8, 1647, *Boyle Correspondence*, 1:59.

37. Seneca, "On the Shortness of Life," 59, 76. See also Montaigne, "On Experience," in *Essays*, III.13, p. 1263, who professes to enjoy all things "twice as much as others."

38. Boyle, *Occasional Reflections*, p. 33; and "The Aretology," in Boyle, *Early Essays*, 9.

39. HP30/4/22A, Ephemerides 1639, part 3. See HP 24/7/14A for Hartlib's copy of a similar view about the appropriate time to train memory versus judgment.

40. HP30/4/22A, Ephemerides 1639, part 3 and HP28/1/53B, Ephemerides 1650, part 2.

41. Boyle, "A Proemial Essay," in *Certain Physiological Essays*, 10.

42. Boyle, "Doctrine of Thinking," in Boyle, *Early Essays*, 198–99, for reference to a "Modell" or plan of meditation.

43. Boyle, *Occasional Reflections*, 37 and 19.

44. Ibid., 26.

45. Ibid., 30.

46. Boyle, "Of the Study of the Book of Nature, For the first Section of my Treatise of Occasional Reflections," 1650s, BP 8, fol. 137^r; also in *Boyle Works*, ed. Hunter and Davis, 13:168; and the editors' comments, 1:xxxi. See Boyle to Lady Katherine Ranelagh, August 31, 1649, *Boyle Correspondence*, 1:82–83, for mention of this work; and Principe, "Virtuous Romance," 393, on the continuity between Boyle's moral and scientific attitudes.

47. Boyle, *Occasional Reflections*, 19.

48. Compare Petrarch's account of a friend's amazing power of recollection, apparently achieved by "personalizing bits of information" (Carruthers, *Book of Memory*, 61). The intensity of Boyle's observational practices is relevant here; on this more generally, see Daston, "On Scientific Observation."

49. Annas, "Aristotle on Memory," 307–10, esp. 308. See Danziger, *Marking the Mind*, 172–73.

50. Lewalski, *Protestant Poetics*, 151–52 and 161–62; also Fisch, "The Scientist as Priest."

51. Petty, *Advice*, 22, 26.

52. Hartlib to Boyle, November 16, 1647, *Boyle Correspondence*, 1:63–64; Evelyn to Wotton, September 12, 1703, cited in Houghton, "The History of Trades," 46.

53. Petty to [Hartlib?], undated, copy in Scribal Hand B, HP7/123/2A. Hartlib's annotation on this letter is: "Mr Pettys Letter in Answer to Mr More." Presumably Petty's letter either responds to, or instigates, some of the comments in Henry More to Hartlib, March 12, 1649, HP18/1/2A–3B. See Turnbull, "Hartlib's Influence," 120.

54. Petty, *Advice*, 22. Wierzbicka, *Experience, Evidence, & Sense*, 34–37, cites Shakespeare—"His years but young, but his experience old"—as an illustration of the second notion, and claims that it was displaced during the 1600s by the new concept of experience as a countable instance, as in the case of an experiments. However, Boyle and others continued to stress the importance of a lifetime of experience in empirical observation.

55. Bacon, *Novum Organum*, Book I, aphorism 97, p. 155.

56. For Boyle's use of Heads, see Hunter, "Robert Boyle and the Early Royal Society."

57. Beale to Boyle, February 23, 1663, *Boyle Correspondence*, 2:62–68.

58. Beale to Boyle, September 28, 1663, ibid., 127, and September 29, 1663, ibid., 129–30.

59. Of course there were other topics, including religious ones. See Wojcik, *Boyle*, 22–23.

60. Beale to Boyle, September 29, 1663, *Boyle Correspondence*, 2:128–42.

61. This point is developed in "Notes upon Mr Hartlib's Accompt of Mr Morleys Art of Memory," BL, Add. MS 4384, fol. 65ʳ.

62. Beale had discussed the art of memory and the prospect of "an Universall character" with Hartlib in a letter of January 9, 1658, HP31/1/61B; he had circulated comments on Caleb Morley's memory treatise on December 23, 1656, HP31/1/7A–8B, and December 2, 1661, HP67/22/13A–14B.

63. Beale to Hartlib, October 4, 1661, copy in scribal hand, HP67/22/11B–12A; see another copy at HP71/6/1A–2B. See also copies of Beale's "Mnemonicall Probleme," BL, Add. MS 4384, fols. 64ʳ and 109ʳ⁻ᵛ (dated "June xi.1663"). These are among John Pell's papers.

64. Beale to Boyle, September 29, 1663, *Boyle Correspondence*, 2:141. A similar confidence underlay Beale's belief that an artificial language of the kind proposed by John Wilkins could be quickly learned, that "the reall Character may be easily taught in few dayes" (Beale to Boyle, June 23, 1682, ibid., 5:301). For a detailed account of Beale's views, see Lewis, "*The Best Mnemonicall Expedient*"; and more generally, his *Language, Mind, and Nature*.

65. Beale to Boyle, February 25, 1665, *Boyle Correspondence*, 2:69. See Beale to Evelyn, September 6, 1662, BL, Add. MS 78683 fol. 49ʳ⁻ᵛ, saying that he told Oldenburg and Boyle about "a Character which came into my head about a yeare agoe. It may seem to belong to that Society as a branch of Mathematic, or the first principle of it. I have only represented it to Mr Boyle & Mr Oldeb & I thinke I have not yet soe far explicated it as if they do thoroughly understand it. Tis impossible, they should apprehend all the uses of it for a Mnemonicall ayde in acquests of languages, words, or sentences or for the draught of a Mother language; naught can I specify these uses into practice with out more laysure than I am allowed in this."

66. Beale to [Hartlib?], December 2, 1661, copy in scribal hand, HP67/22/13B.

67. Beale to Boyle, September 29, 1663, *Boyle Correspondence*, 2:132, 134, and Beale to Boyle, October 2, 1663, 145.

68. Beale to Boyle, February 25, 1663, ibid., 69–70. See *Oldenburg Correspondence*, 1:320–21, for the editors' bad opinion of Beale: "He suffered from total recall and confident reliance upon an unreliable memory."

69. Beale to Boyle, July 30, 1666, *Boyle Correspondence*, 3:196, citing a case from the physician, Guilio Scaliger (1484–1558).

70. Beale to Oldenburg, June 1, 1667, *Oldenburg Correspondence*, 3:425, 427.

71. Beale to Boyle, September 29, 1663, ibid., 2:130.

72. Ibid., 131.

73. Ibid., 132, 133. The more usual point made about erasable tables was that important content should be transferred to a proper commonplace book.

74. Beale to Hartlib, November 4, 1659, BL, Add. MS 15948, fol. 82ʳ. Hartlib passed this on to Evelyn.

75. "A Copy of Mr Beales Letter" [no address or date]; HP67/22/17B. For another copy, see BL, Add. MS 4384, fol. 93ᵛ.

76. Boyle regularly sent Beale copies of his works, the last sent being *Aerial Noctiluca* (1681). Beale replied that "you have obligd me formerly in like manner with all your Volumes." Beale to Boyle, February 16, 1681, *Boyle Correspondence*, 5:239. See also the editors' "Introduction," 1:xii.

77. Beale to Boyle, February 25, 1663, ibid., 2:70. In a subsequent letter Beale added pointing-hand signs (manicules) in the margins; see August 10, 1666, ibid., 3:198–200. For the manicule, see Sherman, *Used Books*, 29–40. See also Carruthers, *Book of Memory*, 107–9, on "notae."

78. Beale to Boyle, July 13, 1666, *Boyle Correspondence*, 3:187.

79. Beale to Boyle, August 10, 1666, ibid., 200. The editors of Boyle's *Works* note that after the peak of his publications in 1666 (that is, after his correspondence with Beale) he produced books resembling "short essays" (*Boyle Works*, ed. Hunter and Davis, 1:xxxvii).

80. Beale to Boyle, April 18, 1666, *Boyle Correspondence*, 3:139–40. Boyle's workdiary (no. 21) entitled "Promiscuous Experiments, Observations, and Notes" (in BP 27, 5–159) dates from the late 1660s. Boyle made entries in these workdiaries (see chap. 6) in sets of 100, thus "centuries." Beale had earlier alluded to the "Sylva of promiscuous Experiments, Upon which you may discharge such of your papers & informations" (Beale to Boyle, September 28, 1663, *Boyle Correspondence*, 2:127).

81. Beale to Boyle, August 10, 1666, ibid., 3:205, 208; italics in original. Beale made similar points in a letter to an unidentified recipient of December 2, 1661, copy in scribal hand, HP67/22/13A–14B.

82. Beale to Boyle, July 13, 1666, *Boyle Correspondence*, 3:187. For Boyle's own remark on "how vast a Disparity there is betwixt experimentall & notionall Learning," see Boyle to Evelyn, May 23, 1657, ibid., 1:214.

83. Beale to Boyle, July 13, 1666, ibid., 3:188.

84. Beale to Boyle, August 10, 1666, ibid., 198. Beale was referring to the opening of Boyle's *Certain Physiological Essays* (1661), but chose to understate Boyle's resistance to systems.

85. Beale to Boyle, August 10, 1666, *Boyle Correspondence*, 3:200 (italics in original). See also editors' comments in *Boyle Works*, ed. Hunter and Davis, 1:lxxxiv.

86. See Beale to Oldenburg, c. June 16, 1671, in *Oldenburg Correspondence*, 8:112, for the suggestion that Theophilus Gale (1628–78) should "methodize and give the substance of Mr Boyle's works."

87. HP1/33/13A, Cyprian Kinner to Hartlib in Latin, June 27, 1647. I use the English translation by W. J. Hitchens, provided in the Hartlib Papers.

88. Ibid. See DeMott, "Science versus Mnemonics"; and Lewis, *Language, Mind, and Nature*, 55–56, for the connection with artificial languages.

89. Beale to Boyle, October 11, 1665, *Boyle Correspondence*, 2:554. For an earlier reference to Jungius in this connection, see Hartlib to Boyle, May 8 or 9, 1654, ibid., 1:172–77.

90. Beale to Boyle, April 28, 1666, ibid., 3:159, 192; see also April 18, 1666, 138, for "Cribo divino."

91. On this conviction in Bacon, see Zargorin, *Francis Bacon*, 104–5; Gaukroger, *Francis Bacon*, 138–48.

92. This assumption fits with the millenarian expectations, shared by some in Hartlib's circle, that a quick installation of recovered prelapsarian knowledge could be achieved. See chap. 4.

93. Hartlib copied this in a letter to John Worthington, August 26, 1661; printed in Worthington, *Diary*, 1: 369; transcript of original letter in BL, Add. MS 32498, fol. 75; a copy in BL, Add. MS 6271, fol. 13ʳ.

94. Boyle, *Excellency of Theology*, 204, 207–8, 219.

Chapter Six

1. Samuel Pepys to John Evelyn, January 9, 1692, *Private Correspondence of Pepys*, 1:51. Pepys added, "besides Mr Evelin," in recognition of Evelyn's comparable status.

2. Gassendi, *Mirrour of True Nobility*, 197 (in Book VI). See my chap. 2 for the dedication of this work to Evelyn. See also Miller, *Peiresc's Europe*.

3. Evelyn to W. Wotton, March 29, 1696, BL, Add. MS 4229, fol. 59ʳ, in Hunter, *Boyle by Himself*, 88; also, a modernized version in Maddison, *Life of Boyle*, 186–88.

4. Wotton to Evelyn, August 8, 1699, cited in Hunter, "Mapping the Mind," 123.

5. Ibid., 123–26; Hunter, Knight, and Littleton, "Robert Boyle's *Paralipomena*," 177–79.

6. Boyle, "Verse mnemonic based on "The Order of my Severall Treatises" (c. 1665), BP 27, pp. 2–3; also in *Boyle Works* 14:331–32. See 1:xxxv–xxxvi for editors' comment.

7. Hunter, "Mapping the Mind," 126–27; Knight, "Organising Natural Knowledge," 99–101. See RS MS 194, fols. 1ʳ, 25ᵛ, and MS 186, fols. 1ᵛ, 2ᵛ, 189ʳ; also Knight, "Robert Boyle et l'organisation du savoir," 157–73.

8. See Evelyn in chap. 2; Hunter, *Science and the Shape of Orthodoxy*, 67–97; Milton, "Locke's *Adversaria*"; Yeo, "John Locke's 'New Method.'"

9. Thomas Dent to Wotton, May 20, 1699, BL, Add. MS 4229, fol. 51ᵛ, in Hunter, *Boyle by Himself*, 105. For examples, see "Information recorded from the Imperial Ambassador" [1677 or 1678], BP 25, pp. 273–78; printed in Principe, *Aspiring Adept*, 296–300.

10. For the introduction of the term "workdiary," see Hunter and Littleton, "The Workdiaries of Boyle," 137.

11. Boyle to Oldenburg, [1670s?], *Boyle Correspondence*, 6:359. The editors suggest this letter was written in the 1670s, before Oldenburg's death in 1677.

12. Robert Boyle, [Boyle's justification for writing on single sheets, c. 1680], BP 36, fol. 6ʳ; also in *Boyle Works*, ed. Hunter and Davis, 12: 360. For piracy and plagiarism, see Boyle, BP 36, fols. 9ᵛ–10ʳ, 15ʳ; Johns, *Nature of the Book*, 504–14.

13. Boyle, *Excellency of Theology*, 82–83; also in Boyle, *Excellencies*, 207, 208.

14. Boyle, "A Proemial Essay," in *Certain Physiological Essays*, 12–14. For the antisystem trope in the prefaces to Boyle's works, see Knight, "Organising Natural Knowledge," esp. 18–19 and chap. 4.

15. Boyle, "An Advertisement of Mr. Boyle about the loss of many of his Writings" [May 1688], BP 36, fols. 50ᵛ–51ʳ; also in *Boyle Works*, ed. Hunter and Davis, 11:169–71. In subsequent references to this "Advertisement" I cite pagination from *Boyle Works*.

16. Boyle to Oldenburg, (1670s?), *Boyle Correspondence*, 6:359; and Boyle, BP 36, fol. 178ʳ, for his mention of "oversight of my writings, Theol[ogical], Philos[ophical], Chymicall, Medic[al] &c."

17. For this practice, Hunter and Davis, "Making of Boyle's *Free Enquiry*," 221–27.

18. Boyle, [Material relating to appendix to *Final Causes*, 1688 or later], BP 9, fol. 37ʳ; also in *Boyle Works*, ed. Hunter and Davis, 14:167. Eleven of the forty-six volumes of the Boyle Papers are available online at www.bbk.ac.uk/boyle/boyle_papers.

19. Boyle, "Advertisement of Mr. Boyle about the loss of many of his Writings," 169. See chap. 1 and below for the notion of "centuries."

20. Boyle, [Sequel to the 1688 Advertisement, 1689?], BP 36, fol. 17ʳ; also in *Boyle Works*, ed. Hunter and Davis, 12:359.

21. Boyle, [his justification], BP 36, fol. 6ʳ; also in *Boyle Works*, ed. Hunter and Davis, 12:360.

22. Boyle to Oldenburg (1670s?), *Boyle Correspondence*, 6:359.

23. "The Publisher to the Ingenious Reader," in Boyle, *New Experiments and Observations Touching Cold*, 205–6.

24. "The Publishers Advertisement to the Reader," in Boyle, *Hydrostatical Paradoxes*, 191.

25. Boyle, preface (September 29, 1682) to *A Free Enquiry into the Vulgarly Receivd Notion of Nature* (London, 1686), in *Boyle Works*, ed. Hunter and Davis, 10: 441; cited in Hunter and Davis, "Making of Boyle's *Free Enquiry*," 222.

26. Boyle, "Of Salts in the Air," Title XI, in Boyle, *History of the Air*, 49; also in *Boyle Works*, ed. Hunter and Davis, 12:36. The material for this work was gathered from the 1660s onwards. See more in chap. 7.

27. Leibniz to G. F. de L'Hospital, mid-March 1693, in Leibniz, *Mathematische Schriften*, 2:227–32; cited and trans. in O'Hara, "A chaos of jottings," 160.

28. Leibniz to G. Wagner, 1696, in Leibniz, *Philosophical Papers*, 465; Placcius, *De arte excerpendi*.

29. Yeo, "Notebooks as Memory Aids," 117–20.

30. Montaigne, "On Experience," in *Essays*, III.13, p. 1240. The Sibylline prophecies were reputedly written, in Greek, on palm-leaves. Montaigne also made notes in copies of books he owned and was able to find and reuse them; see *Essays*, II.32, on his reading of Seneca, Plutarch, and Bodin.

31. Placcius, *De arte excerpendi*, 70–71. Placcius referred to his teacher, Joachim Jungius, who assembled a huge quantity of notes on small slips of paper. See Blair, *Too Much to Know*, 73, 91, 94, 99.

32. HP30/4/46A-B, Ephemerides 1640, part 2; and BL, Add. MS 41846, fols. 194–204.

33. BL, MS Sloane 1466, fol. 165ʳ, dated "1648.12.dec."

34. Placcius, *De arte excerpendi*, 124–49.

35. Malcolm, "Thomas Harrison," 220–21; Clark, "Bureaucratic Plots," 195. For Leibniz's awareness of the choices involved in all classification, see Leibniz, *New Essays*, 523–24.

36. Notes in bound books could be treated as loose by moving them to other notebooks; see Timmermann, "Doctor's Orders." As seen in chap. 1, the transfer of notes was already typical of mercantile double-entry book-keeping. See Soll, *Information Master*, 54–66, on how this method could generate a powerful archive.

37. Hunter, *Boyle Papers*, 527–69.

38. See Boyle, RS MS 186, fol. 1ᵛ; MS 189, fol. 1ᵛ; and MS 194, fol. 1ʳ; also Boyle, *History of the Air*, 249: "This is the Account my Note-Book contains of this Trial."

39. This term was suggested by Michael Hunter and Charles Littleton. I have consulted the originals, but also rely on their report that the early workdiaries used quarto-size leaves apparently torn from a bound notebook, whereas the later ones

comprised folded foolscap paper, creating pages of approximately A4 size. Workdiary 8 from 1652–54 is an early example of this latter kind ("Memorialls Philosopicall," BP 25, pp. 343–46). See Hunter, "Mapping the Mind," 133–34; and Hunter and Littleton, "Workdiaries of Boyle," 141, for workdiaries 8 and 9 as showing signs of wear consistent with being carried in Boyle's pocket.

40. Hunter, *Boyle Papers*, 140. Certainly, Boyle's Geneva notebook (1643), RS MS 44 (approximately 12×8 cm) resembles the commonplace format, as do some other small bound notebooks (approximately 16.5×10.5 cm) such as RS MSS 185–94; later curators have embossed "Commonplace book" on their spines. For Boyle's stock of literary extracts from French romances, see Principe, "Virtuous Romance," 381, and his "Style and Thought of Boyle," 252.

41. Hunter, "How Boyle Became a Scientist," 24–26.

42. These are found respectively as Workdiary 13 (1655) in BP 25, pp. 153–56, 177–83; no. 21 (late 1660s) in BP 27, pp. 5–159; and no. 25 (late 1660s) in BP 27, pp. 219–20.

43. Workdiary 19 (1662–65), BP 22, pp. 1. Several items now lost, but seen by Henry Miles while assisting Thomas Birch with *The Works of the Honourable Robert Boyle*, had titles such as "Promiscuous thôts [thoughts] 1653" and "Promiscuous observat" (Hunter, *Boyle Papers*, 94).

44. Workdiary 6, BP 28, p. 309. See also Workdiary 8, BP 25, p. 343: "Memorialls Philosophicall. Beginning this First day of the Yeare 1651/2."

45. Workdiary 24 (late 1660s), BP 22, pp. 61–73.

46. The first century is in Workdiary 12 (1655), BP 8, fols. 140^r–48^r. Workdiary 38 (1689–1691), BP 21, pp. 219–54, comprises "The XVII Century" and "The XVIII Century." See Boyle, *Certain Physiological Essays*, 17: "And that my intended Centuries might resemble his [Bacon's *Sylva Sylvarum*], to which they were to be annex'd."

47. Boyle, RS MS 44 (his Geneva notebook of 1643) has "Theoremes Arithmetiques" and "Theoremes Geometriques," both of which have 100 points. Principe, "Newly Discovered Boyle Documents," 63, refers to Boyle's lifetime habit of keeping a journal of "sundry peeces."

48. Boyle, *Experimenta et Observationes Physicae* (1691), 419–26; also editors' "Introductory Notes," liii.

49. See Hunter, *Between God and Science*, 53–56.

50. *Boyle Works*, ed. Hunter and Davis, 2:437, 3:534.

51. Ibid., 11:169, cited in Hunter and Littleton, "Workdiaries of Boyle," 139.

52. HP29/3/13B, Ephemerides 1635, part 2, cited in chap. 4.

53. HP31/22/10A–B, Ephemerides 1648, part 1.

54. Boyle, "Dayly Reflection," 222. The editor, J. T. Harwood, says that this diary "has not survived" (his n. 41), but Boyle may have been referring to a set of notes, as in the workdiaries, rather than to a notebook. See Hunter, "How Boyle Became a Scientist," 25.

55. Boyle, *New Experiments about the Preservation of Bodies in Vacuo Boyliano*, in *Tracts containing . . . by the honourable Robert Boyle* (1674–75), in *Boyle Works*, ed. Hunter and Davis, 8:225.

56. Boyle, *Experiments and Notes about the Producibleness of Chemical Principles* (1680), in *Boyle Works*, ed. Hunter and Davis, 9:68.

57. See Boyle, "Cosmicall Suspitions," in *Tracts Written by Boyle* (1670), 304; and Workdiary 21, BP 27, p. 7. For the use of other workdiary entries, see Hunter, *Boyle Papers*, 414–16.

58. Boyle, *Experiments, Notes, &c., about the Mechanical Origin of Qualities* (1675-6), in *Boyle Works*, ed. Hunter and Davis, 8:339.

59. Boyle, *Origine of Formes* (1666), 386.

60. Boyle, *History of the Air*, 249. See "Of the air in reference to the generation, life and health of *Animals*" (Title XL).

61. Hunter, Knight, and Littleton, "Boyle's *Paralipomena*," 198, on the problem of dealing with chronological entries in the absence of thematic heads.

62. Boyle to J. Mallet, September 5, 1655, *Boyle Correspondence*, 1:189, cited in Maddison, *Life of Boyle*, 85, 219; Hunter and Macalpine, "William Harvey and Robert Boyle."

63. Boyle, "To the Reader," in *New Experiments Physico-Mechanical*, sig.A4,3ʳ. This introduction is in the form of a letter to Viscount Dungarvan dated December 20, 1659; also in *Boyle Works*, ed. Hunter and Davis, 1:145. See also Maddison, *Life of Boyle*, 93. Boyle copied "Mr Hartlib's approved remedy for the eyesight" into his notebook of medical receipts: RS MS 41, fol. 13ᵛ.

64. Hunter, "Robert Boyle and His Archive," in Hunter, *Boyle Papers*, 46–52.

65. Evelyn to Wotton, March 29, 1696, BL, Add. MS 4229, fol. 59ʳ, cited in Hunter, *Boyle by Himself*, 88.

66. Gassendi, *Mirrour of True Nobility*, 191 (in Book VI); also chap. 2 above. See Sacchini, *De ratione libros*, 93–94; also in Cevolini, *De arte excerpendi*, 158–59.

67. Boyle, *Occasional Reflections*, 30–31.

68. Bacon, *Advancement of Learning*, 118–19.

69. Bacon, *De augmentis*, 435.

70. Bacon, *Novum Organum*, Book I, aphorism 101, in *Works of Francis Bacon*, ed. Spedding, Ellis, and Heath, 4:96. See also Bacon, *De augmentis*, 435.

71. Bacon, *Novum Organum*, Book I, aphorism 102, pp. 159–61; Book II, aphorism 10, p. 215, and aphorism 26, pp. 286–89.

72. Weeks, "Role of Mechanics," 162–73; Lewis, "A Kind of Sagacity," 171–75.

73. For the term *"Experientia literata,"* see Bacon, *Advancement of Learning*, 111 (where the other stage is *"Interpretatio Naturae"*).

74. Bacon, *Novum Organum*, Book I, aphorism 103, p. 161; also Bacon, "To the Reader," *Sylva Sylvarum*, sig.Aʳ.

75. Bacon, *Parasceve*, 465, 469; and 475–85 for the list of "Titles."

76. Bacon, *Novum Organum*, Book I, aphorism 101, p. 159; also aphorism 102, pp. 159–61 and Book II, aphorism 10, pp. 214–15.

77. Ibid., Book I, aphorism 103, p. 161; and aphorism 92, p. 149.

78. Ibid., Book II, aphorism 10, p. 215. The "form" or "Nature of Heat" was Bacon's exemplar; see 217, 221, 237. See also Malhere, "Bacon's Method of Science," 75–98; Gaukroger, *Bacon and the Transformation of Early-Modern Philosophy*, 138–48.

79. Boyle, "General Heads"; also in *Boyle Works*, ed. Hunter and Davis, 5:508–11.

80. Hunter, "Boyle and the Early Royal Society," 10–12.

81. Bacon advised succinctness and accuracy rather than literary flourish and elaborate organization. Bacon, *Parasceve*, 457, 459.

82. Boyle to Oldenburg, June 13, 1666, BP 25, pp.1–17 (paragraph V, 6–7 and 15). I use

the version, with corrected page order, in Hunter and Anstey, *Text of Robert Boyle's "Designe about Natural History,"* 2–3. See also Anstey and Hunter, "Boyle's 'Designe about Natural History,'" 119–20.

83. Bacon, *Novum Organum*, Book II, aphorism 26, p. 287.

84. Boyle, *Christian Virtuoso. The Second Part*, 463. With many of his contemporaries, Boyle did not believe that there was a satisfactory account of the relationship between classical mnemonic practices and contemporary views of the physical workings of memory.

85. For example, Boyle, "Dayly Reflection," 222.

86. Boyle, "To the Reader," in *New Experiments Physico-Mechanical* (1660), in *Boyle Works*, ed. Hunter and Davis, 1:143.

87. Boyle, *An Essay about the Origine & Virtues of Gems* (1672), in *Boyle Works*, ed. Hunter and Davis, 7:72.

88. Boyle, preface to *Memoirs . . . Humane Blood*, 6; Knight and Hunter, "Boyle's *Memoirs*."

89. Workdiary 21, entry 218, BP 27, p. 11. The "same spirit" here probably refers to "the Spirit of Box-wood" mentioned in *The Sceptical Chymist* (1661), 316; also 288–90. See Boyle, *Experiments and Notes about the Producibleness of Chemical Principles*, in *Boyle Works*, ed. Hunter and Davis, 9:61, for such "Anonymous spirits" which Boyle distinguished from other known acids.

90. Workdiary 21, entry 228, BP 27, p. 14; workdiary 29, entry 286, BP 27, p. 247.

91. Workdiary 19, entry 113, BP 22, p. 53. Boyle used this entry in "Of the Mechanical Origine of Heat and Cold" (1675–76), in *Boyle Works*, ed. Hunter and Davis, 8: 332. See Hunter, *Boyle Papers*, 391, for Boyle's use of other entries from this workdiary.

92. See Knight, "Organising Natural Knowledge," 110–13, for Boyle's retrospective use of titles from the "General Heads" in the margins of workdiary 21.

93. Boyle, *New Experiments and Observations Touching Cold*, 237.

94. Ibid., 233.

95. Boyle, "Of the several degrees or kinds of Natural knowledge," BP 8, fol. 184^r; see another version in RS MS 185, fols. 32^v–37^r. See Hunter, *Boyle Papers*, 330 and 525, for dating of the first manuscript as 1660s–1680s, and the second as late 1680s.

96. Boyle to Oldenburg, December 9, 1665, *Boyle Correspondence*, 2: 598. Boyle also made lists of things to remember, usually opening with "Remember . . ." See BP 9, fol. 22; BP 10, fols. 72 and 102; BP 26, fol. 225; BP 28, fols. 405–6; BP 36, fol. 29.

97. Boyle, RS MS 189, fol. 162^r. See also RS MS 185, fol. 5^r: "Theological Notes of my own To be added under their due Heads to the Passages collected out of other Books, to prove the Truth of the Christian Religion." The latter is a small notebook with no obvious headings.

98. Boyle, "An Advertisement," [curtailing Boyle's willingness to receive visitors]," (c. 1690), two copies in BP 36, fols. 1^r and 2^r; also in *Boyle Works*, ed. Hunter and Davis, 12:363–64.

99. Knight, "Organising Natural Knowledge," 107, observes that "those diaries whose entries remain unnumbered are generally those with the least marginal annotation."

100. For textual excerpts, see Boyle, "Excerpta made out of severall Authors. AD 1663," RS MS 22, and many entries in workdiary 22 (late 1660s and early 1670s), BP 8, fols. 65–116.

101. Boyle, *Greatness of Mind, promoted by Christianity. In a Letter to a Friend* (London: Printed by E. Jones for J. Taylor, 1691), 2; also in *Boyle Works*, ed. Hunter and Davis, 11:347. This work dates from the late 1640s or early 1650s; see Principe, "Virtuous Romance," 387.

102. Boyle, *Occasional Reflections*, 146.

103. Boyle, *Hydrostatical Paradoxes*, 191.

104. Boyle, Workdiary 29, entry 286 in the hand of Frederick Slare, BP 27, p. 247.

105. Boyle, *Essays of Effluviums* (1673), in *Boyle Works*, ed. Hunter and Davis, 7:305.

106. Boyle, *Memoirs . . . Humane Blood*, 79.

107. Boyle, "Advertisement of Mr. Boyle about the loss of many of his Writings," 170.

108. Boyle, "Introductio ad soluta theological nostra Adversaria" [1680s], BP 2, fols. 191–98. See the English translation, "The Introduction to my loose Notes Theological" [1670–1680s], BP 5, fols. 16–34; also in *Boyle Works*, ed. Hunter and Davis, vol. 14, 277–83. Boyle referred to Blaise Pascal's *Pensées* (1670) as indicative of that author's habit of making theological and philosophical points, allocating a separate sheet for each thought. See Pascal, the preface to *Monsieur Pascall's Thoughts*, starting at sig.a4,2^v for this mode of composition; also Pascal, *Pensées*, xvii–xx.

109. Boyle, "Cogitationes Physicae," BP 8, fols. 211^r–212^r.

110. Boyle, "Introduction to my loose Notes," in *Boyle Works*, ed. Hunter and Davis, 14:282–83.

111. Ibid., 280.

112. Carruthers, *Book of Memory*, 253.

113. Descartes, *Rules for the Direction of the Mind*, in *Philosophical Writings*, 1:25, 67 (rules 7 and 16). This was written in Latin and not published until 1701, as *Regulae ad Directionem Ingenii*.

114. Descartes to C. van Hogelande, February 8, 1640, in Descartes, *Philosophical Writings*, 3:144–45. See chap. 4 for Descartes' criticism of Pell's *Idea* in this letter.

115. Boyle, *Some Considerations Touching the Usefulnesse* (1671), 463 (hereafter *Usefulnesse*).

116. "An Intimation of divers Philosophical particulars," *Philosophical Transactions* 6, no. 74 (August 14, 1671): 2216; also in *Boyle Works*, ed. Hunter and Davis, 7:455. A thorough search of the workdiaries might locate this note.

117. See BP 38, fols. 132^r (c. 1670); BP 38, fol. 81^r (1670s); Boyle, *Experimenta et Observationes Physicae*, liii–iv and 377–84; also "Experiments and Notes about the Mechanical Production of Magnetical Qualities," *Boyle Works*, ed. Hunter and Davis, 8:501–8.

118. See BL, MS Sloane 3391, fol. 2^v, which has a summary, including verbatim passages, of Boyle's explanation that the color produced "upon the Mixture of Bodies, is quite different from that of any of the Ingredients." See Boyle, "The Advantages of the Use of Simple Medicines" (1685), in *Boyle Works*, ed. Hunter and Davis, 10: 408. The catalogue of the Sloane manuscripts suggests the maker of these notes was James Petiver (c. 1665–1718), botanist and entomologist.

119. BL, MS Sloane 623, fols. 63^v–71^v. The notebook belonged to Daniel Foote, MD, and is dated "1669/70."

120. Boyle, *New Experiments and Observations Touching Cold*, 210.

121. Boyle, *Usefulnesse* (1671), 400, 402. The passage Boyle refers to is Bacon, *Novum Organum*, Book I, aphorism 103, p. 161: "within one man's knowledge and judgement."

122. Boyle, "Advertisement of Mr. Boyle about the loss of many of his Writings," 171.

123. Boyle, "A Proemial Essay," in *Certain Physiological Essays*, 17–19.

124. Boyle, "Experiments about the Preservation of Bodies in Vacuo Boyliano," in *Boyle Works*, ed. Hunter and Davis, 8:225.

125. Leibniz, *New Essays*, 455–56. For a more sympathetic interpretation stressing Boyle's strong commitment to corpuscular theory, see Boas, "Boyle as a Theoretical Scientist," 264–66.

126. Kuhn, "The Function of Measurement," 192.

127. Daston, "Description by Omission," 16.

128. Boyle, RS MS 198, c. 1680, fol. 104^r; RS MS 189, 1689–90, fols. 27^v-28^r.

129. Bacon, *Novum Organum*, Book I, aphorism 102, p. 159.

130. See Webb, "Art of Note-taking," 369, for this method as an instrument "in actual discoveries." On Hooke's use of loose slips, see Yeo, "Between Memory and Paper-books," 29–30, and chap. 8.

131. Charles Darwin cut pages out of some notebooks and put them with other loose sheets in "portfolios" reserved for different subjects. Darwin, *Autobiography of Darwin*, 53; and his son's account, 102–3.

132. Hunter, "Mapping the Mind," 127.

133. Boyle, *New Experiments and Observations Touching Cold*, 263–64. He made this remark in a letter of February 14, 1662, to Lord Brouncker accompanying the final sections of this work.

Chapter Seven

1. [Locke], "Methode Nouvelle" (1686). Locke's account of these requests is given in the letter to Nicolas Toinard that serves as a preface: "Lettre de Monsieur J. L. de la Société Roiale d'Angleterre, à Monsieur N. T. . . . ." The second edition omitted the mention of the Royal Society. Locke may have feared that this link would disclose his identity.

2. Chambers, *Cyclopaedia*, vol. 1, entry for "Common-Places"; Diderot and d'Alembert, *Encyclopédie*, 13: 868–70, entry for "Recueil (belles lettres)." The latter was copied from Chambers with a reduced image of the Index showing only the letters B, C, D, and E.

3. D'Alembert, *Preliminary Discourse to the Encyclopedia of Diderot*, 83–85.

4. His final catalogue records all but 685 of these. See Laslett, "Locke and His Books," 11–12, 19, for the estimates.

5. For Gessner, see Nelles, "Reading and Memory in the Universal Library"; for Harvey, see Grafton and Jardine, "Studied for Action"; for Drake, see Sharpe, *Reading Revolutions*, 69–89; for Jungius, see Meinel, "Enzyklopädie der Welt," 165–87.

6. Moss, *Printed Commonplace-Books*, 278–79.

7. In the "New Method," English draft, fol. 57^r, Locke recalled that he had first mentioned it to Toinard in Paris, which must have been prior to May 1679. For Toinard's relationship with Locke, see editor's note in *Locke Correspondence*, 1:579–82. Locke opened the correspondence on June 29/July 9, 1678. Ibid., vol. 1, no. 388.

8. Locke to Toinard, February 14/24, 1685, *Locke Correspondence*, vol. 2, no. 811.

9. Locke to Toinard, March 30/April 9, 1685, ibid., no. 818. The copy sent to Toinard is BL, Add. MS 28728, fols. 46–53. Another copy that Locke retained is MS Locke c. 31, fols. 67–78.

10. Locke to Toinard, March 16/26, 1685, *Locke Correspondence*, vol. 2, no. 814. See Bonno, *Les relations intellectuelles*, 151–52, for the suggestion that Toinard was ill in Orleans and not able to arrange publication.

11. Le Clerc, *Life of Locke*, 14; and "Eloge de feu Mr. Locke." For details, see Milton, "Textual Introduction" to the section on the "New Method" in Locke, *Literary and Historical Writings*.

12. In the French publication these two terms were given as *recueuils* and *titre*, being translations from Locke's Latin version, which had "*adversaria*" and "*titulo*"; [Locke], "Methode Nouvelle," 319. The other English translation, "New Method," Greenwood, renders "Adversariorum Methodus" as "the Method of Common Places" (p. 4).

13. Benjamin Furly to Locke, January 30/February 9, 1694, *Locke Correspondence*, vol. 5, no. 1702

14. Locke to Toinard, February 14/24, 1685, *Locke Correspondence*, vol. 2, no. 811 (de Beer's translation of the Latin is used here). The French version included an insertion in italics stressing this point. [Locke], "Methode Nouvelle," 319. This was retained in the English edition of the *Posthumous Works*.

15. Locke uses twenty letters only; he puts *Qu* in the last of the Z cells. For his explanation, see "New Method," English draft, fol.58^r. The translation in "New Method," *Posthumous Works*, 317 reads: "I omit three Letters of the Alphabet as of no use to me, viz. K. Y. W. which are supplied by C. I. U. that are equivalent to them."

16. "New Method," English draft, fol. 58^r.

17. "New Method," *Posthumous Works*, 316. In the French version this passage reads: "Quand je rencontre quelque chose que je croi devoir mettre en mon Recueuil; je cherche d'abord un titre qui soit propre" ("Methode Nouvelle," 321).

18. "New Method," English draft, fol. 58^v; "New Method," *Posthumous Works*, 316.

19. "New Method," English draft, fol. 58^v; "New Method," *Posthumous Works*, 319: "if I see no number in the space."

20. "New Method," English draft, fols. 58^v–59^r. For the different lists in the published versions, see "Methode Nouvelle," 323; "New Method," *Posthumous Works*, 319.

21. "New Method," English draft, fol. 60^r. Here "classes" refers to cells, not to "titles" (Heads): in fact, the number of Heads is not limited to 100 (or 500, with the second vowel) because there can be more than one Head sharing the same letter/vowel combination, as in the examples given. In principle, therefore, Locke's method could deal with at least as many Heads as the 3,300 mentioned in Thomas Harrison's "Arca studiorum" (see chap. 4).

22. See MS Locke d. 10 and d. 11.

23. For Aubrey, see Yale, "With Slips and Scraps"; for Boyle, see chap. 6.

24. "New Method," *Posthumous Works*, 315. In "New Method," English draft, fol. 56^v, Locke says he had used the method "above these twenty years, without haveing found any inconvenience in it or temptation to alter it."

25. Meynell, "Locke's Method of Common-placing," 245: "he often failed to put his principles into practice . . . which was far from consistent."

26. These early variants will be discussed by Milton and Yeo in Locke, *Literary and Historical Writings*.

27. Damaris Cudworth, daughter of the scholar Ralph Cudworth, married Francis Masham in 1685, and had corresponded with Locke since the early 1680s.

28. As the basis for this, Locke used Thomas Hyde's *Catalogus Impressorum Librorum* (1674). He wrote "L" next to those books he owned, and on interleaved pages he added books he owned which were not listed by Hyde. See Harrison and Laslett, *Library of Locke*, plates 3 and 4.

29. In his Will dated April 11, 1704, Locke also directed that "the other Moiety of the said Books . . . I give and bequeath to Francis Cudworth Masham," that is, to Damaris' son. This latter half of the library was subsequently dispersed, many titles being lost. See the Will in John Locke, *The Works of John Locke, Esq.*, 3 vols (London: John Churchill and Sam. Manship, 1714), 1:29–30.

30. See Harrison and Laslett, *Library of Locke*, nos. 23a, 23b, 24, 25, 1712, and 1713.

31. "Adversaria 1661," 30 (I describe the provenance of this notebook below). See also MS Locke c. 25, fol. 18ʳ, for a list including "No. 1 Scritore. Pigeon holed." This follows an "Inventory 75" on fol. 16ᵛ. Locke's desk has ten small pigeon holes at the top, matching his sketch. Formerly in the possession of his relative, Peter King, it was acquired by Christ Church, Oxford, in the 1970s. See *Books and Manuscripts from the Library of Arthur A. Houghton, Jnr.*, Part 1, 1979 (published by Christie's), 248, plate 46.

32. Lord King, preface to *The Life of John Locke*, v.

33. Milton, "John Locke's Medical Notebooks," 138, and also n. 7 for the view that the great majority of the notebooks have survived.

34. This count is based on a sample of entries, of various word lengths, up to p. 326 of the notebook; it excludes the pages of weather records at the end.

35. These are listed and described in Long, *Summary Catalogue of the Lovelace Collection*, but I rely for these estimates on an unpublished list by J. R. Milton. Those counted as commonplace books are usually indexed according to a variant of Locke's "New Method." W. von Leyden estimated "his journal and notebooks" as thirty-eight (Locke, *Essays on the Law of Nature*, 2).

36. MS Locke f. 5, p. 94 (the "59" is not underlined); see p. 93 for Adversaria 60 and 61. See Harrison and Laslett, *Library of Locke*, 270–77 for the full book list of 1681. Milton, "Locke's *Adversaria*," 70–72, suggests that the two missing notebooks might have been renamed "Lemmata Ethica" and "Lemmata Physica" respectively, both of which are listed in the catalogue (nos. 1712 and 1713). See MS Locke f. 19, p. 167, for a cross-reference ("6.Adv.p4") apparently indicating a commonplace book numbered "6" in the series. See also Meynell, "Locke's Method of Common-placing," 266 n. 1.

37. Harrison, "Locke, Physician and Book Collector."

38. Long, "The Mellon Donation." One of these bound manuscripts (MS Locke c. 42) comprised two separate commonplace books, apparently stitched together on Peter King's instructions.

39. These include a notebook belonging to Locke's father containing some entries by Locke (BL, Add. MS 28273), a memorandum book for 1669 (BL, Add. MS 46470), a medical commonplace book (BL, Add. MS 32554), and Locke's journal for 1679 (BL, Add. MS 15642).

40. Fox-Bourne, *Life of John Locke*, 1:145. He did use material connected with Locke in the Shaftesbury Papers. See Milton, "Locke: Modern Biographical Tradition," 90.

41. See Milton, "Locke at Oxford," 46–47; Tully, "Governing Conduct," 189–200, for some entries that relate to the *Essay*.

42. The account book is MS Locke c. 1; the memoranda are MS Locke f. 15 and f. 28,

pp. 2–3 (index). The latter was used in France during 1678–79 and then in the early 1680s in Holland. He did not use the "New Method" index in MS Locke f. 11, f. 12 or f. 13.

43. Masham to Le Clerc, January 12, 1705, in Le Clerc, *Epistolario*, 2:514.

44. MS Locke f. 5, p. 113 (August 18, 1681); MS Locke f. 6, p. 63 (May 30, 1682).

45. BL, Add. MS 46470 at fol. 21^r: "from Oxford to London 47 miles." This is Locke's small memorandum book (97×53 mm) incorporating John Goldsmith's *Almanac* (London: Tho. Ratcliffe and Tho. Daniel, 1669). The relevant section is "The Geographical Description of Wayes from one notable Town to another, all over England, and thereby how to travel from any of them to the City of London" (fol. 20^v).

46. MS Locke f. 5, p. 152 (December 24, 1681); MS Locke f. 6, p. 1 (January 1, 1682).

47. Wood, *Life and Times* (April 23, 1663), 1:472. See also Gunther, *Early Science in Oxford*, 1:23–25.

48. MS Locke f. 25, pp. 100, 136, 154, 173 217, 280, for mention of "Mr Stall" (or "Stal"). Meynell, "Locke, Boyle and Stahl," 185–86, believes Locke later attended a course of lectures in 1666, and that the notes are from that time. Walmsley and Milton, "Locke's Notebook 'Adversaria 4,'" 92–93, provide strong grounds for their being made in 1663. They agree with Meynell about a similar set of topics in Locke's notes and Stahl's lectures.

49. Milton, "John Locke."

50. He began to use this account book at Westminster School; the first page is dated May 26, 1651. See MS Locke f. 11, pp. 10, 11 (for another paperbook for two shillings and three pence).

51. Sanderson, *Logicae Artis Compendium*, appendix II, chap. 3, 321–28. Locke owned the first edition of 1615 (Harrison and Laslett, *Library of John Locke*, no. 2548a). See MS Locke f. 11, p. 10, for purchase of "Sandersons's Logick" for one shilling and six pence. Note the higher cost of the paperbooks. See Mack, *Elizabethan Rhetoric*, 56–59.

52. See [Barlow], *Library for Younger Schollers*, x–xii, for the editors' identification of Barlow as the likely author of the original text on which various manuscripts are based: these are St. John's College, Cambridge MS K.38, pp. 18–96, 163–80; Bodl. MS Rawlinson C. 945; and BL, Harley 2007, fols. 1–9. See also Long, "The Mellon Donation," 188.

53. MS Locke e. 17, pp. 44–71.

54. Ibid., 49; and [Barlow], *Library for Younger Schollers*, 4.

55. [Barlow], *Library for Younger Schollers*, 21 and 50. On the last option, see the discussion in chap. 2.

56. BL, Add. MS 28273, fols. 36–59, 87–101; and recipes at fols. 144–49 for medical complaints. The notebook is approximately 192×145 mm.

57. BL, Add. MS 28273, fols. 125^r and 126^r, respectively.

58. MS Locke e. 4. Some of the content resembles that in BL, Add. MS 28273. The date is written as "[16] 52," but Locke did not start to underline the last two digits of dates until about 1661; see Milton, "Locke's Medical Notebooks," 148.

59. See *Locke Correspondence*, 1:152, n. 1.

60. Locke continued such collections; see University of Glasgow, Special Collections, MS Murray 416, for a notebook used for medical and household matters at Oates between 1691 and 1701. I thank J. R. Milton for this.

61. See Milton, "Date and Significance of Two of Locke's Early Manuscripts," on book lists in MS Locke f. 14, and PRO 30/24/47/30, fols. 42–43 in the Public Record Office, London.

62. MS Locke f. 14, pp. 9, 16, 17, 22–23, 54, 65, 93, 110.

63. Compare BL, Add. MS 32554 (a medical commonplace book) in which the great majority of entries concern concepts, although there are a small number on the reputations of individual physicians (pp. 10, 12, 36, 64, 66) similar to those in MS Locke f. 14.

64. Most entries in MS Locke f. 14 are arranged according to the "New Method," the keyword being the name of the author.

65. On "Physica," see Bylebyl, "The Medical Meaning of 'Physica.'"

66. Milton, "Locke at Oxford," 32, 35–36; Meynell, "A Database for Locke's Medical Notebooks and Medical Reading."

67. "Adversaria Physica" is MS Locke d. 9; it is bound in calf and measures approximately 29×20 cm; the pages are 28.5×18 cm.

68. "Adversaria Ethica" is not in the Lovelace Collection, and thus has no catalogue number. I will refer to it as "Adversaria 1661." The original is now in private ownership. Microfilm copies are held at the Bodleian Library, Oxford, and the Houghton Library, Harvard. It has pages measuring approximately 30×18 cm and consists of 321 numbered pages and 149 blank leaves. A note in the Houghton Library microfilm says 80 leaves are blank after p. 321.

69. Locke also assigned numbers to a series of these adversaria: thus he wrote "Adversaria 5" on the first inside leaf of MS Locke d. 9; but he did not persist with this. See Milton, "Locke's *Adversaria*."

70. See Milton, "The Dating of 'Adversaria 1661.'"

71. BL, Add. MS 28728, fol. 60^v. He later used another two notebooks called "Adversaria 4 Pharmacopaea" (MS Locke f. 25) and "Adversaria Theologica" (MS Locke c. 43).

72. See respectively, "Intellectus" (1671) and "Toleration" (1667), in "Adversaria 1661," 56–89, 94–95; and 106–25, 270–71. For the latter, see Locke, *An Essay Concerning Toleration*.

73. MS Locke d. 9, pp. 30, 218 ("Adversaria Physica"); and MS Locke f. 19, pp. 227, 272–73.

74. Lorraine Daston contrasts the undated pages of Locke's "Adversaria Physica" and the chronological record in Horace-Bénédict de Saussure's journal of 1774. See Daston, "Empire of Observation," 96, 99. This contrast is valid, but it cannot support an inference that Locke generally ignored dating, as I discuss below.

75. MS Locke d. 9, pp. 41, 68.

76. See MS Locke f. 25, p. 33, for "Jun. 1. 66" as the date of an experiment with human blood. Locke referred to this notebook as "Adversaria 4 Pharmacopea." See Walmsley and Milton, "Locke's Notebook 'Adversaria 4.'"

77. See MS Locke d. 9, pp. 87, 236, about poisonous fish in New Providence.

78. These entries start at the back of MS Locke f. 27, a bound notebook measuring 105×75 mm.

79. MS Locke f. 27, p. 1 [fol. 169^v]. For Locke's interest in Helmontian iatrochemistry, see Anstey, "Locke and Helmontian Medicine."

80. Locke's journals contain medical case histories (chronological accounts of symptoms and treatment regime) of his patients and of his own illnesses (often signed JL). See BL, Add. MS 15642, pp. 146–51 for September 5–11, 1679.

81. See chapter 1, n. 3, for the position of the Register in MS Locke d. 9, and the absence of pagination. After some intermittent records in the early 1670s, the Oxford

register resumes in March 1681 and closes on December 3, 1682, except for one final set of readings in late June 1683.

82. MS Locke f. 4, pp. 104–5. Locke noted that one piece of hail was "in circuit 5½ inches almost" and that "Dr. Dan Cox told me he measured one 7½ inches." I have not made a thorough check for weather entries in the journals Locke kept in the Netherlands from October 1683 until his return in early 1689.

83. MS Locke d. 9. This register, titled "Weather at Oates," or variants of this, will be included, with commentary by Anstey and Principe in Locke, *Writings on Natural Philosophy and Medicine*.

84. "A Register kept by Mr Locke, in Oxford," in Boyle, *History of the Air* (1690), 104–32; (the London register is at 116–21); references are from this edition; also in *Boyle Works*, ed. Hunter and Davis, 12:70–89.

85. Stewart, "Locke's Contacts with Boyle," 20–22; Hunter, *Between God and Science*, 94–95, 294.

86. Woolhouse, *Locke*, 60–61, 64. Locke served as Vane's secretary.

87. Locke to Boyle, December 12/22, 1665, *Locke Correspondence*, vol. 1, no. 175.

88. Locke to Boyle, May 5, 1666, ibid., no. 197. These measurements appear in an entry dated April 23, 1666 under "Aeris gravitas" in MS Locke d. 9, p. 41.

89. Boyle must have made this request to Locke in Oxford. When Locke traveled to Pensford in April to check on his property, Boyle sent a barometer there. More generally, see Stewart, "Locke's Professional Contacts," 23.

90. Boyle, "Some Observations and Directions about the Barometer," 184; also in *Boyle Works*, ed. Hunter and Davis, 5:504–7; see also chap. 3.

91. Boyle to Locke, June 2, 1666, *Locke Correspondence*, vol. 1, no. 199.

92. See Gunther, *Early Science in Oxford*, 1:315, for William Merle as "the first man in the world to keep a journal of the weather"—in the mid 1300s. I have already mentioned the daily records in Florence between 1654 and 1670 (see chap. 3).

93. Schove and Reynolds, "Weather in Scotland," 167.

94. For a reprint of the weather remarks, see ibid., 168–75. For the entire diary, see Hay, *Diary*. Earlier parts of the diary are lost.

95. On the latter point, see Golinski, *British Weather*, 81.

96. Nevertheless, in combination with other contemporary observations, it is possible to produce a synoptic pattern on the basis of these comments; see Schove and Reynolds, "Weather in Scotland," 176–77.

97. Wren held the chair of Astronomy at Gresham College, London, between August 1657 and early 1661. On February 5, 1661 he succeeded Seth Ward as Savilian Professor of Astronomy at Oxford.

98. He developed this idea in an address to the Royal Society in 1662. This is printed in Wren, *Parentalia*, 222. *Parentalia* was compiled by Wren's son, Christopher. See Bennett, *Mathematical Science of Wren*, 82–83; and his "A Study of *Parentalia*."

99. Wren, *Parentalia*, 222–23.

100. He reported this to the Royal Society on January 22, 1662; see Birch, *History*, 1:74. It was mentioned by Plot, *Natural History of Oxford-Shire*, 232; and Grew, *Musaeum Regalis Societatis*, 358.

101. Sprat, *History*, 312–13. See Biswas, "Automatic Rain-Gauge"; Bennett, *Mathematical Science of Wren*, 84–86. See Birch, *History*, 3:222, meeting of June 10, 1675, for

mention of Hooke's additions to a "weather-clock . . . which was the more considerable, for that itself records its own effects."

102. Unlike "weather glasses," both these instruments were sealed. Hooke suggested the term "baroscope" in June 1664; Boyle coined "barometer" in 1665. See Middleton, *History of the Barometer*. For an account of how a spirit thermometer registered degrees above and below the "Temperate Point" (not freezing point) for a specific location, see John Warner, *Aeroscopium* (c. 1680), reproduced in Gunther, *Early Science in Oxford*, 12:302–4. This included a printed diary for both barometer and thermometer readings.

103. On the continuing debate about whether the barometer was affected by the weight of an increasing column of air (supplemented by air movements), or by changes in its "specific gravity" due to absorption of "saline substances," see Birch, *History*, 3:509 (November 27, 1679).

104. Gunther, *Early Science in Oxford*, 12:137, citing minutes of the Dublin Society for March 10, 1684. For attempts at comparison between Aberdeen and Oxford, see George Garden to Robert Plot, December 8, 1686, 319.

105. Wren, *Parentalia*, 222.

106. Locke to Sloane, March 15, 1704, *Locke Correspondence*, vol. 8, no. 3489: "tis fit I explain a litle to you some things in the table for the better understanding of it." In this period, the word "chart" was only applied to maps and sea-charts.

107. Hooke, "A Method for Making a History of the Weather," in Sprat, *History*, 173–79; and 179 for the table or "A Scheme" displaying the information. Earlier, on October 7, 1663, Hooke read a paper to the Society entitled "Rules for observing the weather & for an History of it" (RS CP.xx.24, fols. 40^r–41^v). Locke may have heard about this via Boyle's circle in Oxford.

108. Boyle, "Some Observations and Directions about the Barometer," 182–83; also in *Boyle Works*, ed. Hunter and Davis, 5:504–7.

109. Golinski, *British Weather*, 114–20.

110. Hooke, *Micrographia*, scheme XV. Locke mentioned "the turning of a wild Oat-beard, by the insinuation of the Particles of Moisture" in his *Essay* II.ix.11. Wren considered the use of "Lute-strings" but thought that both these and the "Beard of Oats" were unreliable. Wren, *Parentalia*, 224.

111. Locke, "Explication of the foregoing Register," in Boyle, *History of the Air*, 132.

112. There is no evidence that Locke used any kind of rain gauge; but in the early 1680s he began to include the times of day when it rained. See Boyle, *History of the Air*, 126–28.

113. MS Locke d. 9, Register for June 26, 1666.

114. MS Locke d. 9, July 7, 1666; also in Boyle, *History of the Air*, 105. Wren, *Parentalia*, 224, mentions the problem of recording wind at night.

115. Place, of course, was also crucial: this is why Locke kept a separate register for London. See the fragment in figure 1.2 for MS Locke f. 19, p. 394, under "Aer." For the full London observations, see n. 84 above.

116. See MS Locke d. 9, "Explication of my Register of the Air" (last page of notebook) for his note that before April 9, 1682 "the fore and afternoon were marked by a point over or under the hour." But his "Explication of the foregoing Register" (the published version) in Boyle, *History of Air*, 132, does not include this point. In the register at Oates, Locke sometimes used dots around the numeral to indicate fifteen-minute intervals.

117. Boyle, *History of the Air*, 116, 118, 128; and Locke's "Explication" in MS Locke d. 9.

118. "A Register kept by Mr Locke, in Oxford," in Boyle, *History of the Air*, 109–11.

119. RS MS 190, fol. 166^{v-r}; it is headed "weather observations," starting from the back of the notebook.

120. MS Locke d. 9. The first of these annotations is dated December 1691 and he continued to add others through the 1690s. Most of these were made after the publication of the *General History of the Air*, and nothing about this was included in the "Explication" (p. 132) that accompanied the printed version of the earlier registers.

121. But he did not adopt the terms suggested by Hooke in Sprat, *History*, 177. For these, see chap. 8

122. Boyle, *History of the Air*, 115, 120–23. In the original version, Locke had "Thames/Charwell frozen over." MS Locke d. 9, March 8, 1667.

123. Boyle, *History of the Air*, 120, 123. See also *Boyle Works*, ed. Hunter and Davis, 12:80, 82. Locke made this summary even though he did not make any entries between May 21 and August 2, 1681.

124. He made the first of these notices on February 27, 1692.

125. MS Locke d. 9. See also September 4 and 19, 1694, for his intention to compare observations about the swallows' behavior.

126. Sprat, *History*, 312–13; Wren, *Parentalia*, 222–23.

127. The most relevant texts from the corpus were "Epidemics, Books I and III," and "Airs, Waters, Places." See *Hippocratic Writings*, 87–138, 148–69.

128. See Anstey, *Locke and Natural Philosophy*, 54–55.

129. Wren, *Parentalia*, 223.

130. See BL, Add. MS 32554, p. 22; and MS Locke c. 42, pt. I, p. 16, for a note about Hippocrates on the effects of changes in the air.

131. See MS Locke f. 27, Memorandum book [c. 1664–66], p. 3, where Locke ponders whether poor respiration at "great heights" such as the Andes is due to the reduced "pressure of this aer" or to a lack of "those salts" found in the "lower regions" which mix with the blood.

132. Bodl. MS Rawlinson C. 406, fol. 68^r. Dewhurst, *John Locke*, 301, says this copy is in Goodall's hand. For a copy of one response, which includes data from the "Bills of Mortality in Dublin," see fols. 69–82; also Dewhurst, "The Genesis of State Medicine in Ireland."

133. Rusnock, "Hippocrates, Bacon, and Medical Meteorology," 139–41.

134. Locke to Sloane, March 15, 1704, *Locke Correspondence*, vol. 8, no. 3489. See more on this below.

135. Locke had previously accompanied the countess of Northumberland on a three-week visit to Paris in 1668. See Cranston, *John Locke*, 160–83.

136. Laslett, "Locke and His Books," 16, 26, believed that Locke took the large notebook (MS Locke c. 44) to Holland; but his supporting idea—that it served as an exemplar of his method of note-taking when he wrote the "New Method" in 1685—is implausible. Locke needed no reminder of his own method; in any case, he was using it in the small memoranda notebook he carried with him. See MS Locke f. 29 for dates covering his exile in Holland.

137. When Locke copied an entry from a journal, he noted the year of that journal

and the page of the original entry. About the same time, when making a new entry, he began to write in the margin the year in which he made it.

138. Gibbon, *Memoirs*, 135–36.

139. For accounts of Locke's journey, see Lough, ed., *Locke's Travels*, xv–xix; xxxvi–xl; Woolhouse, *Locke*, chap. 4.

140. Locke to Boyle, May 25/June 4, 1677, and Locke to Boyle, July 27/August 6, 1678, *Locke Correspondence*, vol. 1, nos. 335 and 397 respectively.

141. See the judgment on Locke's journals in Matthews, *British Diaries*, 36: "very scrappy and disappointing, but has biographical value."

142. Yeo, "Between Memory and Paperbooks."

143. Anstey, *Locke and Natural Philosophy*, chap. 3.

144. MS Locke f. 3 (March 7–8, 1678), pp. 47–60.

145. Initially, it seems that Locke planned a shorter stay. See Lough, ed., *Locke's Travels*, xvi–xvii.

146. MS Locke f. 1, pp. 37, 49, 69–71, 76–94 (during early 1676).

147. See MS Locke f. 1, p. 151 (March 10, 1676) about the poor state of some vines: "A peasant working in the vineard said they were never the worse."

148. See the mention of "Essay de Intellectu," as a folio, in MS Locke f. 3, p. 183 (July 2, 1678). This manuscript is not the one now known as Draft A, which remained in a large notebook ("Adversaria 1661") in England.

149. See MS Locke f. 1, pp. 325–47 (July 16, 1676). See also Locke, *Essays on the Law of Nature*, 263–72, for transcriptions from Locke's shorthand of some of these.

150. For his reading, see Lough, "Locke's Reading during His Stay in France."

151. Masham to Le Clerc, January 12, 1705, in Le Clerc, *Epistolario*, 2:511. Indeed, Locke confirmed this, saying that talk with jewellers, blacksmiths, gardeners or others skilled in a particular science or art yielded more accurate information than conversation with polite gentlemen (*Essay*, II.xxiii.3).

152. MS Locke f. 1, p. 10 (December 22, 1675). See MS Locke f. 1, p. 37 (January 9, 1676) for his hiring of "a French M^r [Master]." Some of the following entries from Locke's journal are transcribed in Lough, ed., *Locke's Travels*; but he omits most medical and philosophical entries. For selections of the former, see Dewhurst, *John Locke*, 62–151; for the latter, see Locke, *Early Draft of Locke's Essay*, 75–125.

153. MS Locke f. 1, pp. 150–51 (March 8 and 10, 1676); p. 348 (July 18, 1676); and MS Locke f. 2, p. 189 (July 7, 1677).

154. MS Locke f. 1, p. 252 (May 9, 1676); p. 397 (August 9, 1676); and MS Locke f. 3, pp. 161–2 (June 20, 1678).

155. MS Locke f. 1, pp. 252–53 (May 9, 1676); see also MS Locke f. 1, pp. 140–41 (March 5, 1676) for the "Bath at Ballaruc"; see Lough, ed., *Locke's Travels*, 55, on the use of a solution of "powdered gall-nuts" as a test for the presence of iron in water.

156. MS Locke f. 1, p. 92 (February 10, 1676); and pp. 197–98 (April 11, 1676).

157. For the importance of this attitude to the exchange of information, see Lux and Cook, "Closed Circles or Open Networks?" For Locke's contribution to such an informal network, see Harris and Anstey, "John Locke's Seed Lists."

158. Locke, *Essay*, IV.xvi.4. Shapin, *Social History of Truth*, 212–32, places Locke's rules in the context of similar maxims offered by contemporaries such as Boyle.

159. MS Locke f. 2, pp. 292–93 (October 8, 1677).

160. MS Locke f. 1, p. 171 (March 26, 1676), and pp. 264–77 (June 2–9, 1676); for expansion of shorthand passages, see Lough, ed., *Locke's Travels*, 95–100.

161. MS Locke f. 3, p. 256 (August 15, 1678).

162. MS Locke f. 2, p. 83 (March 20, 1677).

163. Ibid., p. 139 (May 19, 1677) and p. 190 (July 7, 1677); and Locke, *Observations upon the Growth and Culture of Vines and Olives*, 7; also in *Locke Works*, 10:323–56. This point is also entered in MS Locke d. 9, p. 264.

164. MS Locke f. 2, p. 68 (March 4, 1677), 68; Lough, ed., *Locke's Travels*, 128, for the part in shorthand. Locke measured the thickness of the vault as 188 of his steps. See Louis de Froideur, *Lettre . . . concernant la relation et la description des travaux qui se font en Languedoc* (Toulouse, 1672), which Locke had read earlier; see MS Locke f. 1, p. 140 (March 4, 1676).

165. For Locke and travel literature, see Laslett, "Locke and His Books," 27–29; Carey, *Locke, Shaftesbury, and Hutcheson*, 76–83; and Talbot, "The Great Ocean of Knowledge."

166. Tully, "Governing Conduct," 189–200, argues that Locke's views on the social conditioning of belief crystallized during this period. See also Yeo, "Locke on Conversation with Friends and Strangers."

167. Although shorthand provided partial secrecy, Locke's use is not predictable: he sometimes used it for weather observations, and sometimes recorded delicate matters in longhand. See Lough, ed., *Locke's Travels*, lxiv–lxv.

168. BL, Add. MS 15642, p. 10 (February 11, 1679, under "Popery"). This entry is not in shorthand, as is the one for February 13 on the "Jesuits" (p. 12).

169. MS Locke f. 2, pp. 93–94 (March 30, 1677); MS Locke f. 3, pp. 260–61 (August 23, 1678) and p. 308 (October 7, 1678).

170. See, for example, on "Turks," MS Locke f. 2, pp. 270–71 (September 19, 1677); also under "Opinion," in BL, Add. MS 15642, p. 101 (June 17, 1679).

171. See MS Locke f. 1, pp. 539–47 ("New Method" index) and pp. 527–32 (alphabetical index). The former, even though incomplete, takes up ten pages, whereas the latter occupies only six.

172. "New Method," English draft, fols. 60^{r-v}; [Locke], "Methode Nouvelle," 325. Of course, he had previously used a second vowel in the two "lemmata" notebooks (MSS Locke d. 10 and d. 11), but these did not require any index, as explained above.

173. See MS Locke f. 2, pp. 86–140. The entry is interrupted by several others on various topics; it concludes in early May 1677. I use Locke's original journal and his pagination; for Axtell's transcription, see Locke, "Of Study," 405–22.

174. See Yeo, "John Locke's 'Of Study' (1677)."

175. See, for example, Grenville to Locke, c. March 6/16–8/18, 1677, *Locke Correspondence*, vol. 1, no. 327. See Locke's entry on "Scrupulosity," MS Locke f. 3, pp. 69–79 (March 20, 1678) and similar themes in Locke to Grenville, March 13/23, 1678, *Locke Correspondence*, vol. 1, no. 374.

176. Nicole, *Essais de Morale*. See Harrison and Laslett, *Library of John Locke*, no. 2040a, for the earliest (1671) of several editions he owned.

177. Nicole, "Traité de la faiblesse de l'homme," para. 33, also paras 48, 67.

178. MS Locke f. 1, pp. 404–5 (August 15, 1676). This entry is in Locke's shorthand; for von Leyden's transcription, see Locke, *Essays on the Laws of Nature*, 257.

179. See Yeo, "Locke and Polite Philosophy," 265–66.

180. MS Locke f. 2, pp. 122–23. See also Locke, *Essay*, IV.xvii.4.

181. MS Locke f. 2, pp. 89 and 123.

182. Ibid., pp. 129, 130.

183. Locke, "Of Study," 128. Locke did not go as far as Montaigne in doubting the use of such classifications; see Montaigne, "On Experience," in *Complete Essays*, III.13, p. 1222.

184. MS Locke f. 2, pp. 122, 128–29.

185. "New Method," English draft, fol. 60^v.

186. MS Locke f. 2, pp. 247–52. He then added a fifth, "Historica physica," which involved the "history of natural causes & effects" that applied both to the "arts" (that is, technology) and "the nature of things" (p. 252).

187. There are another three examples that do not belong to either series; all will be listed in Locke, *Literary and Historical Writings*. I thank John Milton for his transcription and description of these.

188. "Adversaria 1661," first four unnumbered pages and 290–91.

189. MS Locke c. 28, fols. 157^r–58^v. This is undated.

190. MS Locke c. 28, fols. 50–51; MS Locke f. 15, pp. 110, 119–20, 122–23. One possible exception is the undated schema on a loose leaf inserted in "Adversaria 1661" after p. 25. It clearly belongs to type B, but may have been done in a different year.

191. See, for example, MS Locke c. 28, fols. 50–51, and MS Locke f. 15, 119–20.

192. MS Locke c. 28, fol. 51^v.

193. MS Locke f. 2, pp. 250–51(September 4, 1677). See also the definitions in MS Locke c. 28, fols. 50^v and 51^v under "Adversaria" of August 19 and November 12, 1677.

194. See [Locke], "Methode Nouvelle," 326; Locke, *Essay*, IV.xxi.

195. [Locke], "An Introductory Discourse," 511. The Churchill brothers were Locke's publishers, and this fueled the idea of his involvement. It is doubtful, however, that he would have undertaken this very long essay, given his age and declining health.

196. There are also entries in MS Locke d. 1, pp. 1–25, from the journal for 1679. He did not copy anything into "Lemmata Physica" (MS Locke d. 11), presumably because he used "Adversaria Physica" for most scientific material.

197. For examples of these from his time in both France and Holland, see MS Locke c. 33, fols. 11^r–20^v.

198. MS Locke f. 1, pp. 311–12, with the annotation "60, p. 183" indicating that the entry had been copied into "Adversaria Physica." Reciprocally, at p. 268 of this notebook there is a reference to the journal entry as "76 p. 312", where 76 indicates the year of the journal. Locke usually followed this procedure when transferring from the journals.

199. See MS Locke f. 2, pp. 72, 235, 366–68 (hysterica, hypocondriacus, hydrophobia); MS Locke d. 9, pp. 10–11, 182, 268, 300.

200. See BL, Add. MS 15642, p. 1 (January 1, 1679); MS Locke d. 1, pp. 5, 9, 13, 41; Samuel Cottereau Du Clos, *Observations sur les Eaux minérales du plusieurs provinces de France* (Paris, 1675).

201. Locke, *Essay*, IV.xii.13; and Locke to Thomas Molyneux, January 20, 1693, *Locke Correspondence*, vol. 4, no. 1593; also Locke to William Molyneux, June 15, 1697, ibid., vol. 6, no. 2277.

202. "Adversaria Physica" hosted most of the material that Locke used in writing

Observations upon the Growth and Culture of Vines and Olives (not published until 1766). See MS Locke d. 9, pp. 84-85, 110-11, 118, 123, 238, 264.

203. Richard Lilburne to Locke, [August 6] 1674, *Locke Correspondence*, vol. 1, no. 290.

204. Birch, *History*, 3:220 (May 27, 1675).

205. Locke to Oldenburg, May 20, 1675, *Locke Correspondence*, vol. 1, no. 299; and MS Locke d. 9, pp. 87, 236 (under "Pisces") for a copy of his letter to Lilburne, May 12, 1675; also in *Locke Correspondence*, vol. 8, no. 298A.

206. Lilburne to Locke, August 12, 1675, *Locke Correspondence*, vol. 1, no. 300.

207. Sloane to Locke, December 11, 1696, ibid., vol. 5, no. 2160.

208. MS Locke f. 3, pp. 137 and 139 (May 24 and 30, 1678).

209. Dewhurst, *John Locke*, 304.

210. Locke to Boyle, July 27/August 6, 1678, *Locke Correspondence*, vol. 1, no. 397.

211. See Locke to Boyle, June 16, 1679, ibid., vol. 2, no. 478

212. RS CP.xiii.5, fols. 15^r-17^v. This is in a file titled "Monsters: Longevity."

213. Locke to Sloane, March 15, 1697, *Locke Correspondence*, vol. 6, no. 2219.

214. Locke to Sloane, March 22, 1697, ibid., no. 2227.

215. Locke, "An Account of one who had excressencies horns or extraordinary large Nails on his Fingers & Toes, by Mr Locke," *Philosophical Transactions* 19 (1697): 594-96.

216. Locke to Sloane, February 21, 1704, *Locke Correspondence*, vol. 8, no. 3466; see also Sloane to Locke, February 26, 1704, no. 3473, for agreement that copyright remain with John Churchill.

217. This is preserved in BL, MS Sloane 4039, fols. 259-70.

218. Locke to Sloane, March 15, 1704, *Locke Correspondence*, vol. 8, no. 3489. See "A Register of the Weather for the Year 1692, kept at Oates in Essex. By Mr. John Locke," *Philosophical Transactions* 24 (1705): 1017-37. Locke had intended to send records for another year, but this did not happen.

219. See Hunter, *Boyle Papers*, 65-66.

220. Boyle, *Memoirs . . . Humane Blood*, 6. The editors (p. xi) explain that none of the surviving manuscripts relating to this work date from the 1660s.

221. MS Locke f. 19, pp. 272-73; 302 under "Sanguis." For other evidence of these inquiries, see MS Locke f. 25, pp. 33, 275-77; and Dewhurst, "Locke's Contribution." See also MS Locke c. 42, part 1, p. 98 (topics in "the history of Diseases") and pp. 266-67 (a list of ways of producing flames), both based on Boyle's papers.

222. He reported that a local gentleman said, "his Horses are usually short-breathed, which he imputes to the drinking of that Water." Locke to Boyle, May 5, 1666, *Locke Correspondence*, vol. 1, no. 197.

223. Boyle, preface to *History of the Air*, x; also in *Boyle Works*, ed. Hunter and Davis, 12:9. This is evidently a reference to the lost printed list.

224. See, for example, BP 26, fols. 29 and 45^v; for these manuscripts and a full reconstruction of the publishing history, see the editors' Introductory Notes in *Boyle Works*, ed. Hunter and Davis, 12:xi-xliii, especially xi-xiv.

225. MS Locke c. 42, pp. 16-17.

226. See MS Locke c. 37, pp. 64, 66^v, 102-3, for Locke's insertions in his manuscript copy of Boyle's work; cited in Dewhurst, "Locke's Contribution," 205.

227. Locke to Boyle, October 21, 1691, *Locke Correspondence*, vol. 4, no. 1422. For Boyle's tendency to lose papers, see chap. 6.

228. Boyle, preface to *History of the Air*, xi; also in *Boyle Works*, ed. Hunter and Davis, 12:10.

229. [Locke], "Advertisement of the publisher to the reader," in Boyle, *History of the Air*, iii; also in *Boyle Works*, ed. Hunter and Davis, 12:5. The title Boyle appended to the weather records of another contributor nicely corroborates Locke's concerns: "Mr Townly's Register, if I misremember not" (12:69).

230. [Locke], "Advertisement of the publisher to the reader," in Boyle, *History of the Air*, iv; also in *Boyle Works*, ed. Hunter and Davis, 12:5. The editors of *Boyle Works*, ed. Hunter and Davis, 12:6, state that it is not clear to which passage of Bacon Locke is referring. I am not sure either, although it does bear some relation to the advice to Fulke Greville (see chap. 2), presumably not available to Locke.

231. Locke to Boyle, October 21, 1691, *Locke Correspondence*, vol. 4, no. 1422.

232. The last two empty Heads are "Promiscuous experiments and observations of the air" (Title XLVII) and "Desiderata in the history of the air, and proposals towards supplying them" (Title XLVIII).

233. See Boyle, *Medicinal Experiments: or, a Collection of choice remedies* (1692), in *Boyle Works*, ed. Hunter and Davis, 12:187–88; 192 (for three successive entries on contusions).

234. See Stewart, "Locke's Professional Contacts," 39–41. MS Locke c. 44 contains about 200 pages of transcriptions (by one of Locke's amanuenses) from a collection of recipes once in Boyle's papers, but now only listed in BP 36, fols. 102–11. These papers also included chemical processes difficult to understand without the codes Boyle used. See Hunter, *Boyle Papers*, 85, and Newman and Principe, *Alchemy Tried by Fire*, chap. 5.

235. Boyle, *Medicinal Experiments: or, a Collection of choice and safe remedies* (1693). The third volume of 1694 has numbered entries throughout but no other apparent organization. This may have been produced by John Warr, Boyle's servant and executor. See, respectively, *Boyle Works*, ed. Hunter and Davis, 12:207–68 and 269–98.

236. See James Tyrrell to Locke, January 7, 1693, *Locke Correspondence*, vol. 4, no. 1589.

237. Locke to Molyneux, December 26, 1692, ibid., no. 1583.

238. Locke, *Essay*, II.x.8. For other notices of the crucial role of memory in knowledge, see IV.i.8–9 and IV.xi.11.

239. Ibid., II.x.2, 4–8.

240. Locke announced in the *Essay*, I.i.2, that he would not "meddle with the physical consideration of the mind." It is likely, however, that he subscribed to some version of the "trace" theory advanced by Descartes and Hooke. See Sutton, *Philosophy and Memory Traces*, 162–66.

241. MS Locke f. 18, pp. 80, 103 under "Memoria"; and MS Locke e. 6, fol. 12ᵛ (p. 12 in Locke's pagination) under "Memoria/Oblivio" (see chap. 1 for Newton's notes on this).

242. MS Locke f. 2, p. 248 (September 4, 1677).

243. MS Locke e. 6, fol. 12ᵛ cites "Reyn: Pass.c3." This is Reynolds, *Treatise of the Passions*, 13, on the distinction between memory and recollection ("*Reminiscentia*"). Reynolds was dean of Christ Church in 1648–49 and 1659.

244. MS Locke f. 2, p. 122. He was in agreement with his critic, [Sergeant], *Method to Science*, sig.b2ʳ: "Burgersdicus is clearly contriv'd for the Memory onely, and not for the Reason; and he confounds and over-burthens it too, with the Multitude of his Canons, Rules, and Divisions." Locke's copy of Sergeant (Oak Spring 8.226a) has a two-line page

list in ink and some marginal notes. For criticism of neo-scholastic authors, such as Franco Petri Burgersdijck, see Locke, *Some Thoughts*, 157 (sec. 94).

245. Locke, *Some Thoughts*, 232–33 (sec. 176).

246. Coste, "The Character of Mr. Locke," in *Works*, 10:171; also in Locke, *A Collection of Several Pieces of Mr. John Locke*, ed. Des Maizeaux, iv–xxiv. Coste met Locke in 1697 while they were both living in the home of Francis and Damaris Masham in Essex. Here he made the first French translation of the *Essay*.

247. Locke, *Essay*, II.x.1; also II.xix.1.

248. For Hobbes on this, see chap. 1; for Descartes, see chaps. 4 and 6.

249. Bold visited Locke at Oates; see *Locke Correspondence*, 1:xxxiii for background.

250. Samuel Bold to Locke, April 11, 1699, ibid., vol. 6, no. 2567.

251. Locke to Bold, May 16, 1699, ibid., no. 2590. It is not clear what passage Locke had in mind.

252. Ibid.

253. "New Method," English draft, fol. 56^v. See also Locke to Humphry Smith, July 23, 1703; and Locke to Richard King, July 23, 1703, *Locke Correspondence*, vol. 8, nos. 3321 and 3322.

254. Locke, *Conduct of the Understanding*, 262.

255. "New Method," English draft, fol. 58^r.

256. See Yeo, "John Locke's 'New Method,'" 27–29; Allan, *Commonplace Books and Reading*, 61–70.

257. "Epistle Dedicatory" [by the publisher], in Locke, "New Method," Greenwood, sig.A2^r.

258. Le Clerc, *Ars Critica*, vol. 1, part I, chap. v, §§ 8–11, pp. 146–50.

259. Le Clerc, "Monsieur Le Clerc's Character of Mr. Lock's Method, with his Advice about the Use of Common-Places," in Locke, "New Method," Greenwood, iii–v.

260. See earlier discussions of Placcius in chaps. 2, 4, and 6.

261. Placcius, *De arte excerpendi*, 10, referred to the work as "libro Gallico Anonymo"; also 90–91 for a reproduction of Locke's index.

262. Ibid., 89, 92. For another European notice of Locke's contribution, see "Excerpiren," in Zedler, *Grosses vollständiges Universal-Lexicon*, 8: columns 2321-2322. The relevant volume appeared in 1734.

263. Placcius did, however, endorse Locke's way of giving the page number of a citation in relation to the total number pages in the book; he agreed that it worked well enough, even with variation of page numbers across editions (ibid., 28). On this mode of citation, see Yeo, "John Locke's 'New Method,'" 18.

264. John Freke to Locke, November 3/13 [1686], *Locke Correspondence*, vol. 3, no. 874.

265. Bold to Locke, September 9, 1700, ibid., vol. 7, no. 2771.

266. Watts, *Logic*, 72. For a similar deviation, see [Bell], *Bell's Common Place Book*, sig.B^{r-v}.

267. However, in legal commonplace books it made sense to allocate pages for frequently used topics. See, for example, Anon., *A Brief Method of the Law*; also Barber, title page in *Lawyer's Common-Place and Brief book*, which puts Locke in a series of legal authorities who recommended such notebooks: "Fulbec, Roger North, Lord Hale, Phillips, Locke."

268. Soll, *Information Master*.

Chapter Eight

1. Cohen, "Onset of the Scientific Revolution," 10. See also Gaukroger, *Emergence*, 21, on the question of the consolidation "of the scientific enterprise as such."

2. Burke, *Social History of Knowledge*; Yeo, *Encyclopaedic Visions*.

3. The Society first met on November 28, 1660.

4. Sprat, *History*, 61, 44, 64.

5. Ibid., 115.

6. Oldenburg to Daniel Georg Morhof, September 20, 1672, in *Oldenburg Correspondence*, 9:254; and also February 6, 1672/3, 454–56. The reference is to Marquard Gude (1635–89), a German polymath.

7. See the statements of Petty and Ward in chap. 4.

8. [Hartlib], appendix to *Hartlib. His Legacie of Husbandry*. The questions were intended to add information to Gerard Boate's *Irelands Naturall History* (London, 1645), which was composed in 1645. On some of the scant answers Hartlib received, see Coughlan, "Natural History and Historical Nature," 306–7.

9. Hartlib to Boyle, May 8 or 9, 1654, *Boyle Correspondence*, 1:169–70.

10. Rawley, "To the Reader," in Bacon, *Sylva Sylvarum*, sig.A^r. See also *Works of Francis Bacon*, ed. Spedding, Ellis, and Heath, 2:335.

11. Austen, *Observations*, sig.A2^v.

12. Bacon, *Sylva*, 111, 113; Austen, *Observations*, 4, 9.

13. Austen, *Observations*, 5, 47–48, on "Experiment 411" in *Sylva*. Some of Henry Power's notebooks reveal a similar engagement with Bacon's *Sylva*, in this case with experiments concerning the properties of bodies and fluids; see BL, MS Sloane 1334, fols. 3–43^r.

14. Childrey, *Britannia Baconica*, preface, sig.B4,3^v.

15. Bacon, *Parasceve*, 465.

16. Joshua Childrey to Oldenburg, July 12, 1669, in *Oldenburg Correspondence*, 6:108.

17. Wood, *Athenae Oxonienses*, vol. 3, col. 904.

18. [Ray], preface to *A Collection of English Proverbs* (1670), sig.A2^r, sig.A3,1^r, 32–39.

19. Keynes, *John Ray*, 24, 42. More generally, see Raven, *John Ray Naturalist*, 167–69.

20. Ray, *Collection of English Proverbs* (1678), sig.A2^r.

21. [Ray], *Collection of English Proverbs* (1670), sig.A3^v.

22. Ray, *Collection of English Proverbs* (1678), sig.A3^v–A3,2^v.

23. See "Proverbs and Proverbial Observations concerning Husbandry, Weather and the Seasons of the Year," in [Ray], *Collection of English Proverbs* (1670), 40–45; and *Collection of English Proverbs* (1678), 43–53, with the addition of some Italian proverbs; here there are nineteen caveats.

24. [Ray], *Collection of English Proverbs* (1670), 45, 46–47.

25. Ray, *Collection of English Proverbs* (1678), sig.A2^r. See Beale's offer to Hartlib of eight "Oeconomicall Aphorisms" from Herefordshire (HP62/15/2B, Beale to Hartlib, April 8, 1657). The letter is found in Hartlib Papers, Bundle 62, but incorrectly labeled 61/15/2B.

26. Ray, *Collection of English Proverbs* (1678), sig.A2^r; A3^r.

27. Sprat, *History*, 95, 252.

28. Birch, *History*, 1:8–10, for "Questions propounded and agreed to be sent to Teneriffe by the Lord Brouncker and Mr. Boyle."

29. Boyle, "General Heads." See Hunter, "Robert Boyle and the Early Royal Society."

30. Oldenburg to Boyle, March 24, 1666, *Boyle Correspondence*, 3:126. Oldenburg relayed this extract to Boyle. I cite the editors' translation of the French.

31. Sprat, *History*, 155–56. On the problem of "coordinated observation," see Daston, "Empire of Observation," 87–91.

32. Sprat, *History*, 158–78. See Carey, "Compiling Nature's History"; Shapiro, *Culture of Fact*, chaps. 3–6; Daston and Park, *Wonders and the Order of Nature*, chap. 6.

33. Oldenburg to René Sluse, April 2, 1669, in *Oldenburg Correspondence*, 5:469–70; Oldenburg's Latin here is "Adversaria Philosophica." See Hunter, *Science and Society*, 53. Two of Oldenburg's notebooks are held in the Royal Society: RS MS MM I ("Liber Epistolaris") and MM XXII.

34. RS CP.xx.50b, fol. 110^r; printed in Hunter, *Establishing the New Science*, 337–38. Hooke held this lectureship from 1664 until his death.

35. "The Preface," *Philosophical Transactions* 13 (1683): 2, quoted in Hunter, *Science and Society*, 52.

36. RS CP.xx.50b, fol. 110^r, transcribed in Hunter, *Establishing the New Science*, 338. See also Oldenburg to Johannes Hevelius, January 31, 1667/8, in *Oldenburg Correspondence*, 4:137, promising that "the names of all of you who dedicated your efforts to improving the researches of the Royal Society will be handed down to posterity." On accurate records allowing priority claims, see Iliffe, "In the Warehouse."

37. Beale to Oldenburg, September 24, 1666, in *Oldenburg Correspondence*, 3:232; Beale, "Some Promiscuous Observations, made in Somersetshire," *Philosophical Transactions* 1 (1666): 323.

38. Lister to Oldenburg, May 30, 1671, in *Oldenburg Correspondence*, 8:69.

39. See the three-page "Alphabetical Table" in *Philosophical Transactions* 5 (1670), after p. 2083. See *Oldenburg Correspondence*, editors' comment, vol. 7, n. 5, p. 441.

40. Oldenburg to Boyle, January 27, 1666, in *Boyle Correspondence*, 3:46. He was probably referring to Hooke's paper in RS CP.xx.50a, fols. 99^r–109^v, entitled "Lectures of things requisite to a Ntral History" (on fol. 109^v); printed in Oldroyd, "Some Writings of Robert Hooke," transcription at 151–59.

41. Boyle to Oldenburg, June 13, 1666, *Boyle Correspondence*, 3:170–75; and the recent edition of this letter and associated manuscripts in Hunter and Anstey, *Text of Robert Boyle's "Designe about Natural History."*

42. Boyle, "General Heads"; and Boyle to Oldenburg, June 13, 1666, *Boyle Correspondence*, 3:174.

43. Boyle to Oldenburg, June 13, 1666, *Boyle Correspondence*, 3:171.

44. Boyle, *New Experiments and Observations Touching Cold*, 210, 221. See Rawley, "To the Reader," in Bacon, *Sylva*, sig. A2^v, on the reason for not putting "these Particulars into any exact *Method*."

45. Boyle to Oldenburg, *Boyle Correspondence*, June 13, 1666, 3:171–72.

46. Sprat, *History*, 156, 177; and 179 for "A Scheme at one View representing to the Eye the Observations of the Weather for a Month." There were nine columns. For Hooke's manuscript, see RS CP.xx.2, fols. 2^r–3^v. Despite the detail included there was no mention of how to "standardize and calibrate instruments." See Golinski, *British Weather*, 83–84.

47. See Birch, *History*, 1:311, for minutes of October 7, 1663. See also the mention of "an hygroscope made of the beard of a wild oat, with an index."

48. Boyle, "Some Observations and Directions about the Barometer"; also in *Boyle Works*, ed. Hunter and Davis, 5:504–7. See chap. 3.

49. Childrey to Oldenburg, July 12, 1669, in *Oldenburg Correspondence*, 6:108. See Golinski, *British Weather*, 100, for Childrey's reform of astrological weather forecasting.

50. This letter was included in Boyle, *History of the Air*, Title XIII, 76, as if written by him to Hartlib. For the correct identification, see *Boyle Works*, ed. Hunter and Davis, 12:xiv; and 48–56 for the letter. On the astrological context, see Clericuzio, "New Light on Worsley's Natural Philosophy," 236–46.

51. [Robert Plot], *Quaer's to be propounded to the most ingenious of each County*. Column 3 asks that people "are desired that whatsoever you meet with (worthy notice) in your reading, or converse in the World, upon any of these subjects, to take notice of, and transmit it." Plot ends by saying that "all persons shall receive due acknowledgment according to the merits of the information."

52. Plot, *Natural History of Oxford-Shire*, 5–6. See John Wallis, "A Relation of an Accident by Thunder and Lightning, at Oxford," *Philosophical Transactions* 1 (1666): 222–26.

53. Plot, *Natural History of Oxford-Shire*, 6. See Bodl. MS Digby 176; [Beale], "The Copy of a Letter from Somersetshire, concerning a Strange Frost," *Philosophical Transactions* 8 (1673): 5140–41.

54. See Oldenburg to Boyle, March 27, 1666, *Boyle Correspondence* 3:128: "I wish, Dr Beale had digested his owne sense for you, and not commissioned me to cull it out of his letters here and there." See also his letter of March 17, 1666, 3:114–15, relaying a Beale letter; and Boyle to Oldenburg, March 21, 1666, *Boyle Correspondence*, 3:118.

55. Beale, *Herefordshire Orchards*, 13. Beale later sent a note on apple and pear trees to Ray who printed it in *Collection of English Proverbs* (1678), 53.

56. Beale, "Some Promiscuous Observations"; and in his letter of September 24, 1666, cited in n. 37 above.

57. "Extract of more Barometricall Observations [by John Beale]"; enclosure with Oldenburg's letter to Boyle, January 16, 1665/6, in *Oldenburg Correspondence*, 3:21.

58. Beale to Oldenburg, March 19, 1665/6, in *Oldenburg Correspondence*, 3:64.

59. Beale to Oldenburg, January 13, 1672/3, ibid., 9:406–7. An edited version of this letter was printed in the *Philosophical Transactions* 7 (January 1672/3): 5138–42.

60. Beale to Oldenburg, January 13, 1672/3, in *Oldenburg Correspondence*, 9:407.

61. John Wallis and John Beale, "Some Observations concerning the Baroscope and Thermoscope," *Philosophical Transactions* 4 (1669): 1114–15. (This is drawn from Beale's letters of December 18 and 29, 1669, and January 3, 1669/70.)

62. See Petty on the case of weather records (chap. 7 above); also Golinski, *British Weather*, chap. 3.

63. Ray, preface to *Ornithology*, sig.A2,3ʳ, sig.A2,2ᵛ.

64. Ogilvie, *Science of Describing*, 181, 262–64.

65. Johannes Faber, *Rerum medicarum Novae Hispaniae thesaurus* (Rome: printed by Vitalis Mascardi, 1651), 540, cited and trans. in De Renzi, "Writing and Talking of Exotic Animals," 161. See also Ogilvie, "Many Books of Nature," 33–35.

66. Tancred Robinson to Ray, April 18, 1684, in Ray, *Philosophical Letters*, 153–54. See

also Ray, *Correspondence*, 141. Robinson was knighted in 1714 and appointed physician-in-ordinary to George II. According to an inventory made by William Derham, Ray wrote at least 250 letters to Robinson, many of which are lost. See Ray, *Further Correspondence*, 285–306.

67. Ray to Robinson, April 29, 1685, in Ray, *Philosophical Letters*, 180–81; Ray, *Correspondence*, 165–66.

68. Ray, *Observations Topological, Moral, & Physiological*; Willughby's "A Relation of a Voyage made through a great part of Spain," is at 466–99.

69. [Willughby], "Observations Anatom Patavii 1663," Nottingham University Library, Mi LM 15/2 (I thank the staff of this library for providing a microfilm copy). This manuscript, found within the covers of Willughby's commonplace book (Mi LM 15/1), is in two unidentified hands, with marginal additions by Willughby. The notes cover the period December 10, 1663–January 2, 1664. See Welch, "Francis Willoughby," 73, 83.

70. Ray, *Collection of Curious Travels*, sig.a^{r-v}.

71. See Roos, *Web of Nature*.

72. Raven, *John Ray Naturalist*, 138–39; Ray, *Further Correspondence*, 110–11.

73. John Ray, *Catalogus plantarum circa Cantabrigiam nascentium* (Cambridge: J. Field, 1660). Ray to Lister, June 18, 1667, in Ray, *Correspondence*, 14. Lister, *Historiae animalium Anglicae*; the first tract concerns spiders.

74. Lister to Ray, December 22, 1669, in Ray, *Correspondence*, 48–49. For the earlier letter in Latin, 11, 19.

75. All in Ray, *Correspondence*: Lister to Ray, December 22, 1669, 49; Ray to Lister, February 13, 1669, 52; Lister to Ray, December 22, 1669, 49; Ray to Lister, February 13, 1669, 52, and April 28, 1670, 55.

76. Lister to Ray, March 21, 1670, in ibid., 83. For a similar declaration, see Lister to Oldenburg, April 28, 1671, in *Oldenburg Correspondence*, 8:43.

77. Robinson to Ray, August 29, 1684, in Ray, *Correspondence*, 151.

78. Ray to Lister, June 18, 1667, in Ray, *Further Correspondence*, 112.

79. Ray to Lister, November 15, 1669, in Ray, *Correspondence*, 43.

80. Ray to Lister, December 10, 1669, in ibid., 47; Lister to Ray, December 22, 1669, in ibid., 51.

81. Lister to Ray, June 29, 1669, in ibid., 60.

82. Lister to Ray, October 8, and December 22, 1670, in ibid., 65, 74.

83. Ray to Lister, March 3, 1670, in ibid., 81. In Willughby's "Commonplace book" covering the period 1658–65 (Nottingham University Library: Mi LM 15/10), at p. 433 there is an entry on rising sap in birch trees, dated March 1664/5. Ray had access to this notebook sometime after Willughby died on July 3, 1672.

84. Lister to Ray, February 8, 1670, in Ray, *Correspondence*, 79. For the context, see Hartley, "Exploring Knowledge of Trees."

85. Lister to Ray, June 20, 1673, in Ray, *Philosophical Letters*, 116; Ray, *Correspondence*, 103.

86. See Welch, "Francis Willoughby," for publications. The will provided Ray with a sixty-pound annuity. One result was Ray's *Historia Plantarum*, 3 vols. (London: Henry Fairthorne, 1686, 1688, 1704).

87. Robinson to Ray, April 18, 1684, in Ray, *Correspondence*, 141.

88. Lister to Ray, undated, c. 1669, in Ray, *Correspondence*, 53, 54.

89. Lister to Ray, December 22, 1670, in ibid., 73.

90. Lister to Ray, February 8, 1670, in ibid., 78. These exchanges continued despite a cooling in their relationship at the time of the priority dispute between Lister and Edward Hulse over the description of the gossamer threads of spiders. Ray sought to clarify his own position in *Philosophical Transactions* 5 (1670): 2103–5. After Willughby's death, Lister invited Ray to live with him in York. See Raven, *John Ray Naturalist*, 139, 165.

91. Lister to Oldenburg, April 28, 1671, in *Oldenburg Correspondence*, 8:44.

92. Ray to Aubrey, August 24, 1692, Bodl. MS Aubrey 13, fol. 176$^{r, v}$, printed in Ray, *Further Correspondence*, 175.

93. Ray to Aubrey, October 20, 1692, Bodl. MS Aubrey 13, fol. 177, printed in Ray, *Further Correspondence*, 179.

94. Sprat, *History*, 94. William Croone, a founding member, was the first to occupy this role. See Birch, *History*, 2:223–24, for the "Letter-book appointed by statute for that purpose" (December 5, 1667); and 3:343 for a decision on September 24, 1677, about "the officiating secretary taking short notes of all that passes." For the list of formal "Books," see Hall, "A Note on Sources," in her *Library and Archives of the Royal Society*, 221–22, on "Journal Books" (minutes of meetings); Council Minutes 1662; and "Register Books" (copies of important papers read to the meetings).

95. Hunter, *Establishing the New Science*, 4; Feingold, "Of Records and Grandeur," 173.

96. Oldenburg to Boyle, July 4, 1665, in *Oldenburg Correspondence*, 2:430, quoted in Feingold, "Of Records and Grandeur," 175.

97. The best account is in Hunter, *Establishing the New Science*, 123–55. He cites (at 125) Birch, *History*, 1:316, for the first use of this term in the minutes of the council meeting of October 19, 1663, which directed that "Mr Hooke have the keeping of the repository of the society, for which the west gallery of Gresham College was appointed." See also "Minutes of the Committee for taking care of the Repository," February 3, 1676 (RS DM/16/39).

98. Oldenburg to Moray, November 7, 1665, in *Oldenburg Correspondence*, 2:592. This answers Moray to Oldenburg, November 5, 1665, 590–91.

99. Oldenburg to Boyle, December 17, 1667, in *Oldenburg Correspondence*, 4:58.

100. Birch, *History*, 1:123 (November 5), 127 (November 19). See chap. 4 for Haak's knowledge of Harrison's indexes.

101. See RS MM 4.72, signed "A. B" and dated "Oct 19. 1674"; transcription in Hunter, *Establishing the New Science*, 230.

102. Birch, *History*, 3:343 (September 24, 1677). A German traveler later reported that the "library is locked away in small cupboards" (von Uffenbach, *London in 1710*, 101–2).

103. Hoskyns to Aubrey, March 22, 1674, Bodl. MS Aubrey 12, fol. 214^{r}, cited in Hunter, *Science and Society*, 67. See Grew, *Musaeum regalis societatis*. This choice of the word "museum" for a collection of objects was relatively novel in England. See also Tradescant, *Musaeum Tradescantianum*; and *OED*.

104. We naturally want to say the "archive" of the Society, yet the members did not use this term, a relatively new one in English. Phillips, *The New World of English Words*, defines "archive" as "a place where ancient Records are kept"; and "repository" as "a storehouse . . . built for the laying up of rarities either in picture or other arts." But for

the Latin term, see Oldenburg to Johannes Hevelius, January 31, 1667/8, in *Oldenburg Correspondence*, 4:135 and 137: "Tradentur ea Omnia Regiae Societatis Archivis" ("all these things have been deposited in the Royal Society's archives").

105. Sprat, *History*, 115–16, also 95, 318–19; for the contrast with Hooke, see Wood, "Methodology and Apologetics," especially 7–8.

106. Glanvill, *Plus Ultra*, 90–91.

107. Boyle, *Tracts Written by Boyle* (1670), 261.

108. "The Publisher to the Ingenious Reader," in Boyle, *New Experiments and Observations Touching Cold*, 205.

109. Boyle, *Usefulnesse* (1671), 400.

110. These notes and papers occupy the first 100 or so pages of a large folio book, bound together with Hooke's own rough notes for the minutes (the next 400 pages) he took in his new role as secretary, succeeding Oldenburg. See RS MS 847.

111. Adams and Jardine, "The Return of the Hooke Folio," 236, 238. citing RS MS 847.

112. Waller, "Life of Dr. Robert Hooke," in *Posthumous Works of Hooke*, iii.

113. Hooke, "The Preface," in Knox, *An Historical Relation of the Island Ceylon*, sig.a2^r, sig.a^r.

114. On September 11, 1677, Hooke made a diary entry about treating "a bad memory and severall other distempers [by swallowing] . . . very fine filings of the best refined silver." Hooke, *Diary*, 311. For the full range of self-experimentation, see Jardine, "Hooke the Man," 163–206.

115. Hooke, *Diary*, September 6, 1672, 7: "bought August *Transactions*, Streets book of mnemonick verses, both 1sh." See Mulligan, "Robert Hooke's 'Memoranda'"; and her "Self-scrutiny."

116. Aubrey, *Brief Lives*, 1:411.

117. Hooke, "General Scheme," 5. Hesse, "Hooke's Philosophical Algebra," 68, dates its composition as 1666; but see Pugliese, "Scientific Achievement of Hooke," 10, for the likely date of 1668.

118. Waller, "Life of Dr. Robert Hooke," in *Posthumous Works of Hooke*, i, quotes this from "a small Pocket-Diary." I do not attempt to treat in detail Hooke's use of his diaries, but see Henderson, "Unpublished Material from the Memorandum Book of Robert Hooke."

119. Hooke, preface to *Micrographia*, sig.b1^v, alludes to "another Discourse," which must be "A General Scheme."

120. Hooke, Preface to *Micrographia*, sig.a1^r; also sig.b2^v. On the perceived effects of original sin, see Harrison, *Fall of Man*.

121. Hooke, preface to *Micrographia*, sig.b2^r. On this point, see Inwood, *The Man Who Knew Too Much*, 64–65. Hooke used various synonyms for reason, almost interchangeably: thus in a manuscript of April 21, 1692, he refers to "the Intellect and the mind & judgment." See the transcription of Trinity College, Cambridge, MS o.11a.1^{14} in Oldroyd, "Some 'Philosophicall Scribbles.'" Note that "Philosophicall Scribbles" (MS o.11a.1^{28}) is the item that Oldroyd transcribes in full.

122. Hooke, Preface to *Micrographia*, sig.b1^v.

123. Hooke, "General Scheme," 34.

124. Ibid., 18. On the close links with Bacon's method, see Hunter, "Hooke the Natural Philosopher."

125. Birch, *History*, 2:283, for May 9, 1668; see Lewis, "Efforts of the Aubrey Correspondence Group"; and his *Language, Mind, and Nature*, 194–200.

126. Hooke to Andrew Paschall, March 1679, RS Early Letters H3/61. Paschall was using Seth Ward's idea. See also Hooke, "Some Observations . . . Concerning the Chinese Characters."

127. Hooke to Leibniz, May 15, 1681, RS Letter Book, EL/H3/64. Hooke went on to call this "the Algebra of Algebras or the Science of methods." He envisaged a real character, based on a reduction of things and notions to primitive natural kinds, as a part of this yet to be achieved "Generall method." Hooke's letter to Leibniz on "the farther usefulnes of the philosophical language and character was read" on July 15, 1680; Birch, *History*, 4:47. See Lewis, "Hooke's Two Buckets," 356, n. 61, on Haak as the intermediary for this correspondence.

128. See the table in Hooke, "General Scheme," 22.

129. Hooke, "Lectures of Light." This was given on June 21, 1682.

130. See Hooke, *Posthumous Works*, 138–39, for Waller's attempt at suggesting the possible link.

131. Hooke, "Lectures of Light," sec. VII, 144. For this hypothesis, see Singer, "Robert Hooke on Memory"; Kassler, *Inner Music*, chap. 3; Richards, *Mental Machinery*, 67–69; Draaisma, *Metaphors of Memory*, 56–61; Draaisma, "Hooke on Memory."

132. Hooke was challenged on this point when he presented this part of the lecture at the request of some members of the Royal Society. To the objection that his account tended "to prove the soul mechanical," Hooke replied that "no such thing was hinted," only that "the soul forms for its own use certain corporeal ideas" (Birch, *History*, 4:153–54, for minutes of the meetings of June 21 and 28, 1682). For Evelyn's positive reaction in his diary of June 20, 1682, see Inwood, *The Man Who Knew Too Much*, 331.

133. Hooke, "Lectures of Light," 143–44. See also Draaisma, "Hooke on Memory," 117; and Landauer, "How Much Do People Remember."

134. Ibid., 140. On Hooke's weather clock as an analog of memory, see Wilding, "Graphic Technologies," 124.

135. Hooke, "Lectures of Light," 140, 146.

136. Hooke, "General Scheme," 6. Unlike Hobbes, Hooke did not specifically mention association of ideas as one of the problems, although this is implied here. In 1650 Hobbes gave the following example: "from St. Andrew the mind runneth to St. Peter, because their names are read together; from St. Peter to a stone, because we see them together; and for the same cause, from foundation to church, from church to people, and from people to tumult." Hobbes, *Human Nature*, 20. See also Hobbes, *Leviathan*, vol. 2, chap. 3.

137. See the undated manuscript in Hooke's hand entitled "Lectures of things requisite to a Ntral History," RS CP.xx.50a, fols. 99^r–109^v. This is transcribed in Oldroyd, "Some Writings of Hooke," 155.

138. Hooke, "General Scheme," 21. Compare Locke's allusion to Bacon's advice; see chap. 7.

139. Ibid., 62.

140. Hooke, "Proposalls for the Good of the R.S" [no date, possibly 1673], RS CP.xx.50, fol. 86^r. See also fol. 94^r for the means of attaining knowledge: "to wit in three places. first in bookes. 2dly in men. 3dly in the things themselves." For an account of how this worked, see Johns, "Reading and Experiment."

141. Hooke, *Micrographia*, sig.a2^r.

142. Hooke, "General Scheme," 8.

143. RS CP.xx.50, fol. 92^r; printed in Hunter, *Establishing the New Science*, 236.

144. Hooke, Trinity College Library, MS 0.11a.1 [undated]; printed in Oldroyd, "Some 'Philosophicall Scribbles,'" 18. Given the speculation in this manuscript on the physical basis of memory, Oldroyd suggests a date of 1681–82, around the time of the lecture on light.

145. "Dr Hook's Discourse concerning Telescopes and Microscopes," in Hooke, *Philosophical Experiments*, 263–64.

146. Hooke, "General Scheme," 5–6.

147. BL, Sloane MS 1039, fols. 1^r and 2^r. This is probably a draft of Hooke's preface to Pitt's work, which did not appear in the published version. See Pitt to Hooke, April 25, 1680, Sloane MS 1039, fol. 5^r asking for this.

148. Sprat, *History*, 175; and 179 for the table.

149. Hooke, "Rules for observing the weather & for an History of it" (RS CP.xx.24, fol. 41^r.). This is marked "read Oct:7: 63."

150. "Mr. Hooke's Analysis of the whole Businesse of Navigation under the Title of Hydrographie," Bodl. MS Rawlinson, A. 171, fols. 245^v–246^r. [c. 1686]. For a related document on "Hydrography," dated March 24, 1685, see RS CP.xx.70, fol. 156.

151. Hooke, "General Scheme," 64. For relevant discussion, see Harwood, "Rhetoric and Graphics in *Micrographia*," 134–47.

152. Hooke, "General Scheme," 21, 63. For a similar rejection of "Needless philology, avoyding also the citations of all kinds of authors & opinions," see Hooke, "Lectures of things requisite to a natural history," in Oldroyd, "Some Writings of Hooke," 155. Hooke followed Bacon in alleging that textual scholarship was verbose.

153. Hooke, "General Scheme," 21, 42, 62, 63–64.

154. Ibid., 18. In an undated paper by Hooke, entitled, by the editor, "Dr. Hook's method of making experiments," there is a prescription "to register the whole Process of the Proposal, Design, Experiment, Success, or Failure" (Hooke, *Philosophical Experiments*, 27).

155. Hooke, "General Scheme," 64.

156. Ibid., 7, 61. Waller interpolated: "This I think Dr. Hook never wrote," 5–6. For a reference to "mechanical algebra," see Christopher Wren to Matthew Wren, c. 1651–54, BL, Add. MS 25071, fols. 42–43, printed in Bennett, "A Study of *Parentalia*," 144–45.

157. Whewell, *Philosophy of the Inductive Sciences*, 2:267–68.

158. Hooke, "General Scheme," 6–7.

159. RS CP.xx.50a, fol. 94^r, c. 1670s, in Hunter, *Establishing the New Science*, 232.

160. Hooke, Guidhall, London MS1757, fols. 103^v, 104^r.

161. Hooke, "General Scheme," 7.

162. Hooke set out the reasoning behind such a notion in "Mathematical Language," RS CP.xx.72, fols. 160^r–161^v (undated); see also Slaughter, *Universal Languages*, 183. Hooke did not use term "Philosophical Algebra" in this text.

163. Hesse, "Hooke's Philosophical Algebra," 75–77; Oldroyd, "Robert Hooke's Methodology of Science."

164. Oldenburg to Boyle, January 27, 1666, *Boyle Correspondence*, 3:46. He asked Boyle that "I may not be named" for mentioning this.

165. Boyle, BP 9, fol. 73^r [1670s–1680s]; transcription in Hunter and Anstey, *Text of Robert Boyle's "Designe about Natural History,"* 7; also Anstey and Hunter, "Boyle's 'Designe about Natural History,'" 118, for comment on the relation to Hooke.

166. Hooke, *Micrographia*, sig.d1^r.

167. See Lewis, "Hooke's Two Buckets," 355–58.

168. See "Proposals concerning the arrangement &c of the publications of the R. Society," RS CP.xx.50, fol. 97 (not in Hooke's hand). For the degree of control Hooke envisaged, see Johns, *Nature of the Book*, 484–89; Wilding, "Graphic Technologies," 131–33. Of course, Hooke was not alone in these concerns. See Feingold, "Of Records and Grandeur."

169. See Holmes, Renn, and Rheinberger, *Reworking the Bench*, on the later development of laboratory notebooks.

170. Hooke, "General Scheme," 63–65.

171. Ibid., 64–65.

172. Hooke, "Dr Hook's method of making experiments," in Hooke, *Philosophical Experiments*, 26–28.

173. RS CP.xx.50a, fol. 100^r; in Oldroyd, "Some Writings of Hooke,"152.

174. RS CP.xx.50b, fol. 110^v; printed in Hunter, *Establishing the New Science*, 338. See the similar promise in Hooke, *Micrographia*, sig.b1^v.

175. Hooke, *Micrographia*, sig.b2^r. For Hooke's architectural metaphor, see "Lectures of Light," 144–45; "Discourse of Earthquakes," 330. For notices of this, see Oldroyd, "Robert Hooke's Methodology of Science," 115; Hunter, "The Theory of the Impression," 183.

176. Hooke, "General Scheme," 20–21.

177. Hooke, "To the Reader," in *An Attempt to Prove the Motion of the Earth*, sig.A2, 2$^{r–v}$. He continued that "the greatest part of human Invention being but a luckey hit of chance, for the most part not in our power." Hesse contrasts this comment with the "somewhat extravagant claims Hooke was making for the infallibility of his method up to 1666" ("Hooke's Philosophical Algebra," 83).

178. Hooke, preface to *Micrographia*, sig.b1^r, sig.d1^r

179. Bacon, *Advancement of Learning*, 85; also *De augmentis*, 361 on uniting axioms.

180. Hooke, preface to *Micrographia*, sig.g2^v. He also professed to be content "if I have contributed the *meanest foundations* whereon others may raise noble *Superstructures.*" (sig.b1$^{r–v}$). For his famous drawing of a microscopically enlarged mite, see *Micrographia*, scheme XXXVI.

Chapter Nine

1. Helmholtz, *Popular Lectures on Scientific Subjects*, 12.

2. Huxley, "On the Advisableness of Improving Natural Knowledge," in *Fortnightly Review*, vol. 3, ed. George Henry Lewes (London: Chapman and Hall, 1866), 628.

3. Gibbon, *Essay*, 12 and 13–14 (in French ed., 1761, pp. 11, 13). Gibbon went on to defend "the Belles Lettres," pointing out that Descartes, Newton, and Gassendi did not confine themselves to mathematics (13–16).

4. Gibbon, *Essay*, 4. See Pocock, *Barbarism and Religion* (1999), 1:215, for Gibbon's inference that since other subjects have fallen from the throne, so too might physics and mathematics.

5. D'Alembert, *Preliminary Discourse to the Encyclopedia of Diderot*, 150, 163; and the

"Système Figuré" in "Observations sur la Division des Sciences du Chancellor Bacon," in Diderot and d'Alembert, *Encyclopédie*, 1:li–lii, facing lii.

6. Bacon, *New Atlantis*, in *Francis Bacon: Major Works*, ed. Vickers, 486.

7. Beale to Hartlib, as conveyed by Hartlib to John Worthington, August 26, 1661, in Worthington, *Diary*, 1:369. See discussion in chap. 4.

8. Oldenburg to René Sluse, October 23, 1667, in *Oldenburg Correspondence*, 3:537.

9. See chap. 3 for Evelyn to Beale, July 11, 1679, BL, Add. MS 78299, fol. 2ᵛ.

10. For a recent debate, stimulated in part by methods of data analysis not available to the virtuosi, see Leonelli, "Making Sense of Data-driven Research."

11. Weber, "Science as a Vocation," 12. This was given as a lecture in 1917 and published in 1922.

[Manuscript Sources]

Major manuscripts are listed here; other manuscripts are identified in the notes.

University of Sheffield Library

Samuel Hartlib Papers

British Library, London

MS Sloane
- 623 Notebook of Daniel Foote, c. 1670
- 1039 Papers of Robert Hooke
- 1334 Henry Power's notebook, c. 1650–1660s
- 1466 Thomas Harrison material
- 2891–2900 Abraham Hill's commonplace books
- 2903 William Petty papers
- 3391 Notes by James Petiver on Boyle
- 4039 John Locke's weather register, 1691–92, sent to Hans Sloane in March 1704

Additional MSS
- 4384 Correspondence about mnemonics, c. 1661–63, collected by Thomas Birch
- 6038 Sir Julius Caesar's commonplace book on the template of John Foxe's *Pandectae locorum communium* (1572)
- 15642 John Locke's journal, 1679
- 15948, 78221, 78298–99, 78683 John Evelyn correspondence
- 15950 John Evelyn's notebook
- 28273 Notebook originally belonging to John Locke Snr.
- 28728 Letters and papers of John Locke and Nicolas Toinard
- 32554 John Locke's medical commonplace book, c. 1660–c. 1667

41846 Thomas Harrison's "Arca Studiorum"

46470 John Locke's pocket memorandum book, 1669

78323–78325 John Evelyn's "Kalendarium"

78327 John Evelyn's "Vade Mecum"

78329–31 John Evelyn's commonplace books

78337, 78340 John Evelyn's collection of recipes

Bodleian Library, University of Oxford

MS Locke—Lovelace Collection of John Locke's Papers

c. 28 Papers on philosophy, religion, and medicine c. 1662–c. 1694

c. 33 Locke's notes on his reading when abroad

c. 37 Copy of Boyle's *General History of the Air*

c. 42 Commonplace books: Part I, Medical, and Part II, Philosophical and ethical

d. 1 Commonplace book, 1679

d. 9 "Adversaria Physica," a medical and scientific commonplace book

d. 10 "Lemmata Ethica"

d. 11 "Lemmata Physica"

e. 4 Medical notebook, 1650s

e. 6 Notebook c. 1660–64

e. 17 Commonplace book, including book list by Thomas Barlow c. 1650s

f. 1 Journal, November 1675–December 1676

f. 2 Journal, 1677

f. 3 Journal, 1678

f. 4 Journal, 1680

f. 5 Journal, 1681

f. 6 Journal, 1682

f. 11 Pocket memorandum book, 1649–66

f. 14 Commonplace book, c. 1659–c. 1667

f. 15 Pocket memorandum book, 1677

f. 18 Medical commonplace book, 1659–60

f. 19 Medical commonplace book, c. 1662–c. 1669

f. 25 Chemical notebook, 1663–67

f. 27 Pocket memorandum book, 1664–66

f. 28 Pocket memorandum book, 1678–85

f. 29 Pocket memorandum book, 1683–1702

MS Aubrey

4, 6, 10, 13 John Aubrey Papers

MS Rawlinson

A. 171, fols 245^v–246^r, "Mr. Hooke's Analysis of the whole Businesse of Navigation under the Title of Hydrographie"

Beinecke Library, Yale University

MS Osborn

b112 Robert Southwell's commonplace book, c. 1660s

b182 Richard Cromleholme Bury's notebook, 1681

MS Osborn Files

    14242 Memorandum recording the advice of Matthew Hale, in the possession of Robert Southwell, 1664

The Royal Society of London

BP (Boyle Papers)

    2, 5, 7, 8, 9, 10, 21, 22, 25, 26, 27, 28, 36, 38. These papers include workdiaries, papers, drafts, loose notes.

RS MS

    22 Robert Boyle's commonplace book

    41 Medical commonplace book used by Robert Boyle and Lady Ranelagh

    185–191 Robert Boyle's notebooks

    193 Robert Boyle's workdiary 18

    194 Robert Boyle's workdiary 33

    198 Robert Boyle's notebook

    847 Robert Hooke Folio

    MM I, MM XXII Henry Oldenburg's notebooks

RS Classified Papers

    RS CP.xx Robert Hooke papers

    RS CP.xiii Contains John Locke's material about a boy with long nails

RS DM/16/39 Domestic Manuscripts

RS Miscellaneous Manuscripts 4.72

Houghton Library, Harvard University

MS Eng.992.7 John Evelyn's commonplace book, 1690

Fitzwilliam Museum, Cambridge

MS 1936 Isaac Newton's Fitzwilliam notebook

Cambridge University Library

Add. 6986 "Dr. [James] Duport's Rules"

Add. 3996 Isaac Newton's commonplace book containing "Questiones quaedam Philosophicae"

Trinity College, Cambridge, Library

MS 0.10A.33 James Duport, "Rules to be observed by young scholars in the University"

MS 0.11A.1 Robert Hooke, "Some Philosophicall Scribbles."

Guildhall Library, London

MS 1757 Robert Hooke's notes for a lecture, c. 1665–66

[Bibliography]

Aczel, Amir D. *Descartes' Secret Notebook: A True Tale of Mathematics, Mysticism, and the Quest to Understand the Universe.* New York: Broadway Books, 2005.

Adams, Frank. *Writing Tablets with a Kalender for xxiii Yeres.* London, 1581.

Adams, Robyn, and Lisa Jardine. "The Return of the Hooke Folio." *Notes and Records of the Royal Society of London* 60 (2006): 235–39.

Agrippa, Cornelius. *The Uncertainty and Vanity of the Arts and Sciences.* London: Henry Wykes, 1569. First published in 1531 as *De incertitudine et vanitate scientiarum et Artium.*

Aho, James. *Confession and Bookkeeping: The Religious, Moral and Rhetorical Roots of Accounting.* New York: State University of New York Press, 2005.

———. "Rhetoric and the Invention of Double Entry Bookkeeping." *Rhetorica* 3 (1985): 21–43.

Algazi, Gadi. "Food for Thought: Hieronymus Wolf Grapples with the Scholarly Habitus." In Dekker, *Egodocuments and History*, 21–43.

Allan, David. *Commonplace Books and Reading in Georgian England.* Cambridge: Cambridge University Press, 2010.

Annas, Julia. "Aristotle on Memory and Self." In *Essays on Aristotle's De Anima*, ed. M. C. Nussbaum and A. O. Rorty, 297–311. Oxford: Oxford University Press, 1992.

Anon. *Ad Herennium. De ratione dicendi.* Translated by Harry Caplan. Cambridge, MA: Harvard University Press, 1954.

———. *A Brief Method of the Law. Being an exact Alphabetical Disposition of all the Heads necessary for a Perfect Common-Place.* London: Printed by the Assignees of Richard and Edward Atkins esquires, for John Kidgell, 1680.

———. *A Brief Vindication of the Royal Society: From the late Invectives and Misrepresentations of Mr. Henry Stubbe.* London: Printed for John Martin, 1670.

———. "Observazioni meterologiche fatte in Firenze (December 1654–March 1670)." In

V. Antinori, *Archivo meterologico centrale Italiano*, nell I. E. R. Museo di Fisica e Storia Naturali di Firenze. Firenze: Società Tipografica sulla Legge del Grano, 1858.

———. *The Polite Gentleman; or, Reflections upon the several Kinds of Wit*. London: R. Basset, 1700.

Anstey, Peter R. "Experimental versus Speculative Natural Philosophy." In *The Science of Nature in the Seventeenth Century: Patterns of Change in Early Modern Natural Philosophy*, ed. Peter Anstey and John Schuster, 215–42. Dordrecht: Springer, 2005.

———. "John Locke and Helmontian Medicine." In *The Body as Object*, ed. Wolfe and Gal, 93–117.

———. *John Locke and Natural Philosophy*. Oxford: Oxford University Press, 2011.

———. "Literary Responses to Robert Boyle's Natural Philosophy." In *Science, Literature and Rhetoric in Early Modern England*, ed. Juliet Cummins and David Burchell, 145–62. Aldershot: Ashgate, 2007.

———. *The Philosophy of Robert Boyle*. London: Routledge, 2000.

Anstey, Peter R., and John Burrows. "John Locke, Thomas Sydenham, and the Authorship of Two Medical Essays." *Electronic British Library Journal* (2009), article 3.

Anstey, Peter R., and Michael Hunter. "Robert Boyle's 'Designe about Natural History.'" *Early Science and Medicine* 13 (2008): 83–126.

Aristotle. *Aristotle: The Complete Works, the Revised Oxford Translation*. 2 vols. Edited by Jonathan Barnes. Princeton, NJ: Princeton University Press, 1984.

Aubrey, John. *Aubrey on Education: A Hitherto Unpublished Manuscript by the Author of "Brief Lives."* Edited by J. E. Stephens. London: Routledge, 1972.

———. *"Brief Lives" chiefly of Contemporaries, set down by John Aubrey, between the Years 1669 & 1696*. 2 vols. Edited by Andrew Clark. Oxford: Clarendon Press, 1898.

Augustine. *Confessions*. Translated by R. S. Pine-Coffin. London: Penguin Classics, 1961.

Austen, Ralph. *Observations upon some part of Sir Francis Bacon's Naturall History, as it concernes Fruit-trees, Fruits, and Flowers: Especially the Fifth, Sixth, and Seventh Centuries, Improving the Experiments mentioned, to the best Advantage*. Oxford: Printed by Hen. Hall for Thomas Robinson, 1658.

Avramov, Iordan, Michael Hunter, and Hideyuki Yoshimoto. *Boyle's Books: The Evidence of His Citations*. Occasional Papers 4. London: Robert Boyle Project, 2010.

Bacon, Francis. *The Advancement of Learning*. Vol. 4 of *The Oxford Francis Bacon*, ed. Michael Kiernan. Oxford: Oxford University Press, 2000.

———. "Advice to Fulke Greville on his Studies." In *Francis Bacon: The Major Works*, ed. Brian Vickers, 102–6.

———. "[Advice to the Earl of Rutland on his Travels]." In *Works of Francis Bacon*, ed. Spedding, Ellis, and Heath, 9:6–20.

———. *Baconiana, Or Certain Genuine Remains of Sr. Francis Bacon*. London: J. D. for Richard Chiswell, 1679.

———. "Comentarius solutus sive pandecta, sive ancilla memoriae." In *Works of Francis Bacon*, ed. Spedding, Ellis, and Heath, 11:39–95.

———. *De dignitate et augmentis scientiarum* (called *De augmentis*). In *Works of Francis Bacon*, ed. Spedding, Ellis, and Heath, 4:275–498.

———. *Descriptio globi intellectualis*. In *Philosophical Studies, c. 1611–c. 1619*. Vol. 6 of *The Oxford Francis Bacon*, ed. Graham Rees, 95–169. Oxford: Oxford University Press, 1996.

———. "A Description of the Intellectual Globe." In *Works of Francis Bacon*, ed. Spedding, Ellis, and Heath, 5:503–44.

———. *The Essayes or Counsels, Civill and Morall*. Vol. 15 of *The Oxford Francis Bacon*, ed. Michael Kiernan. Oxford: Clarendon Press, 1985.

———. *Francis Bacon: The Major Works*. Edited by Brian Vickers. Oxford: Oxford University Press, 2008.

———. *The Instauratio Magna, Part II: Novum Organum and Associated Texts*. Vol. 11 of *The Oxford Francis Bacon*, ed. Graham Rees. Oxford: Clarendon Press, 2004.

———. *The Instauratio Magna, Part III: Historia naturalis et experimentalis; historia ventorum and historia vitae & mortis*. Vol. 12 of *The Oxford Francis Bacon*, ed. G. Rees and M. Wakely. Oxford: Clarendon Press, 2007.

———. "A Letter, and Discourse, to Sir Henry Savill, touching helps, for the Intellectual Powers." In *Resuscitatio, Or, Bringing into Publick Light Severall Pieces, of the Works*, ed. W. Rawley, 225–31. London, 1657.

———. *Novum Organum*. In *The Oxford Francis Bacon*, 11:50–447.

———. *Of the Advancement and Proficience of Learning*. Translation of *De augmentis* by Gilbert Watts. Oxford, 1640.

———. *Parasceve, ad historiam naturalem, et experimentalem*. Translated as *Preparative to a Natural and Experimental History*. In vol. 11 of *The Oxford Francis Bacon*, ed. Graham Rees, 448–85.

———. *Sylva Sylvarum or a Naturall Historie, in Ten Centuries, published after the Author's death by William Rawley*. London: Printed by J. H. for William Lee, 1627.

———. *The Works of Francis Bacon*. 14 vols. Edited by J. Spedding, R. L. Ellis, and D. Heath. Stuttgart: F. Frommann Verlag G. Holzboog, 1963. Reprint of 1857–1874 edition.

Barber, G. L. *Lawyers' Common-Place and Brief Book with an Explanation of the Method of Common-Placing, an Alphabetical Index of about Three Thousand Titles and Subjects, Together with an Index to Cases*. Chicago: E. B. Myers, n.d.

[Barlow, Thomas.] A *"Library for Younger Schollers": Compiled by an English Scholar-Priest about 1655*. Edited by Alma De Jordy and Harris Francis Fletcher. Urbana: University of Illinois Press, 1961.

Bayle, Pierre. *The Dictionary Historical and Critical of Mr Peter Bayle. The second edition, . . . to which is prefixed, the life of the author, revised, corrected and enlarged, by Mr Des Maizeaux*. 5 vols. London: J. J. and P. Knapton, D. Midwinter et al., 1734–38.

———. *Dictionnaire Historique et Critique*. 3rd ed. 4 vols. Rotterdam: M. Bohm, 1720.

Beadle, John. *The Journal or Diary of a Thankful Christian: Presented in some Meditations upon Numb. 33:2*. London: Printed by E. Cotes for Tho. Parkhurst, 1656.

Beal, Peter. "Notions in Garrison: The Seventeenth-Century Commonplace Book." In *New Ways of Looking at Old Texts: Papers of the Renaissance Text Society*, ed. W. Speed Hill, 131–47. Tempe, AZ: ACMRS, 1993.

Beale, John. *Herefordshire Orchards, a Pattern for all England. Written in an Epistolary Address to Samuel Hartlib Esq*. London: Roger Daniel, 1657.

[Bell, John]. *Bell's Common Place Book, for the Pocket; formed generally upon the Principles Recommended and Practised by Mr. Locke*. London: Printed for John Bell, 1770.

Benedict, Barbara M. *Curiosity: A Cultural History of Early Modern Inquiry*. Chicago: University of Chicago Press, 2001.

Bennett, J. A. *The Mathematical Science of Christopher Wren*. Cambridge: Cambridge University Press, 1982.

———. "A Study of *Parentalia*." *Annals of Science* 30 (1973): 129–47.

Bennett, Jim, Michael Cooper, Michael Hunter, and Lisa Jardine, eds. *London's Leonardo: The Life and Work of Robert Hooke*. Oxford: Oxford University Press, 2003.

Bennett, Jim, and Scott Mandelbrote, eds. *The Garden, the Ark, the Tower, the Temple: Biblical Metaphors of Knowledge in Early Modern Europe*. Oxford: Museum of the History of Science, 1998.

Bennett, Michael. "Note-Taking and Data-Sharing: Edward Jenner and the Global Vaccination Network." *Intellectual History Review* 20 (2010): 415–32.

Bernecker, Sven. *Memory: A Philosophical Study*. Oxford: Oxford University Press, 2010.

Birch, Thomas. *The History of the Royal Society of London*. 4 vols. London: A. Millar, 1756–57.

Biswas, Asit K. "The Automatic Rain-Gauge of Sir Christopher Wren, F.R.S." *Notes and Records of the Royal Society of London* 22 (1967): 94–104.

Blair, Ann. "Humanist Methods in Natural Philosophy: The Commonplace Book." *Journal of the History of Ideas* 53 (1992): 541–51.

———. "Note Taking as Transmission." *Critical Inquiry* 31 (2004): 85–107.

———. "The Rise of Note-Taking in Early Modern Europe." *Intellectual History Review* 20 (2010): 303–16.

———. *The Theater of Nature: Jean Bodin and Renaissance Science*. Princeton, NJ: Princeton University Press, 1997.

———. *Too Much to Know: Managing Scholarly Information before the Modern Age*. New Haven, CT: Yale University Press, 2010.

Blair, Ann, and Richard Yeo, eds. *Note-Taking in Early Modern Europe*. Special issue of *Intellectual History Review* 20 (2010): 301–432.

Bloch, David. *Aristotle on Memory and Recollection: Text, Translation, Interpretation and Reception in Western Scholasticism*. Leiden: Brill, 2007.

Blount, Thomas. *Glossographia; or, A Dictionary, Interpreting All Such Hard Words . . . Now Used in Our Refined English Tongue* (1656). Menston: Scolar Press, 1969.

Boas, Marie. "Boyle as a Theoretical Scientist." *Isis* 41 (1950): 261–68.

Bolgar, Robert Ralph. *The Classical Heritage and Its Beneficiaries*. Cambridge: Cambridge University Press, 1963.

Bonno, Gabriel. *Les relations intellectuelles de Locke avec la France*. Berkeley: University of California Press, 1955.

Boswell, James. *The Life of Samuel Johnson*. Edited by David Womersley. London: Penguin Classics, 2008.

Boyle, Robert. *Certain Physiological Essays and other Tracts* (1661; 2nd ed., 1669). In *Boyle Works*, ed. Hunter and Davis, 2:3–203.

———. *The Christian Virtuoso, The First Part* (1690–91). In *Boyle Works*, ed. Hunter and Davis, 11:281–344.

———. *The Christian Virtuoso: The Second Part* (1744). In *Boyle Works*, ed. Hunter and Davis, 12:428–30.

———. *The Correspondence of Robert Boyle, 1636–1691*. 6 vols. Edited by Michael Hunter, Antonia Clericuzio, and Lawrence M. Principe. London: Pickering and Chatto, 2001.

————. *The Early Essays and Ethics of Robert Boyle.* Edited by John T. Harwood. Carbondale: Southern Illinois University Press, 1991.

————. *The Excellencies of Robert Boyle.* Edited by J. J. MacIntosh. Claremont, Canada: Broadview Editions, 2008.

————. *The Excellency of Theology, compared with Natural Philosophy* (1674). In *Boyle Works,* ed. Hunter and Davis, 8:3–98.

————. *Experimenta et Observationes Physicae* (1691). In *Boyle Works,* ed. Hunter and Davis, 11:367–426.

————. "General Heads for a Natural History of a Countrey, great or small, imparted likewise by Mr. Boyle." *Philosophical Transactions* 1 (1666): 186–89.

————. *The General History of the Air Designed and Begun by the Honorable Robert Boyle.* London: Printed for Awnsham and John Churchill, 1692.

————. *Hydrostatical Paradoxes, made out by new Experiments* (1666). In *Boyle Works,* ed. Hunter and Davis, 5:189–279.

————. "An Invitation to free and generous Communication." In Hartlib, *Chymical, Medicinal, and Chyrurgical Addresses,* 113–50.

————. *Memoirs for the Natural History of Humane Blood* (1684). In *Boyle Works,* ed. Hunter and Davis, 10:3–101.

————. *New Experiments and Observations Touching Cold* (1665). In *Boyle Works,* ed. Hunter and Davis, 4:203–458.

————. *New Experiments Physico-Mechanical, touching the Spring of the Air.* Oxford: T. Robinson, 1660.

————. *Occasional Reflections upon several Subjects. Whereto is premis'd A Discourse about such kind of Thoughts* (1665). In *Boyle Works,* ed. Hunter and Davis, 5:3–187.

————. *The Origine of Formes and Qualities, According to the "Corpuscular Philosophy"* (1666). In *Boyle Works,* ed. Hunter and Davis, 5:281–491.

————. *Robert Boyle by Himself and His Friends.* Edited by Michael Hunter. London: William Pickering, 1994.

————. *The Sceptical Chymist* (1661). In *Boyle Works,* ed. Hunter and Davis, 2:205–378.

————. *Some Considerations Touching the Usefulnesse of Experimental Naturall Philosophy* (1663). In *Boyle Works,* ed. Hunter and Davis, 3:189–290 (First Part); 3:291–560 (Second Part).

————. *Some Considerations Touching the Usefulnesse of Experimental Naturall Philosophy, . . . the Second Tome* (1671). In *Boyle Works,* ed. Hunter and Davis, 6:389–540.

————. "Some Observations and Directions about the Barometer." *Philosophical Transactions* 1 (1666): 181–85.

————. *Three Tracts Written by the Honourable Robert Boyle* (1670). In *Boyle Works,* ed. Hunter and Davis, 6:317–64.

————. *Tracts Written by the Honourable Robert Boyle* (1670). In *Boyle Works,* ed. Hunter and Davis, 6:259–315.

————. *The Works of the Honourable Robert Boyle.* 6 vols. Edited by T. Birch. London: J. and F. Rivington et al., 1772.

————. *The Works of Robert Boyle.* 14 vols. Edited by Michael Hunter and Edward B. Davis. London: Pickering and Chatto, 1999–2000.

Braddick, M. J., and M. Greengrass, eds. *The Letters of Sir Cheney Culpeper (1641–57).* In

Camden Miscellany 33, Camden Fifth Series, 7:105–402. Cambridge: Cambridge University Press, 1996.

Brinsley, John. *Ludus literarius: or, the Grammar Schoole* (1612). Menston: Scolar Press, 1968.

Brown, John Seeley, and Paul Duguid. *The Social Life of Information*. Cambridge, MA: Harvard Business School Press, 2000.

Browne, Thomas. *Pseudodoxia Epidemica; or, Enquiries into very many received Tenents, and commonly presumed Truths*. London: T. H. for Edward Dodd, 1646.

Burke, Peter. *A Social History of Knowledge: From Gutenberg to Diderot*. Malden, MA: Blackwell, 2000.

Burton, Robert. *The Anatomy of Melancholy* (1621). 3 vols. Edited by Thomas C. Faulkner, Nicholas K. Kiessling, and Rhonda L. Blair. Oxford: Oxford University Press, 1989.

Bylebyl, Jerome J. "The Medical Meaning of 'Physica.'" *Osiris* 6 (1990): 16–41.

Cagnolati, Anotella. *Il circolo di Hartlib: Riforme educative e diffusione del sapere, Inghilterra 1630–1660*. Bologna: Clueb, 2001.

Cantor, David, ed. *Reinventing Hippocrates*. Aldershot: Ashgate, 2002.

Čapková, Dagmar. "Comenius and His Ideals: Escape from the Labyrinth." In *Samuel Hartlib and Universal Reformation*, ed. Greengrass, Leslie, and Raylor, 75–91.

Carey, Daniel. "Compiling Nature's History: Travellers and Travel Narratives in the Early Royal Society." *Annals of Science* 54 (1997): 269–92.

———. *Locke, Shaftesbury, and Hutcheson: Contesting Diversity in the Enlightenment and Beyond*. Cambridge: Cambridge University Press, 2006.

Carruthers, Mary. *The Book of Memory: A Study of Memory in Medieval Culture*. Cambridge: Cambridge University Press, 1990.

———. *The Craft of Thought: Meditation, Rhetoric, and the Making of Images, 400–1200*. Cambridge: Cambridge University Press, 1998.

Casaubon, Meric. *Generall Learning: A Seventeenth-Century Treatise on the Formation of the General Scholar*. Edited by Richard Serjeantson. Cambridge: RTM Publications, 1999.

Cave, Terence. *The Cornucopian Text. Problems of Writing in the French Renaissance*. Oxford: Oxford University Press, 1979.

Cevolini, Alberto. *De arte excerpendi. Imparare a dimenticare nella modernità*. Florence: L. S. Olschki Editore, 2006.

Chambers, Ephraim. *Cyclopaedia: Or, an Universal Dictionary of Arts and Sciences*. 2 vols. London: J. and L. Knapton, J. Darby, D. Midwinter et al., 1728.

Chartier, Roger. *Inscription and Erasure: Literature and Written Culture from the Eleventh to the Eighteenth Century*. Translated by Arthur Goldhammer. Philadelphia: University of Pennsylvania Press, 2009.

Chartier, Roger, Frank Mowery, Peter Stallybrass, and Heather Wolfe. "Hamlet's Tables and the Technologies of Writing in Renaissance England." *Shakespeare Quarterly* 55 (2004): 379–419.

Childrey, Joshua. *Britannia Baconica: Or, the Natural Rarities of England, Scotland, & Wales . . . Historically related, according to the Precepts of Lord Bacon; Methodically digested*. London: Printed for the author, 1661.

Cicero, Marcus Tullius. *De oratore*. 2 vols. Translated by E. W. Sutton and with an introduction by H. Rackham. London: Heineman, 1959.

————. *The Orations of Marcus Tullius Cicero*. Translated by C. D. Yonge. London: Henry G. Bohn, 1856.

Clanchy, Michael T. *From Memory to Written Record: England 1066–1307*. 2nd ed. Oxford: Blackwell, 1993.

Clark, Andrew, ed. *The Life and Times of Anthony Wood*. 5 vols. Oxford: Oxford Historical Society, 1891–1900.

Clark, Andy, and David Chalmers. "The Extended Mind." *Analysis* 58 (1998): 10–23.

Clark, W. "On the Bureaucratic Plots of the Research Library." In *Books and the Sciences in History*, ed. Marina Frasca-Spada and Nick Jardine, 190–206. Cambridge: Cambridge University Press, 2000.

Clarke, Desmond. *Descartes' Theory of the Mind*. Oxford: Clarendon Press, 2003.

Clarke, Samuel. *The Lives of Thirty-Two Divines*. London: Printed for William Birch, 1677.

Clavell, Robert. *A Catalogue of Books Printed in England Since the Dreadful Fire of London in 1666*. London: Printed for R. Clavel and B. Tooke, 1696.

Clericuzio, Antonio. "New Light on Benjamin Worsley's Natural Philosophy." In *Samuel Hartlib and Universal Reformation*, ed. Greengrass, Leslie, and Raylor, 236–46.

Clucas, Stephen. "The Correspondence of a Seventeenth-Century 'Chymicall Gentleman': Sir Cheney Culpeper and the Chemical Interests of the Hartlib Circle." *Ambix* 40 (1993): 147–70.

————. "In Search of *The True Logick*: Methodological Eclecticism among the 'Baconian reformers.'" In *Samuel Hartlib and Universal Reformation*, ed. Greengrass, Leslie, and Raylor, 51–74.

————. "'A Knowledge Broken': Francis Bacon's Aphoristic Style and the Crisis of Scholastic and Humanist Knowledge-Systems." In *English Renaissance Prose*, ed. N. Rhodes, 147–72. Tempe, AZ: ACMRS, 1997.

————. "Samuel Hartlib's *Ephemerides*, 1635–59, and the Pursuit of Scientific and Philosophical Manuscripts: The Religious Ethos of an Intelligencer." *The Seventeenth Century* 6 (1991): 33–55.

————. "*Scientia* and *Inductio Scientifica* in the *Logica Hamburgensis* of Joachim Jungius." In *"Scientia" in Early Modern Philosophy: Seventeenth-Century Thinkers on Demonstrative Knowledge from First Principles*, ed. Tom Sorell, G. A. J. Rogers, and Jill Kraye, 53–70. Dordrecht: Springer, 2010.

Cohen, H. Floris. "The Onset of the Scientific Revolution: Three Near Simultaneous Transformations." In *The Science of Nature in the Seventeenth Century: Patterns of Change in Early Modern Natural Philosophy*, ed. P. R. Anstey and J. A. Schuster, 9–33. Dordrecht: Springer, 2005.

Cohen, I. Bernard. *The Newtonian Revolution: With Illustrations of the Transformation of Scientific Ideas*. Cambridge: Cambridge University Press, 1980.

Coleman, Janet. *Ancient and Medieval Memories: Studies in the Reconstruction of the Past*. Cambridge: Cambridge University Press, 1992.

Collinson, Patrick. "The Coherence of the Text: How It Hangeth Together; The Bible in Reformation England." In *The Bible, the Reformation and the Church*, ed. W. P. Stephens, 84–108. Sheffield: Sheffield Academic Press, 1995.

Comenius, John Amos. *Orbis sensualium pictus*. Translated by Charles Hoole. Facsimile of the Third London Edition 1672. With an introduction by James Bowen. Sydney: Sydney University Press, 1967.

————. *A Patterne of Universall Knowledge, in a Plaine and true Draught.* Translated by Jeremy Collier. London: Printed for T. H. and J. O. Collins, 1651.

Cook, Harold J. *Matters of Exchange: Commerce, Medicine and Science in the Dutch Golden Age.* New Haven, CT: Yale University Press, 2007.

————. *Trials of an Ordinary Doctor: Joannes Groenevelt in Seventeenth-Century London.* Baltimore, MD: Johns Hopkins University Press, 1994.

Cooper, Michael, and Michael Hunter, eds. *Robert Hooke: Tercentenary Studies.* Burlington, VT: Ashgate, 2006.

Corneanu, Sorana. *Regimens of the Mind: Boyle, Locke, and the Early Modern Cultura Animi Tradition.* Chicago: University of Chicago Press, 2011.

Coste, Pierre. "The Character of Mr. Locke." In *A Collection of Several Pieces,* ed. Pierre Desmaizeau. London: R. Francklin, 1720.

Costello, William T. *The Scholastic Curriculum at Early Seventeenth-Century Cambridge.* Cambridge, MA: Harvard University Press, 1958.

Coughlan, Patricia. "Natural History and Historical Nature: The Project for a Natural History of Ireland." In *Samuel Hartlib and Universal Reformation,* ed. Greengrass, Leslie, and Raylor, 298–319.

Cranston, Maurice. *John Locke: A Biography.* New York: Macmillan, 1957.

da Costa, Palmira Fontes. *The Singular and the Making of Knowledge at the Royal Society of London in the Eighteenth Century.* Newcastle upon Tyne: Cambridge Scholars Publishing, 2009.

d'Alembert, Jean Le Rond. *Preliminary Discourse to the Encyclopedia of Diderot.* Translated by Richard N. Schwab with Walter E. Rex. Chicago: University of Chicago Press, 1995.

Danziger, Kurt. *Marking the Mind: A History of Memory.* Cambridge: Cambridge University Press, 2008.

Darley, Gillian. *John Evelyn: Living for Ingenuity.* New Haven, CT: Yale University Press, 2007.

Darwin, Francis, ed. *The Autobiography of Charles Darwin and Selected Letters.* New York: Dover Publications, 1958.

Daston, Lorraine. "Baconian Facts, Academic Civility, and the Prehistory of Objectivity." *Annals of Scholarship* 8 (1991): 337–63.

————. "Description by Omission: Nature Enlightened and Obscured." In *Regimes of Description: In the Archive of the Eighteenth Century,* ed. John Bender and Michael Marrinan, 11–24. Stanford, CA: Stanford University Press, 2005.

————. "The Empire of Observation, 1600–1800." In *Histories of Scientific Observation,* ed. Lorraine Daston and Elizabeth Lunbeck, 81–113. Chicago: University of Chicago Press, 2011.

————. "Perché i fatti sono brevi?" *Quaderni Storici* 108 (2001): 745–70.

————. "On Scientific Observation." *Isis* 99 (2008): 97–110.

————. "Taking Note(s)." *Isis* 95 (2004): 443–48.

Daston, Lorraine, and Katherine Park. *Wonders and the Order of Nature, 1150–1750.* New York: Zone Books, 2001.

Davies, Roy. "The Creation of New Knowledge by Information Retrieval and Classification." *Journal of Documentation* 45 (1989): 273–301.

Dear, Peter. *Discipline and Experience.* Chicago: University of Chicago Press, 1995.

———. "The Meanings of Experience." In *The Cambridge History of Science*, vol. 3: *Early Modern Science*, ed. Katherine Park and Lorraine Daston, 106–31. Cambridge: Cambridge University Press, 2006.

———. "Totius in Verba: Rhetoric and Authority in the Early Royal Society." *Isis* 76 (1985): 145–61.

Décultot, Elizabeth. "Introduction: L'art de l'extrait, définition, évolution, enjeux." In *Lire, Copier, Écrire: Les Bibliothèques Manuscrites et leurs usages au XVIIIe siècle*, ed. Elizabeth Décultot, 7–28. Paris: Centre National de la Recherche Scientifique, 2003.

Dekker, Rudolf, ed. *Egodocuments and History: Autobiographical Writing in Its Social Context since the Middle Ages*. Hilversum: Verloren, 2002.

De la Bédoyère, Guy, ed. *Particular Friends: The Correspondence of Samuel Pepys and John Evelyn*. Woodbridge: Boydell Press, 2005.

De Maria, Robert. *Johnson and the Life of Reading*. Baltimore, MD: Johns Hopkins University Press, 1997.

DeMott, B. "Science versus Mnemonics: Notes on John Ray and on John Wilkins' 'Essay toward a real Character, and a Philosophical Language.'" *Isis* 48 (1957): 3–12.

De Renzi, Silvia. "Writing and Talking of Exotic Animals." In *Books and Sciences in History*, ed. Marina Frasca-Spada and Nick Jardine, 151–67. Cambridge: Cambridge University Press, 2000.

Descartes, Rene. *The Correspondence*. Vol. 3 of *The Philosophical Writings of Descartes*. Translated by John Cottingham, Robert Stoothoff, Dugald Murdoch, and Anthony Kenny. Cambridge: Cambridge University Press, 1985.

———. *A Discourse on the Method*. Translated by Ian Maclean. New York: Oxford University Press, 2006.

———. *The Philosophical Writings of Descartes*. Vol. 1. Translated by John Cottingham, Robert Stoothoff, and Dugald Murdoch. Cambridge: Cambridge University Press, 1985.

Des Maizeaux, Pierre. *A Collection of Several Pieces of Mr. John Locke*. Printed by J. Bettenham for R. Francklin, 1720.

De Vivo, Filippo. *Information and Communication in Venice: Rethinking Early Modern Politics*. Oxford: Oxford University Press, 2007.

Dewhurst, Kenneth. *Dr. Thomas Sydenham (1624–1689): His Life and Original Writings*. London: The Wellcome Historical Medical Library, 1966.

———. "The Genesis of State Medicine in Ireland." *Irish Journal of Medical Science* 31 (1956): 365–84.

———. *John Locke, 1632–1704, Physician and Philosopher: A Medical Biography with an Edition of the Medical Notes*. London: Wellcome Historical Medical Library, 1963.

———. "Locke's Contribution to Boyle's Researches on the Air and on Human Blood." *Notes and Records of the Royal Society of London* 17 (1962): 198–206.

———, ed. *Thomas Willis's Oxford Lectures*. Oxford: Sandford Publication, 1980.

Diderot, Denis, and Jean Le Rond d'Alembert, eds. *Encyclopédie, ou dictionnaire raisonné des sciences, des arts et des métiers*. 17 vols. Paris: Briasson, 1751–65.

Dircks, Henry. *A Biographical Memoir of Samuel Hartlib, Milton's Familiar Friend*. London: J. Russell, [1865].

Donald, Merlin. *A Mind So Rare: The Evolution of Human Consciousness*. New York: W. W. Norton, 2001.

———. *Origins of the Modern Mind: Three Stages in the Evolution of Culture and Cognition*. Cambridge, MA: Harvard University Press, 1991.

Doody, Aude. *Pliny's Encyclopedia: The Reception of the "Natural History."* New York: Cambridge University Press, 2010.

Dooley, Brendan, and Sabrina Baron, eds. *The Politics of Information in Early Modern Europe*. London: Routledge, 2001.

Draaisma, Douwe. "Hooke on Memory and the Memory of Hooke." In *Robert Hooke*, ed. Cooper and Hunter, 111–21.

———. *Metaphors of Memory: A History of Ideas about the Mind*. Translated by Paul Vincent. Cambridge: Cambridge University Press, 2000.

Drexel, Jeremias. *Aurifodina artium et scientiarum omnium: excerpendi solertia amantibus monstrata*. Antwerp: Apud viduam Ioannis Cnobbari, 1638.

Dryden, John. *Of Dramatic Poesy, an Essay* (1668). In *Essays of John Dryden*, ed. W. P. Ker, 1:21–108. New York: Russell and Russell, 1961.

Dury [or Durie], J. *The Reformed Librarie-Keeper, with a supplement to the Reformed School . . . wherunto is added 1. An Idea of Mathematicks*. London: W. Du-Gard, 1650.

Eisenstein, Elizabeth. *The Printing Press as an Agent of Change: Communications and Cultural Transformations in Early Modern Europe*. 2 vols. Cambridge: Cambridge University Press, 1979.

Erasmus, Desiderius. *De copia* (1512). In Erasmus, *Literary and Educational Writings 2*, 279–659.

———. *De ratione studii* (1512). In Erasmus, *Literary and Educational Writings 2*, 661–91.

———. *Literary and Educational Writings 2: De copia /De ratione Studii*. Vol. 24 of *Collected Works of Erasmus*. Edited by Craig R. Thomson. Translated by Betty I. Knott. Toronto: University of Toronto Press, 1978.

Essex, Robert (Earl of), and Philip Sidney. *Profitable Instructions*. London: B. Foster, 1633.

Evelyn, John. *The Diary of John Evelyn* (1955). 6 vols. Edited by E. S. de Beer. Oxford: Clarendon Press, 2000.

———. *Memoires for My Grand-son*. Transcribed and furnished with a preface and notes by Geoffrey Keynes. Oxford: Nonesuch Press, 1926.

———. *Sylva, or a Discourse of Forest-Trees and the Propagation of Timber*. London, 1664.

———. *Sylva, or a Discourse of Forest-Trees and the Propagation of Timber*. 3rd ed. London: Printed for John Martyn, 1679.

Farnaby, Thomas. *Index rhetoricus* (1625). Menston: Scolar Press, 1970.

Feingold, Mordechai. "English Ramism: A Reinterpretation." In *The Influence of Petrus Ramus: Studies in Sixteenth-and Seventeenth-century Philosophy and Sciences*, ed. Mordechai Feingold, Joseph S. Freedman, and Wolfgang Rother, 127–76. Basel: Schwabe, 2001.

———. "Of Records and Grandeur: The Archive of the Royal Society." In *Archives of the Scientific Revolution*, ed. Hunter, 171–84.

———. "Science as a Calling? The Early Modern Dilemma." *Science in Context* 15 (2002): 79–119.

Feisenberger, H. A. "The Libraries of Newton, Hooke and Boyle." *Notes and Records of the Royal Society* 20 (1965): 42–55.

Fisch, Harold. "The Scientist as Priest: A Note on Robert Boyle's Natural Theology." *Isis* 43 (1953): 252–65.

Fischer, K. *Francis Bacon of Verulam: Realistic Philosophy and Its Age.* Translated from the German by John Oxenford. London: Longmans and Roberts, 1857.

Fitzmaurice, Lord Edmond. *The Life of Sir William Petty.* London: John Murray, 1895.

Flemming, Rebecca. "Commentary." In *The Cambridge Companion to Galen,* ed. R. J. Hankinson, 323–54. Cambridge: Cambridge University Press, 2008.

Fletcher, Harris. F. *The Intellectual Development of John Milton.* 2 vols. Urbana: University of Illinois Press, 1956–61.

Floridi, Luciano. *Information: A Very Short Introduction.* Oxford: Oxford University Press, 2010.

Foster, Jonathan K. *Memory: A Very Short Introduction.* Oxford: Oxford University Press, 2009.

Fothergill, Robert A. *Private Chronicles: A Study of English Diaries.* London: Oxford University Press, 1974.

Fox-Bourne, H. R. *The Life of John Locke.* 2 vols. London: H. S. King, 1876.

Foxe, John. *Acts and Monuments.* Edited by S. R. Cattley. London: Seeley and Burnside, 1838.

———. *Locorum communium tituli et ordines centum quinquaginta, ad seriem predicamentorum decem descripti.* Basel: Apud Joannem Oporinum, 1557.

———. *Pandectae locorum communium.* London: Johannes Dayus, 1572.

Freud, Sigmund. "A Note upon the 'Mystic Writing-Pad' (Notiz über den Wunderblock)." In *The Complete Psychological Works of Sigmund Freud,* translated under the general editorship of J. Strachey, 19:227–32. London: Hogarth Press, 1961.

Fuller, Thomas. *The Holy State.* Cambridge: Roger Daniel for John Williams, 1642.

Galen. *Galen: On the Therapeutic Method, Books I and II.* Translated and with an introduction and commentary by R. J. Hankinson. Oxford: Clarendon Press, 1991.

———. *Three Treatises on the Nature of Science.* Translated by Richard Walzer and Michael Frede. Indianapolis: Hackett Publishing, 1985.

Galton, Francis. *The Art of Travel: Or, Shifts and Contrivances Available in Wild Countries.* Introduction by D. Middleton. London: Phoenix Press, 1971. Reprint of 5th edition, 1872.

Garberson, Eric. "Libraries, Memory, and the Space of Knowledge." *Journal of the History of Collections* 18 (2006): 105–36.

Garrett, Jeffrey. "The Legacy of the Baroque in Virtual Representations of Library Space." *Library Quarterly* 74 (2004): 42–62.

Gascoigne, John. *Joseph Banks and the English Enlightenment: Useful Knowledge and Polite Culture.* Cambridge: Cambridge University Press, 1994.

Gassendi, Pierre. *The Mirrour of True Nobility and Gentility: Being the life of the renowned Nicolaus Claudius Fabricus Lord of Peiresk, Englished by W. Rand.* London: J. Streater for Humphrey Moseley, 1657. First published 1641 in Latin.

Gaukroger, Stephen. *Descartes' System of Natural Philosophy.* Cambridge: Cambridge University Press, 2002.

———. *The Emergence of a Scientific Culture: Science and the Shaping of Modernity, 1210–1685.* Oxford: Clarendon Press, 2006.

———. *Francis Bacon and the Transformation of Early-Modern Philosophy*. Cambridge: Cambridge University Press, 2001.

Geijsbeek, J. B. *Ancient Double-Entry Bookkeeping. Lucas Pacioli's Treatise* (1914). Houston, TX: Scholars Book Co., 1974. Reprint of 1914 edition.

Gessner, Conrad. *Pandectae sive partitionum universalium*. Zurich, 1548.

Gibbon, Edward. *An Essay on the Study of Literature*. London: T. Becket, 1764.

———. *Memoirs of My Life* (1796). Edited from the manuscripts by Georges A. Bonnard. New York: Funk and Wagnalls, 1966.

Glanvill, Joseph. *Plus Ultra, or, The Progress and Advancement of Knowledge since the Days of Aristotle*. London: Printed for James Collins, 1668.

Golinski, Jan. *British Weather and the Climate of Enlightenment*. Chicago: University of Chicago Press, 2007.

Grafton, Anthony. *Defenders of the Text: The Traditions of Scholarship in an Age of Science, 1450–1800*. Cambridge, MA: Harvard University Press, 1991.

———. *The Footnote: A Curious History*. London: Faber and Faber, 1997.

———. "Martin Crusius Reads his Homer." *Princeton University Library Chronicle* 64 (2002): 63–86.

———. "Where Was Salomon's House? Ecclesiastical History and the Intellectual Origins of Bacon's *New Atlantis*." In *Die europäische Gelehrtenrepublik im Zeitalter des Konfessionalismus*, ed. Herbert Jaumann, 21–38. Wiesbaden: Harrassowitz, 2001.

———. "The World of Polyhistors: Humanism and Encyclopaedism." *Central European History* 18 (1985): 31–47.

———. *Worlds Made by Words: Scholarship and Community in the Modern West*. Cambridge, MA: Harvard University Press, 2009.

Grafton, Anthony, and Lisa Jardine. "'Studied for Action': How Gabriel Harvey Read His Livy." *Past and Present* 129 (1990): 30–78.

Greengrass, Mark. "Archive Refractions: Hartlib's Papers and the Workings of an Intelligencer." In *Archives of the Scientific Revolution*, ed. Hunter, 35–48.

———. "An 'Intelligencer's Workshop': Samuel Hartlib's Ephemerides." *Studia Comeniana et Historica* 26 (1996): 48–62.

———. "Samuel Hartlib and the Commonwealth of Learning." In *The Cambridge History of the Book in Britain*, vol. 4, ed. John Barnard and D. F. McKenzie, 304–22. Cambridge: Cambridge University Press, 2002.

———. "Samuel Hartlib and Scribal Communication." *Acta Comeniana* 12 (1997): 47–62.

Greengrass, Mark, Michael Leslie, and Timothy Raylor, eds. *Samuel Hartlib and Universal Reformation: Studies in Intellectual Communication*. Cambridge: Cambridge University Press, 1994.

Grew, Nehemiah. *Musaeum regalis societatis*. London: Printed by W. Rawlins for the author, 1681.

Gunther, Robert W. T. *Early Science in Oxford*. 15 vols. Oxford: Oxford University Press, 1923–45.

Halbwachs, Maurice. *On Collective Memory* (1952). Edited and translated by Lewis A. Coser. Chicago: University of Chicago Press, 1992.

Hall, A. R. "Sir Isaac Newton's Note-Book, 1661–65." *Cambridge Historical Journal* 9 (1948): 239–50.

Hall, Marie Boas. *The Library and Archives of the Royal Society, 1660–1990*. London: The Royal Society, 1992.

———. "Samuel Hartlib." In *Dictionary of Scientific Biography*, vol. 6, ed. Charles C. Gillispie, 140–42. New York: Charles Scribner's Sons, 1972.

Hanford, James Holly. "The Chronology of Milton's Private Studies." *Publications of the Modern Language Association* 36 (1921): 251–314.

Hanson, Craig A. *The English Virtuoso: Art, Medicine and Antiquarianism in the Age of Empiricism*. Chicago: University of Chicago Press, 2009.

Harkness, Deborah E. *The Jewel House: Elizabethan London and the Scientific Revolution*. New Haven, CT: Yale University Press, 2007.

Harris, Stephen, and Peter Anstey. "John Locke's Seed Lists: A Case Study in Botanical Exchange." *Studies in History and Philosophy of Biological and Biomedical Sciences* 40 (2009): 256–64.

Harrison, John. "John Locke, Physician and Book Collector." *Journal of the History of Medicine and Allied Sciences* 19 (1964): 70–71.

———. *The Library of Isaac Newton*. Cambridge: Cambridge University Press, 1978.

Harrison, John, and Peter Laslett. *The Library of John Locke*. Oxford: Oxford University Press for the Oxford Bibliographical Society, 1965.

Harrison, Peter. *The Fall of Man and the Foundations of Science*. Cambridge: Cambridge University Press, 2007.

———. "Natural Philosophy." In *Wrestling with Nature: From Omens to Science*, ed. Peter Harrison, Ronald L. Number, and Michael H. Shank, 117–48. Chicago: University of Chicago Press, 2011.

Hartley, Beryl. "Exploring and Communicating Knowledge of Trees in the Early Royal Society." *Notes and Records of the Royal Society of London* 64 (2010): 229–50.

Hartlib, Samuel. *A Brief Discourse Concerning the Accomplishment of our Reformation*. London, 1647.

———. *Chymical, Medicinal, and Chyrurgical Addresses made to Samuel Hartlib*. London: G. Dawson, 1655.

———. *A Further Discoverie of The Office of Publick Addresse for Accommodations*. London: n.p., 1648.

———. *The Hartlib Papers: A Complete Text and Image Database of the Papers of Samuel Hartlib (c. 1600–1662)*. 2nd ed. Sheffield: HROnline, Humanities Research Institute, 2002. Electronic resource held in Sheffield University Library, Sheffield, UK.

[———]. *The Reformed Common-Wealth of Bees. Presented in Several Letters and Observations to Samuel Hartlib Esq*. London: Giles Calvert, 1655.

[———]. *Samuel Hartlib. His Legacie; or an enlargement of the Discourse of Husbandry used in Brabant & Flaunders*. 2nd ed. London: printed for R. Wodnothe, 1652.

Harvey, E. Ruth. *The Inward Wits: Psychological Theory in the Middle Ages and the Renaissance*. London: Warburg Institute, 1975.

Harvey, William. *Anatomical Exercitations, Concerning the Generation of Living Creatures*. London: James Young, 1653. First published in 1651 in Latin.

Harwood, John T. "Rhetoric and Graphics in *Micrographia*." In *Robert Hooke: New Studies*, ed. Michael Hunter and Simon Schaffer, 119–47. Woodbridge: The Boydell Press, 1989.

Havens, Earle. *Commonplace Books: A History of Manuscripts and Printed Books from Antiquity to the Twentieth Century*. New Haven, CT: Beinecke Rare Book and Manuscript Library, 2001.

———, ed. *"Of Common Places, or Memorial Books." A Seventeenth-Century Manuscript from the James Marshall and Marie-Louise Osborn Collection*. New Haven, CT: Yale University Library, 2001.

Hay, Andrew, *The Diary of Andrew Hay of Craignethan, 1659–1660*. Edited by Alexander G. Reid. Edinburgh: Scottish History Society, 1901.

Heinsius, Daniel. "Funeral Oration." In *Autobiography of Joseph Scaliger*, edited and translated by George W. Robinson. Cambridge, MA: Harvard University Press, 1927.

Helmholtz, Hermann von. *Popular Lectures on Scientific Subjects*. Translated by Edmund Atkinson. London: Longmans, Green and Co., 1873.

Henderson, Felicity. "Unpublished Material from the Memorandum Book of Robert Hooke, Guildhall Library MS 1758." *Notes and Records of the Royal Society* 61 (2007): 129–75.

Henry, John. "The Origins of Modern Science: Henry Oldenburg's Contribution." *British Journal for the History of Science* 21 (1988): 103–10.

Hesse, Mary B. "Hooke's Philosophical Algebra." *Isis* 57 (1966): 67–83.

Heyworth, P. L., ed. *Letters of Humfrey Wanley: Palaeographer, Anglo-Saxonist, Librarian, 1672–1726*. Oxford: Clarendon Press, 1989.

[Hill, Abraham]. *Familiar Letters which passed between Abraham Hill, Esq. and several eminent and ingenious persons of the last century. Transcribed from the Original Letters*. London: W. Johnston, 1767.

Hippocrates. *The Aphorismes of Hippocrates, Prince of Physitians*. London: H. Moseley, 1655.

———. *Hippocratic Writings*. Translated by J. Chadwick and W. N. Mann. Edited by G. E. R. Lloyd. Harmondsworth: Penguin, 1978.

Hobart, M. E., and Z. S. Schiffman. *Information Ages: Literacy, Numeracy, and the Computer Revolution*. Baltimore, MD: Johns Hopkins University Press, 1998.

Hobbes, Thomas. *Human Nature*. In *The Elements of Law, Natural and Politic*, ed. Ferdinand Tönnies, 3–73. Cambridge: Cambridge University Press, 1928.

———. *Leviathan*. 3 vols. Edited by Noel Malcolm. Oxford: Clarendon Press, 2012.

Holdsworth, Richard. "Directions for a Student in the Universitie." In H. F. Fletcher, *The Intellectual Development of John Milton*, 2:623–55.

Holmes, F., J. Renn, and H.-J. Rheinberger, eds. *Reworking the Bench: Research Notebooks in the History of Science*. Dordrecht: Kluwer Academic Publishers, 2003.

Hooke, Robert. *An Attempt to Prove the Motion of the Earth from Observations*. London: Printed by T. R. for John Martyn, 1674.

———. *The Diary of Robert Hooke, 1672–1680*. Edited by Henry W. Robinson and Walter Adams. London: Wykeham Publications, 1968.

———. "A Discourse of Earthquakes." In *Posthumous Works of Hooke*, ed. Waller, 279–450.

———. "A General Scheme, or Idea of the present state of Natural Philosophy." In *Posthumous Works of Hooke*, ed. Waller, 1–70.

———. "Lectures of Light, explicating its nature, properties, and effects." In *Posthumous Works of Hooke*, ed. Waller, 71–148.

———. *Micrographia, or, some Physiological Descriptions of Minute Bodies made by Magnifying Glasses*. London: Printed by Jo. Martyn and Ja. Allestry, 1665.

———. *Philosophical Experiments and Observations of the Late Eminent Dr. Robert Hooke*. Edited by W. Derham. London: Printed by W. and J. Innys, printers to The Royal Society, 1726.

———. *The Posthumous Works of Robert Hooke*. Edited by Richard Waller. London: Printed by S. Smith and B. Walford, 1705.

———. "Some Observations and Conjectures Concerning the Chinese Characters." *Philosophical Transactions of the Royal Society of London* 16 (1686): 63–78.

Hoole, Charles. *A New Discovery of the Old Art of Teaching Schoole* (1660). Menston: Scolar Press, 1969.

Hotson, Howard. *Commonplace Learning: Ramism and Its German Ramifications, 1543–1630*. Oxford: Oxford University Press, 2007.

———. *Johann Heinrich Alsted (1588–1638): Between Renaissance, Reformation, and Universal Reform*. Oxford: Clarendon Press, 2000.

Houghton, Walter E. "The English Virtuoso in the Seventeenth Century." Parts I and II. *Journal of the History of Ideas* 3 (1942): 51–73, and 190–219.

———. "The History of Trades: Its Relation to Seventeenth-Century Thought." *Journal of the History of Ideas* 2 (1941): 33–60.

Howarth, David. *Lord Arundel and His Circle*. New Haven, CT: Yale University Press, 1985.

Huarte, Juan. *The Examination of Mens Wits. Translated by M. Camillio Camilli, Englished by Richard Carew*. London, 1594.

Hunter, Ian M. "Lengthy Verbatim Recall: The Role of Text." In *Progress in the Psychology of Language*, ed. A. Ellis, 1:207–35. Hillsdale: Erlbaum, 1985.

Hunter, Matthew. "The Theory of the Impression According to Robert Hooke." In *Printed Images in Early Modern Britain*, ed. Michael Hunter, 167–90. Aldershot: Ashgate, 2010.

Hunter, Michael, ed. *Archives of the Scientific Revolution: The Formation and Exchange of Ideas in Seventeenth-Century Europe*. Woodbridge: The Boydell Press, 1998.

———. *Boyle: Between God and Science*. New Haven, CT: Yale University Press, 2009.

———. *The Boyle Papers: Understanding the Manuscripts of Robert Boyle*. Aldershot: Ashgate, 2007.

———. "Boyle versus the Galenists: A Suppressed Critique of Seventeenth-Century Medical Practice and Its Significance." In Hunter, *Robert Boyle: Scrupulosity and Science*, 157–201.

———. "The British Library and the Library of John Evelyn." In *John Evelyn in the British Library*, 82–102. London: British Library, 1995.

———. *Establishing the New Science: The Experience of the Early Royal Society*. Woodbridge: The Boydell Press, 1989.

———. "Hooke the Natural Philosopher." In *London's Leonardo*, ed. Bennett et al., 119–22.

———. "How Boyle Became a Scientist." In Hunter, *Robert Boyle: Scrupulosity and Science*, 15–57.

———. *John Aubrey and the Realm of Learning*. London: Duckworth, 1975.

———. "Mapping the Mind of Robert Boyle: The Evidence of the Boyle Papers." In *Archives of the Scientific Revolution*, ed. Hunter, 121–36.

————. *Robert Boyle (1627-91): Scrupulosity and Science*. Woodbridge: The Boydell Press, 2000.

————. "Robert Boyle and the Early Royal Society: A Reciprocal Exchange in the Making of Baconian Science." *British Journal for the History of Science* 40 (2007): 1-23.

————. *Science and Society in Restoration England*. Cambridge: Cambridge University Press, 1981.

————. *Science and the Shape of Orthodoxy: Intellectual Change in Late Seventeenth-Century Britain*. Woodbridge: The Boydell Press, 1995.

Hunter, Michael, and Peter Anstey, eds. *The Text of Robert Boyle's "Designe about Natural History."* Occasional Papers no. 3. London: Robert Boyle Project, 2008.

Hunter, Michael, and E. B. Davis. "The Making of Robert Boyle's *Free Enquiry into the Vulgarly Receiv'd Notion of Nature* (1686)." In Hunter, *Boyle Papers*, 219-76.

Hunter, Michael, Harriet Knight, and Charles Littleton. "Robert Boyle's *Paralipomena*: An Analysis and Reconstruction." In Hunter, *Boyle Papers*, 177-218.

Hunter, Michael, and C. Littleton, "The Workdiaries of Robert Boyle: A Newly Discovered Source and Its Internet Publication." In Hunter, *Boyle Papers*, 137-76.

Hunter, Michael, and Paul B. Wood. "Towards Solomon's House: Rival Strategies for Reforming the Early Royal Society." *History of Science* 24 (1986): 49-108.

Hunter, Richard, and Ida Macalpine. "William Harvey and Robert Boyle." *Notes and Records of the Royal Society of London* 13 (1958): 115-27.

Hurley, Susan. *Consciousness in Action*. London: Harvard University Press, 1998.

Hutchins, Edwin. *Cognition in the Wild*. Cambridge, MA: MIT Press, 1995.

Hyde, Thomas. *Catalogus impressorum librorum Bibliothecae Bodleianae in Academia Oxoniensi*. Oxonii: E Theatro Sheldoniano, 1674.

Iliffe, Rob. "'In the Warehouse': Privacy, Property and Priority in the Early Royal Society." *History of Science* 30 (1992): 29-68.

Inwood, Stephen. *The Man Who Knew Too Much: The Strange and Inventive Life of Robert Hooke, 1635-1703*. London: Pan Macmillan, 2003.

Irving, Sarah. *Natural Science and the Origins of the British Empire*. London: Pickering and Chatto, 2008.

Jardine, Lisa. "Hooke the Man: His Diary and His Health." In *London's Leonardo*, ed. Bennett et al., 163-206.

Johns, Adrian. *The Nature of the Book: Print and Knowledge in the Making*. Chicago: University of Chicago Press, 1998.

————. "Reading and Experiment in the Early Royal Society." In *Reading, Society, and Politics in Early Modern England*, ed. Kevin Sharpe and Steven Zwicker, 244-71. Cambridge: Cambridge University Press, 2003.

Johnson, Samuel. *A Dictionary of the English Language*. 2nd ed. 2 vols. London: Printed for W. Strachan, 1755.

Johnson, Thomas. *Quaestiones philosophicae in justi systematis ordinem dispositae*. 2nd ed. Cambridge: G. Thurlbourn, 1735.

Jones, Matthew. *The Good Life in the Scientific Revolution: Descartes, Pascal, Leibniz, and the Cultivation of Virtue*. Chicago: University of Chicago Press, 2006.

Jonson, Ben. "Timber, or, Discoveries, made upon Men and Matter." In *Ben Jonson*, ed. Ian Donaldson, 521-94. Oxford: Oxford University Press, 1985.

Kassler, Jamie C. *The Honourable Roger North, 1651–1734: On Life, Morality, Law and Tradition*. Farnham: Ashgate, 2009.

———. *Inner Music: Hobbes, Hooke and North on Internal Character*. Madison, NJ: Fairleigh Dickinson University Press, 1995.

Kearney, Hugh. *Scholars and Gentlemen: Universities and Society in Pre-Industrial Britain, 1500–1700*. Ithaca, NY: Cornell University Press, 1970.

Keller, Vera. "Accounting for Invention: Guido Pancirolli's Lost and Found Things and the Development of *Desiderata*." *Journal of the History of Ideas* 73 (2012): 223–45.

Kenny, Neil. *The Uses of Curiosity in Early Modern France and Germany*. Oxford: Oxford University Press, 2004.

Keynes, Geoffrey. *John Ray: A Bibliography*. London: Faber and Faber, 1951.

King, Lord Peter. *The Life of John Locke, with Extracts from his Correspondence, Journals, and Common-Place Books*. London: H. Colburn, 1829.

Kliegel, M., M. A. McDaniel, and G. O. Einstein, eds. *Prospective Memory: Cognitive, Neuroscience, Developmental, and Applied Perspectives*. Mahwah, NJ: L. Erlbaum, 2008.

Knight, Harriet. "Organising Natural Knowledge in the Seventeenth Century: The Works of Robert Boyle." PhD diss., University of London, 2003.

———. "Robert Boyle et l'organisation du Savoir." In *La Philosophie Naturelle de Robert Boyle*, ed. M. Dennehy and R. Charles, 157–73. Paris: Vrin, 2009.

Knight, Harriet, and Michael Hunter. "Robert Boyle's *Memoirs for the Natural History of Human Blood* (1684): Print, Manuscript and the Impact of Baconianism in Seventeenth-Century Medical Science." *Medical History* 51 (2007): 145–64.

Knox, Robert. *An Historical Relation of the Island Ceylon, in the East-Indies*. London: R. Chiswell, 1681.

Krajewski, Markus. *Paper Machines: About Cards and Catalogs, 1548–1929*. Translated by Peter Krapp. Cambridge, MA: MIT Press, 2011.

Krzysztof, Pomian. *Collectors and Curiosities: Paris and Venice, 1500–1800*. Translated by Elizabeth Wiles-Portier. Oxford: Polity, 1990. Original French edition 1987.

Kuhn, T. S. "The Function of Measurement in Modern Physical Science." *Isis* 52 (1961): 161–93.

Kusukawa, Sachiko. "Bacon's Classification of Knowledge." In *The Cambridge Companion to Bacon*, ed. Markku Peltonen, 47–74. Cambridge: Cambridge University Press, 1996.

Landauer, T. K. "How Much Do People Remember? Some Estimates of the Quantity of Learned Information in Long-Term Memory." *Cognitive Science* 10 (1986): 477–93.

Lashley, Karl S. "In Search of the Engram." In *The Neuropsychology of Lashley: Selected Papers*, ed. Frank A. Beach, Clifford T. Morgan, and Henry W. Nissen, 478–505. New York: McGraw Hill, 1960.

Laslett, Peter. "Locke and His Books." In Harrison and Laslett, *Library of John Locke*, 1–65.

Le Clerc, Jean. *Ars critica*. 3 vols. Amsterdam: George Gallet, 1697.

———. "Eloge de feu Mr. Locke." *Bibliothèque Choisie* 6 (1705): 342–411.

———. *Epistolario*. 4 vols. Edited by Maria Grazia and Mario Sina. Firenze: L. S. Olschki, 1987–94.

————. *The Life and Character of John Locke*. London: John Clarke, 1706.

Lechner, Joan Marie. *Renaissance Concepts of the Commonplaces*. Westport, CT: Greenwood Press, 1962.

Leibniz, G. W. "An Introduction to a Secret Encyclopaedia" (1679). In *Leibniz: Philosophical Writings*, ed. G. H. R. Parkinson, translated by M. Morris and G. H. R. Parkinson, 5–9. London: J. M. Dent, 1995.

————. *Leibniz: Selections*. Edited by Philip P. Wiener. New York: Scribner's Sons, 1951.

————. *Mathematische Schriften*. 7 vols. Edited by C. I. Gerhardt. Berlin, 1849–63. Reprint, Hildsheim: Olms, 1962.

————. *New Essays on Human Understanding*. Trans. and ed. Peter Remnant and Jonathan Bennett. Cambridge: Cambridge University Press, 1981.

————. *Philosophical Papers and Letters*. 2nd ed. Edited by L. E. Loemker. Dordrecht: Reidel, 1969.

Leonelli, S. "Introduction: Making Sense of Data-driven Research in the Biological and Biomedical Sciences." *Studies in History and Philosophy of Biological and Biomedical Sciences* 43 (2012): 1–3.

Levine, Joseph M. *Between the Ancients and the Moderns: Baroque Culture in Restoration England*. New Haven, CT: Yale University Press, 1999.

Levy, F. J. "How Information Spread among the Gentry, 1550–1640." *Journal of British Studies* 21 (1982): 11–34.

Lewalski, B. K. *Protestant Poetics and the Seventeenth-Century Religious Lyric*. Princeton, NJ: Princeton University Press, 1979.

Lewis, C. T., and C. Short. *A Latin Dictionary* (1879). Oxford: Clarendon Press, 1966.

Lewis, Rhodri. "*The Best Mnemonicall Expedient*: John Beale's Art of Memory and Its Uses." *The Seventeenth Century* 20 (2005):113–44.

————. "The Efforts of the Aubrey Correspondence Group to Revise John Wilkins' *Essay* (1668) and Their Context." *Historiographia Linguistica* 28 (2001): 331–63.

————. "Hooke's Two Buckets: Memory, Mnemotechnique and Knowledge in the Early Royal Society." In *Ars Reminiscendi: Mind and Memory in Renaissance Culture*, ed. D. Beecher, and G. Williams, 339–63. Toronto: Centre for Reformation and Renaissance Studies Publications, 2009.

————. "A Kind of Sagacity: Francis Bacon, the *ars memoriae* and the Pursuit of Natural Knowledge." *Intellectual History Review* 19 (2009): 155–75.

————. *Language, Mind, and Nature: Artificial Languages in England from Bacon to Locke*. Cambridge: Cambridge University Press, 2007.

————. "Robert Hooke at 371." *Perspectives on Science* 14 (2006): 558–73.

————, ed. *William Petty on the Order of Nature: An Unpublished Manuscript Treatise*. Tempe, AZ: ACMRS, 2012.

Lister, Martin. *Historiae animalium anglicae tres tractatus*. London: J. Martyn, 1678.

Lloyd, Claude. "Shadwell and the Virtuosi." *Publications of the Modern Languages Association of America* 44 (1929): 472–94.

Locke, John. *A Collection of Several Pieces of Mr. John Locke*. Edited by [Pierre Des Maizeaux]. London: R. Francklin, 1720.

————. *The Correspondence of John Locke*. 9 vols. Edited by E. S. de Beer. Oxford: Clarendon Press, 1976–89.

———. *An Early Draft of Locke's Essay Together with Excerpts from His Journals.* Edited by Richard I. Aaron and Jocelyn Gibb. Oxford: Clarendon Press, 1936.

———. *An Essay Concerning Human Understanding.* Edited by Peter H. Nidditch. Oxford: Clarendon Press, 1975.

———. *An Essay Concerning Toleration: And Other Writings on Law and Politics, 1667–1683.* Edited by J. R. Milton and Philip Milton. Oxford: Clarendon Press, 2006.

———. *Essays on the Law of Nature and Associated Writings.* Edited by W. von Leyden. Oxford: Clarendon Press, 2002.

[———]. "An Introductory Discourse, containing the whole History of Navigation from its Original to this Time." In *Works of John Locke,* 10: 359–572.

———. *Literary and Historical Writings.* Edited by J. R. Milton. Oxford: Clarendon Press, forthcoming.

[———]. "Methode nouvelle de dresser des recueuils. Communiquée par l'Auteur." *Bibliothèque Universelle et Historique* 2 (1686): 315–40.

———. "A New Method of a Common-Place-Book. Translated out of French from the second volume of the Bibliothèque Universelle." In *Posthumous Works of Mr. John Locke,* 311–36.

———. *A New Method of Making Common-Place-Books; Written by the late Learned Mr Lock, author of the Essay concerning Humane Understanding. Translated from the French. To which is added something from Monsieur Le Clerc, relating to the subject.* London: J. Greenwood, 1706.

———. *Observations upon the Growth and Culture of Vines and Olives: The Production of Silk: The Preservation of Fruits.* London: W. Sandby, 1766.

———. "Of Study" (1677). In *The Educational Writings of John Locke: A Critical Edition with Introduction and Notes,* ed. James L. Axtell, 405–22. Cambridge: Cambridge University Press, 1969.

———. *Of the Conduct of the Understanding.* In *Works of John Locke,* 3: 205–89.

———. *Posthumous Works of Mr. John Locke.* Edited by Peter King and Anthony Collins. London: W. B. for A. and J. Churchill, 1706.

———. *Some Thoughts Concerning Education* (1693). Edited by John W. Yolton and Jean S. Yolton. Oxford: Clarendon Press, 1989.

———. *The Works of John Locke.* 10 vols. A new Edition, Corrected. London: T. Tegg et al., 1823. Reprint, Scientia Verlag Aalen: Germany, 1963.

———. *Writings on Natural Philosophy and Medicine.* Edited by Peter R. Anstey and Lawrence M. Principe. Oxford: Clarendon Press, forthcoming.

Long, P. "The Mellon Donation of Additional Manuscripts of John Locke from the Lovelace Collection." *Bodleian Library Record* 7 (1964): 185–93.

———. *A Summary Catalogue of the Lovelace Collection of the Papers of John Locke in the Bodleian Library.* Oxford: Printed for the Society at the University Press, 1959.

Lough, John. "Locke's Reading during His Stay in France (1675–79)." *The Library* 8 (1953): 229–58.

———, ed. *Locke's Travels in France 1675–1679: As Related in his Journals, Correspondence and other Papers.* Cambridge: Cambridge University Press, 1953.

Love, Harold. *The Culture and Commerce of Texts: Scribal Publication in Seventeenth-Century England.* Amherst: University of Massachusetts Press, 1998.

Lund, Roger D. "Wit, Judgment, and the Misprisions of Similitude." *Journal of the History of Ideas* 68 (2004): 53–74.

Luria, Aleksandr R. *The Mind of a Mnemonist: A Little Book about a Vast Memory.* Translated by L. Solotaroff. Harmondsworth: Penguin, 1975.

Lux, David S., and Harold J. Cook. "Closed Circles or Open Networks? Communicating Science at a Distance during the Scientific Revolution." *History of Science* 36 (1998): 179–211.

Mack, P. *Elizabethan Rhetoric: Theory and Practice.* Cambridge: Cambridge University Press, 2002.

Maclean, Ian. *Logic, Signs, and Nature in the Renaissance.* Cambridge: Cambridge University Press, 2002.

Maddison, R. E. W. "Abraham Hill, F.R.S. (1635–1722)." *Notes and Records of the Royal Society of London* 15 (1960): 173–82.

———. *The Life of the Honourable Robert Boyle F.R.S.* London: Taylor and Francis, 1969.

———. "A Tentative Index of the Correspondence of the Honourable Robert Boyle, F.R.S." *Notes and Records of the Royal Society of London* 13 (1958): 128–201.

Malcolm, Noel. "The Library of Henry Oldenburg." *Electronic British Library Journal* (2005), article 7.

———. "The Life of John Pell." In N. Malcolm and J. Stedall, *John Pell (1611–1685) and His Correspondence with Sir Charles Cavendish: The Mental World of an Early Modern Mathematician,* 12–244. Oxford: Oxford University Press, 2005.

———. "Thomas Harrison and His 'Ark of Studies': An Episode in the History of the Organization of Knowledge." *The Seventeenth Century* 19 (2004): 196–232.

Malham, John, ed. *The Harleian Miscellany: Or, a Collection of scarce, curious, and entertaining Tracts,* vol. 6. London: Printed for Robert Dutton, 1810.

Malhere, M. "Bacon's Method of Science." In *The Cambridge Companion to Bacon,* ed. M. Peltonen, 75–98. Cambridge: Cambridge University Press, 1996.

Marbeck [or Merbecke], John. *A Booke of Notes and Common places, with their expositions, collected and gathered out of the workes of divers singular writers, and brought alphabetically into order.* London: Thomas East, 1581.

Mattern, Susan, P. *Galen and the Rhetoric of Healing.* Baltimore, MD: Johns Hopkins University Press, 2008.

Matthews, William. *British Diaries: An Annotated Bibliography of British Diaries Written between 1442 and 1942.* Gloucester, MA: P. Smith, 1967.

McGuire, J. E., and Martin Tamny. *Certain Philosophical Questions: Newton's Trinity Notebook.* Cambridge: Cambridge University Press, 1983.

Meinel, Christof. "Enzyklopädie der Welt und Verzettelung des Wissens: Aporien der Empirie bei Joachim Jungius." In *Enzyklopädien der frühen Neuzeit: Beiträge zu ihrer Forschung,* ed. Franz M. Eybl, 162–87. Tübingen: Neimeyer, 1995.

Menary, Richard, ed. *The Extended Mind.* Cambridge, MA: MIT Press, 2010.

Meynell, G. G. "A Database for John Locke's Medical Notebooks and Medical Reading." *Medical History* 42 (1997): 473–86.

———. "John Locke's Method of Common-placing, as Seen in His Drafts and His Medical Notebooks, Bodleian MSS Locke d. 9, f. 21, f. 23." *The Seventeenth Century* 8 (1993): 245–67.

———. "Locke, Boyle and Stahl." *Notes and Records of the Royal Society* 49 (1995): 185–92.

Middleton, W. E. Knowles, *The History of the Barometer*. Baltimore, MD: Johns Hopkins University Press, 1964.

———, ed. and trans. *Lorenzo Magalotti at the Court of Charles II*. Waterloo, Canada: Wilfred Laurier University Press, 1980.

Miller, Peter N. *Peiresc's Europe: Learning and Virtue in the Seventeenth Century*. New Haven, CT: Yale University Press, 2000.

Milton, J. R. "The Date and Significance of Two of Locke's Early Manuscripts." *Locke Newsletter* 19 (1988): 47–89.

———. "The Dating of 'Adversaria 1661.'" *Locke Newsletter* 29 (1998): 105–17.

———. "John Locke." In *Oxford Dictionary of National Biography*. 61 vols. Edited by H. C. G. Matthew and Brian Harrison, 34:216–29. Oxford: Oxford University Press, 2004–12.

———. "John Locke: The Modern Biographical Tradition." *Eighteenth-Century Thought* 3 (2007): 89–109.

———. "John Locke's Medical Notebooks." *Locke Newsletter* 28 (1997): 135–56.

———. "Locke at Oxford." In *Locke's Philosophy: Content and Context*, ed. G. A. J. Rogers, 29–47. Oxford: Clarendon Press, 1994.

———. "Locke's *Adversaria*." *Locke Newsletter* 18 (1987): 63–75.

Milton, John. *A Common-Place Book of John Milton*. In the possession of Sir F. U. Graham and edited from the original MSS by Alfred J. Horwood. London: Printed for The Camden Society, 1876.

Mohl, Ruth. *John Milton and His Commonplace Book*. New York: Frederick Ungar, 1969.

Monk, J. H. "Memoir of Dr. James Duport." *Museum Criticum* 2 (1826): 672–98.

Montaigne, Michel de. *The Complete Essays*. Translated and edited by M. A. Screech. London: Penguin Books, 1987.

Morhof, Georg Daniel. *Polyhistor sive de notitia autorum et rerum commentarii* (1688). 3rd ed. Lübeck: Petrus Boeckmannus, 1732.

Moss, Ann. *Printed Commonplace-Books and the Structuring of Renaissance Thought*. Oxford: Clarendon Press, 1996.

Mulligan, Lotte. "Robert Hooke's 'Memoranda': Memory and Natural History." *Annals of Science* 49 (1992): 47–61.

———. "Self-Scrutiny and the Study of Nature: Robert Hooke's Diary as Natural History." *Journal of British Studies* 35 (1996): 311–42.

Murdock, Bennet B., Jr. "The Serial Position Effect of Free Recall." *Journal of Experimental Psychology* 64 (1962): 482–88.

Murphy, Trevor. *Pliny the Elder's Natural History: The Empire in the Encyclopedia*. Oxford: Oxford University Press, 2004.

Naudé, Gabriel. *Instructions Concerning Erecting of a Library . . . Interpreted by J. Evelyn*. London: Printed for G. Bedle and T. Collins, 1661.

Nelles, Paul. "Note-taking Techniques and the Role of Student Notebooks in the Early Jesuit colleges." *Archivum Historicum Societatis Iesu* 76 (2007): 75–112.

———. "Reading and Memory in the Universal Library: Conrad Gessner and the Renaissance Book." In *Ars Reminiscendi: Mind and Memory in Renaissance Culture*, ed. Donald Beecher and Grant Williams, 147–69. Toronto: Centre for Reformation and Renaissance Studies Publications, 2009.

———. "Seeing and Writing: The Art of Observation in the Early Jesuit Missions." *Intellectual History Review* 20 (2010): 317–33.

Newman, William R., and Lawrence M. Principe. *Alchemy Tried by Fire: Starkey, Boyle, and the Fate of Helmontian Chymistry*. Chicago: University of Chicago Press, 2002.

Nicole, Pierre. *Essais de morale, contenus en divers traittez sur plusiers devoirs importans.* 4 vols. Paris: Charles Savreux, 1672–78.

———. "Traité de la faiblesse de l'homme." In *John Locke as Translator: Three of the Essais of Pierre Nicole in French and English*, ed. Jean Yolton, 43–113. Oxford: Voltaire Foundation, 2000.

North, Roger. *General Preface and Life of Dr John North*. Edited by P. Millard. Toronto: University of Toronto Press, 1984.

———. *Notes of Me: The Autobiography of Roger North*. Edited by Peter Millard. Toronto: University of Toronto Press, 2000.

Nunberg, Geoffrey. "Farewell to the Information Age." In *The Future of the Book*, ed. Geoffrey Nunberg, 103–37. Berkeley: University of California Press, 1996.

O'Brien, J. J. "Samuel Hartlib's Influence on Robert Boyle's Scientific Development." *Annals of Science* 21 (1965): 257–76.

Ogilvie, Brian W. "The Many Books of Nature: Renaissance Naturalists and Information Overload." *Journal of the History of Ideas* 64 (2003): 29–40.

———. *The Science of Describing: Natural History in Renaissance Europe*. Chicago: University of Chicago Press, 2006.

O'Hara, J. G. "'A chaos of jottings that I do not have the leisure to arrange and mark with headings': Leibniz's Manuscript Papers and Their Repository." In *Archives of the Scientific Revolution*, ed. Hunter, 159–70.

Oldenburg, Henry. *The Correspondence of Henry Oldenburg*. 13 vols. Edited and translated by A. R. Hall and M. B. Hall. Madison: University of Wisconsin Press, 1965–86.

Oldroyd, D. R. "Robert Hooke's Methodology of Science as Exemplified in His 'Discourse of Earthquakes.'" *British Journal for the History of Science* 6 (1972): 109–30.

———. "Some 'Philosophicall Scribbles' attributed to Robert Hooke." *Notes and Records of the Royal Society of London* 35 (1980): 17–32.

———. "Some Writings of Robert Hooke on Procedures for the Prosecution of Scientific Enquiry." *Notes and Records of the Royal Society of London* 41 (1987): 145–67.

Ong, Walter J. *Ramus, Method, and the Decay of Dialogue: From the Art of Discourse to the Art of Reason*. Chicago: University of Chicago Press, 1958.

———. "Typographic Rhapsody: Ravisius Textor, Zwinger, and Shakespeare." In *An Ong Reader: Challenges for Further Inquiry*, ed. Thomas J. Farrell and Paul A. Soukup, 429–63. Creskill, NJ: Hampton Press, 2002.

Pascal, Blaise. *Monsieur Pascall's Thoughts*. Translated by Jos. Walker. London, 1688.

———. *Pensées*. Translated by A. J. Krailsheimer. Harmondsworth: Penguin, 1995.

———. *Thoughts, Letters, Minor Works*. New York: P. Collier and Sons, 1910.

Passmore, John. *A Hundred Years of Philosophy*. Harmondsworth: Pelican, 1970.

Pattison, Mark. "Diary of Casaubon." *Quarterly Review* 93 (1853): 462–500.

———. *Isaac Casaubon, 1559–1614*. 2nd ed. Oxford: Clarendon Press, 1892.

Peacham, Henry. *The Compleat Gentleman*. London: F. Constable, 1634.

[Pell, John]. "An Idea of Mathematicks, long since written by Dr. John Pell." In *Philosophical Collections*, no. 5, ed. Robert Hooke, 127–34. London, 1681/82.

————. "An Idea of Mathematics, written by Mr Joh. Pell to Samuel Hartlib." In Dury, *The Reformed Librarie-Keeper*, 33–46.

[————]. *Tabula numerorum quadratorum decies millium, . . .* London: Thomas Ratcliffe, 1672.

Pepys, Samuel. *The Diary of Samuel Pepys*. 11 vols. Edited by Robert Latham and William Matthews. London: Bell and Hymann, 1970–83.

————. *Private Correspondence and Miscellaneous Papers of Samuel Pepys, 1679–1703.* 2 vols. Edited by J. R. Tanner. London: G. Bell and Sons, 1926.

Petrarch, Francesco. *The Secret by Francesco Petrarch with related Documents*. Edited by Carol E. Quillen. Boston: Bedford/St. Martins, 2003.

Petty, William. *The Advice of W. P. To Mr. Samuel Hartlib for the Advancement of some Particular Parts of Learning*. London, 1647.

————. *The Petty Papers: Some Unpublished Writings of Sir William Petty*. 2 vols. Edited from the Bowood Papers by the Marquis of Lansdowne. London: Constable, 1927.

————. *The Petty-Southwell Correspondence, 1676–1687* (1928). Edited from the Bowood Papers by the Marquis of Landsdowne. New York: A. M. Kelley, 1967.

Phillips, Edward. *The New World of English Words: Or, a general Dictionary containing the Interpretations of such hard Words as are derived from other Languages*. London: Printed by E. Tyler, 1658.

Phillips, William. *Studii legalis ratio; or, Directions for the Study of the Law*. London, 1675.

Pinkus, Steven. *1688: The First Modern Revolution*. New Haven, CT: Yale University Press, 2009.

Placcius, Vincent. *De arte excerpendi, vom Gelährten Buchhalten liber singularis quo genera et praecepta excerpendi*. Hamburg: G. Liebezeit, 1689.

Plato. *Plato's Phaedrus*. Translated by R. Hackworth. Cambridge: Cambridge University Press, 1972.

Platte, Hugh. *The Jewell House of Arts and Nature*. London: P. Short, 1594.

Plattes, Gabriel. "A Caveat for Alchymists." In Hartlib, *Chymical, Medicinal, and Chyrurgical Addresses*, 49–88.

[————]. *A Description of the Famous Kingdome of Macaria*. London: Printed for Francis Constable, 1641.

Playford, John. *Vade Mecum: Or, the Necessary Pocket Companion* (1679). 7th ed. London: J. Sawbridge, 1699.

Pliny the Elder. *Naturalis Historia*. 10 vols. Edited by H. Rackham. London: W. Heinemann, 1974.

Pliny the Younger. *The Letters of the Younger Pliny*. Translated by Betty Radice. Harmondsworth: Penguin, 1963.

Plot, Robert. *The Natural History of Oxford-Shire*. Oxford: Printed at the Theater, 1677.

[————]. *Quaer's to be propounded to the most ingenious of each County in my Travels through England*. Oxford, [1674].

Pocock, J. G. A. *Barbarism and Religion*. 5 vols. Cambridge: Cambridge University Press, 1999–2011.

Pomata, Gianna. "Observatio ovvero Historia. Note su empirisimo e storia in etá moderna." *Quaderni Storici* 31 (1996): 173–98.

————. "*Praxis Historialis*: The Uses of *Historia* in Early Modern Medicine." In *Historia*, ed. Pomata and Siraisi, 105–46.

———. "Sharing Cases: The *Observationes* in Early Modern Medicine." *Early Science and Medicine* 15 (2010): 193–236.

Pomata, Gianna, and Nancy G. Siraisi, eds. *Historia: Empiricism and Erudition in Early Modern Europe*. Cambridge, MA: MIT Press, 2005.

Poole, William. *Francis Lodwick (1619–1694): A Country Not Named (MS. Sloane 913, fols 1ʳ–33ʳ)*. Tempe, AZ: ACMRS, 2007.

———. "The Genres of Milton's Commonplace Book." In *The Oxford Handbook of Milton*, ed. Nicholas McDowell and Nigel Smith, 367–81. Oxford: Oxford University Press, 2009.

———. *John Aubrey and the Advancement of Learning*. Oxford: Bodleian Library, 2010.

———. *The World Makers: Scientists of the Restoration and the Search for the Origins of the Earth*. Oxford: Peter Lang, 2010.

Poovey, Mary. *A History of the Modern Fact: Problems of Knowledge in the Sciences of Wealth and Society*. Chicago: University of Chicago Press, 1998.

Pope, Walter. *The Life of the Right Reverend Father in God Seth, Lord Bishop of Salisbury* (1697). Edited by J. B. Bamborough. Oxford: Published for the Luttrell Society by B. Blackwell, 1961.

Power, Henry. *Experimental Philosophy*. London: T. Roycroft, 1663–64.

Principe, Lawrence. *The Aspiring Adept: Robert Boyle and his Alchemical Quest*. Princeton, NJ: Princeton University Press, 1998.

———. "Newly Discovered Boyle Documents in the Royal Society Archive: Alchemical Tracts and His Student Notebook." *Notes and Records of the Royal Society of London* 49 (1995): 57–70.

———. "Style and Thought of the Early Boyle: Discovery of the 1648 Manuscript of *Seraphic Love*." *Isis* 85 (1994): 247–60.

———. "Virtuous Romance and Romantic Virtuoso: The Shaping of Robert Boyle's Literary Style." *Journal of the History of Ideas* 56 (1995): 377–97.

Pugliese, Patri Jones. "The Scientific Achievement of Robert Hooke: Method and Mechanics." PhD diss., Harvard University, 1982.

Purchas, Samuel. *Purchas His Pilgrimage. Or relations of the World*. London, 1613.

Purver, Margery. *The Royal Society: Concept and Creation*. Cambridge, MA: MIT Press, 1967.

Quattrone, Paul. "Accounting for God: Accounting and Accountability Practices in the Society of Jesus (Italy, XVI–XVII Centuries)." *Accounting, Organizations and Society* 29 (2004): 647–83.

Quillen, Carol E. *Rereading the Renaissance: Petrarch, Augustine, and the Language of Humanism*. Ann Arbor: University of Michigan Press, 1998.

Quintilian. *The Institutio Oratoria of Quintilian*. 4 vols. Translated by H. E. Butler. The Loeb Classical Library. Cambridge, MA: Harvard University Press, 1922.

Raven, Charles E. *John Ray Naturalist. His Life and Work*. 2nd ed. Cambridge: Cambridge University Press, 1950.

Ray, John. *A Collection of Curious Travels and Voyages*. 2 vols. London: The Royal Society, 1693.

[———]. *A Collection of English Proverbs digested into a Convenient Method for the speedy Finding upon any one Occasion*. Cambridge: John Hayes, 1670.

————. *A Collection of English Proverbs digested into a Convenient Method for the speedy Finding upon any one Occasion*. Cambridge: John Hayes, 1678.

————. *The Correspondence of John Ray*. Edited by Edwin Lankester. London: Printed for the Ray Society, 1848.

————. *Further Correspondence of John Ray*. Edited by Robert W. T. Gunther. London: Printed for the Ray Society, 1928.

————. *Observations Topological, Moral, & Physiological*. London: J. Martyn, 1673.

————. *The Ornithology of Francis Willughby*. London: Printed by A. C. for John Martyn, 1678.

————. *Philosophical Letters between the late Learned Mr Ray . . . to which are added those of Francis Willughby*. Edited by William Derham. London: The Royal Society, 1718.

Rayward, W. Boyd. "Some Schemes for Restructuring and Mobilising Information in Documents: Historical Perspective." *Information Processing and Management* 30 (1994): 163–75.

Rechtien, John G. "John Foxe's Comprehensive Collection of Commonplaces: A Renaissance Memory System for Students and Theologians." *Sixteenth Century Journal* 9 (1978): 82–89.

Renaissance Commonplace Books from the Huntington Library. Microfilm, 4 reels. Marlborough: Adam Matthew Publications, 1994.

Reynoldes, Edward. *A Treatise of the Passions and Faculties of the Soule of Man*. London: R. H[earne] and John Norton for Robert Bostock, 1640.

Richards, Graham. *Mental Machinery: The Origins and Consequences of Psychological Ideas, part 1: 1600–1850*. London: The Athlone Press, 1992.

Richter, Irma A., ed. *The Notebooks of Leonardo da Vinci*. Oxford: Oxford University Press, 1998.

Rigaud, Stephen Peter, ed. *Correspondence of Scientific Men of the Seventeenth Century*. 2 vols. 1841; reprint, Hildesheim: L. G. Olms, 1965.

Robinson, H. W. "An Unpublished Letter of Dr Seth Ward." *Notes and Records of the Royal Society of London* 7 (1949): 68–70.

Roos, Anna Marie. *The Salt of the Earth: Natural Philosophy, Medicine, and Chymistry in England, 1650–1750*. Leiden: Brill, 2007.

————. *Web of Nature: Martin Lister (1639–1712), the First Arachnologist*. Leiden: Brill, 2011.

Rosenberg, Daniel. "Early Modern Information Overload." *Journal of the History of Ideas* 64 (2003): 1–9.

Rossi, Paolo. *Francis Bacon: From Magic to Science*. Translated by Sacha Rabinovitch. Chicago: University of Chicago Press, 1968.

Rostenberg, Leona. *The Library of Robert Hooke: The Scientific Book Trade of Restoration England*. Santa Monica, CA: Modoc Press, 1989.

Rowbottom, M. E. "The Earliest Published Writing of Robert Boyle." *Annals of Science* 6 (1950): 376–89.

Rusnock, Andrea. "Hippocrates, Bacon, and Medical Meteorology at the Royal Society, 1700–1750." In Cantor, *Reinventing Hippocrates*, 136–53.

Sacchini, Francesco. *De ratione libros cum profectu legendi libellus . . .* (1614). Sammieli: F. du Bois, 1615.

Sanderson, Robert. *Logicae artis compendium*. 8th ed. Oxford: H. Hall, 1672.

Sankey, Margaret. "Writing the Voyage of Scientific Exploration: The Logbooks, Journals and Notes of the Baudin Expedition (1800–1804)." *Intellectual History Review* 20 (2010): 401–13.

Scaliger, Joseph. *Scaligerana ou bons mots, rencontres agréables, et remarques judicieuses & sçavantes de J. Scaliger*. Cologne, 1695.

Schacter, Daniel. *Searching for Memory: The Brain, the Mind, and the Past*. New York: Basic Books, 1996.

Schaffer, Simon. "Newton on the Beach: The Information Order of *Principia Mathematica*." *History of Science* 47 (2009): 243–76.

Schiebinger, Londa, and Claudia Swan, eds. *Colonial Botany: Science, Commerce and Politics in the Early Modern World*. Philadelphia: University of Pennsylvania Press, 2005.

Schmidt-Biggemann, Wilhelm. *Topica Universalis: Eine Modellgeschichte Humanistischer und Barocker Wissenschaft*. Hamburg: F. Meiner, 1983.

Schönpflug, W., and K. B. Esser, "Memory and Its *Graeculi*: Metamemory and Control in Extended Memory Systems." In *Discourse Comprehension: Essays in Honor of Walter Kintsch*, ed. Charles A. Weaver, Suzanne Mannes, and Charles R. Fletcher, 245–55. Hillsdale, NJ: L. Erlbaum, 1995.

Schove, D. J., and David Reynolds. "Weather in Scotland, 1659–1660: The Diary of Andrew Hay." *Annals of Science* 30 (1973): 165–77.

Selden, John. *The Historie of Tithes*. London, 1618.

———. *Table-Talk: Being the Discourses of John Selden Esq*. London, 1698.

Semon, Richard. *The Mneme*. London: George Allen and Unwin, 1921.

Seneca, Lucius Annaeus. *Dialogues and Letters*. Edited and translated by C. D. N. Costa. Harmondsworth: Penguin, 1997.

———. "On Gathering Ideas." Letter no. 84 in *Ad Lucilium epistulae morales*, 3 vols., trans. Richard M. Gummere, 2:277–85. London: Heineman, 1917–25.

[Sergeant, John S]. *The Method to Science*. London: W. Redmayne, 1696.

Shadwell, Thomas. *The Virtuoso*. Edited by Marjorie Hope Nicolson and David Stuart Rodes. Lincoln: University of Nebraska Press, 1966.

Shapin, Steven. *A Social History of Truth: Civility and Science in Seventeenth-Century England*. Chicago: University of Chicago Press, 1994.

Shapin, Steven, and Simon Schaffer. *Leviathan and the Air Pump: Hobbes, Boyle, and the Experimental Life*. Princeton, NJ: Princeton University Press, 1985.

Shapiro, Barbara J. *A Culture of Fact: England 1550–1720*. Ithaca, NY: Cornell University Press, 2000.

Shapiro, Barbara, and Robert G. Frank Jr. *English Scientific Virtuosi in the 16th and 17th Centuries*. Los Angeles: William Andrews Clark Library, 1979.

Sharpe, Kevin. *Reading Revolutions: The Politics of Reading in Early Modern England*. New Haven, CT: Yale University Press, 2000.

Sherman, Stuart. *Telling Time: Clocks, Diaries, and English Diurnal Form, 1660–1785*. Chicago: University of Chicago Press, 1996.

Sherman, William H. *John Dee: The Politics of Reading and Writing in the English Renaissance*. Amherst: University of Massachusetts Press, 1995.

———. *Used Books: Marking Readers in Renaissance England*. Philadelphia: University of Pennsylvania Press, 2008.

Singer, B. R. "Robert Hooke on Memory, Association and Time Perception." *Notes and Records of the Royal Society of London* 31 (1976): 115–31.

Siraisi, Nancy G. *History, Medicine, and the Traditions of Renaissance Learning*. Ann Arbor: University of Michigan Press, 2007.

Skinner, Quentin. *Reason and Rhetoric in the Philosophy of Hobbes*. Cambridge: Cambridge University Press, 1996.

Slaughter, Mary. *Universal Languages and Scientific Taxonomy in the Seventeenth Century*. Cambridge: Cambridge University Press, 1982.

Small, Jocelyn. *Wax Tablets of the Mind: Cognitive Studies of Memory and Literacy in Classical Antiquity*. London: Routledge, 1997.

Smith, A. Mark. "Picturing the Mind: The Representation of Thought in the Middle Ages and Renaissance." *Philosophical Topics* 20 (1992): 149–70.

Smith, Pamela H., and Paula Findlen, eds. *Merchants and Marvels: Commerce, Science and Art in Early Modern Europe*. London: Routledge, 2002.

Snow, C. P. *The Two Cultures and the Scientific Revolution*. Cambridge: Cambridge University Press, 1959.

Snow, Vernon. "Francis Bacon's Advice to Fulke Greville on Research Techniques." *Huntington Library Quarterly* 23 (1960): 369–78.

Soll, Jacob. "Amelot de la Houssaye (1634–1706) Annotates Tacitus." *Journal of the History of Ideas* 61 (2000): 167–87.

———. "From Note-Taking to Data Banks: Personal and Institutional Information Management in Early Modern Europe." *Intellectual History Review* 20 (2010): 355–75.

———. *The Information Master: Jean-Baptiste Colbert's Secret State Intelligence System*. Ann Arbor: University of Michigan Press, 2009.

Solomon, H. M. *Public Welfare, Science and Propaganda in Seventeenth-Century France: The Innovations of Théophraste Renaudot*. Princeton, NJ: Princeton University Press, 1972.

Sorabji, Richard, ed. *Aristotle on Memory*. London: Duckworth, 1972.

Sorell, T., G. A. J. Rogers, and J. Kraye, eds. *Scientia in Early Modern Philosophy: Seventeenth-Century Thinkers on Demonstrative Knowledge from First Principles*. Dordrecht: Springer, 2010.

South, Robert. *Sermons Preached upon Several Occasions*. 7 vols. Oxford: Clarendon Press, 1823.

Spence, Jonathan D. *The Memory Palace of Matteo Ricci*. New York: Viking Penguin, 1984.

Sprat, Thomas. *The History of the Royal Society of London, for the Improving of Natural Knowledge*. London: The Royal Society, 1667.

Stagl, J. *A History of Curiosity: The Theory of Travel, 1550–1800*. Chur, Switzerland: Harwood, 1995.

Stedall, J. "The Mathematics of John Pell." In N. Malcolm and J. Stedall, *John Pell (1611–1685) and His Correspondence with Sir Charles Cavendish: The Mental World of an Early Modern Mathematician*, 245–328. Oxford: Oxford University Press, 2005.

Stewart, M. A. "Locke's Professional Contacts with Robert Boyle." *Locke Newsletter* 12 (1981): 19–44.

Stimson, Dorothy. "Hartlib, Haak, and Oldenburg: Intelligencers." *Isis* 31 (1940): 309–26.

———. *Scientists and Amateurs: A History of the Royal Society*. New York: Greenwood Press, 1968.

Stubbe, Henry. *A Censure upon Certain Passages Contained in the History of the Royal Society . . . Destructive to Established Religion and the Church of England*. Oxford, 1670.

Stubbs, Mayling. "John Beale, Philosophical Gardener of Herefordshire. Part I: Prelude to the Royal Society (1608–1663)." *Annals of Science* 39 (1982): 463–89.

Sutton, John. "Distributed Cognition: Domains and Dimensions." *Pragmatics and Cognition* 14 (2006): 235–47.

———. *Philosophy and Memory Traces: Descartes to Connectionism*. Cambridge: Cambridge University Press, 1998.

———. "Porous Memory and the Cognitive Life of Things." In *Prefiguring Cyberculture: An Intellectual History*, ed. Darren Tofts, 130–41. Cambridge, MA: MIT Press, 2002.

Swann, Marjorie. *Curiosities and Texts: The Culture of Collecting in Early Modern England*. Philadelphia: University of Pennsylvania Press, 2001.

Swanson, Don. "Undiscovered Public Knowledge." *Library Quarterly* 56 (1986): 93–118.

Takarangi, Melanie K. T., and Garry Loftus. "Dear Diary, Is Plastic Better than Paper? I Can't Remember." *Psychological Methods* 11 (2006): 119–22.

Talbot, Ann. *"The Great Ocean of Knowledge": The Influence of Travel Literature on the Work of John Locke*. Leiden: Brill, 2010.

Thomas, Keith. "Literacy in Early Modern England." In *The Written Word: Literacy in Transition*, ed. Gerd Baumann, 97–131. Oxford: Clarendon Press, 1986.

Timmermann, A. "Doctor's Orders: An Early Modern Doctor's Alchemical Notebooks." *Early Science and Medicine* 13 (2008): 25–52.

Todd, Margo. *Christian Humanism and the Puritan Social Order*. Cambridge: Cambridge University Press, 1987.

Tradescant, John. *Musaeum Tradescantianum; or, a Collection of Rarities, preserved at South Lambeth, neer London*. London: Printed by John Grismond, 1656.

Trentman, John A. "The Authorship of *Directions for a Student in the Universitie*." *Transactions of the Cambridge Bibliographical Society* 7 (1978): 170–83.

Trevor-Roper, Hugh. *"Three Foreigners*: The Philosophers of the Puritan Revolution." In Trevor-Roper, *Religion, the Reformation and Social Change, and Other Essays*, 2nd ed., 237–93. London: Macmillan, 1972.

Tribble, Evelyn. "Distributing Cognition in the Globe." *Shakespeare Quarterly* 56 (2005): 135–55.

Tribble, Evelyn, and Nicholas Keene. *Cognitive Ecologies and the History of Remembering: Religion, Education and Memory in Early Modern England*. Chippenham: Palgrave Macmillan, 2011.

Tully, James. "Governing Conduct: Locke on the Reform of Thought and Behaviour." In Tully, *An Approach to Political Philosophy: Locke in Contexts*, 179–241. Cambridge: Cambridge University Press, 1993.

Turnbull, G. H. *Hartlib, Dury and Comenius: Gleanings from Hartlib's Papers*. London: Hodder and Stoughton, 1947.

———. *Samuel Hartlib. A Sketch of his Life and His Relations to J. A. Comenius*. Oxford: Oxford University Press, 1920.

———. "Samuel Hartlib's Influence on the Early History of the Royal Society." *Notes and Records of the Royal Society of London* 10 (1953): 101–30.

Turner, A. J. "Andrew Paschall's Tables of Plants for the Universal Language, 1678." *Bodleian Library Record* 9 (1978): 346–50.

van Fraassen, Bas. C. *The Empirical Stance*. New Haven, CT: Yale University Press, 2002.

Verney, Frances Parthenhope, and M. M. Verney. *Memoirs of the Verney Family during the Civil War*. 4 vols. London: Longmans, Green and Co, 1892–99.

Vickers, Brian. "The Authenticity of Bacon's Earliest Writings." *Studies in Philology* 94 (1997): 248–96.

Vine, Angus. "Commercial Commonplacing: Francis Bacon, the Waste-Book, and the Ledger." In *Manuscripts Miscellanies c.1450–1700*, ed. Richard Beadle and Colin Burrow, 197–218. Chicago: University of Chicago Press, 2011.

von Uffenbach, Zacharias. *London in 1710 from the Travels of Zacharias Conrad von Uffenbach*. Translated and edited by W. H. Quarrell and M. Mare. London: Faber and Faber, 1934.

Waldow, Anik. "Empiricism and Its Roots in the Ancient Medical Tradition." In *The Body as Object*, ed. Wolfe and Gal, 287–308.

Walker, Obadiah. *Of Education Especially of Young Gentlemen. In Two Parts. The Third Impression with Additions* (1673). Oxford: Printed at the Theatre, 1677.

Wallis, P. J. "An Early Mathematical Manifesto: John Pell's *Idea of Mathematics*." *Durham Research Review* 18 (1967): 139–48.

Walmsley, J. C., and J. R. Milton. "Locke's Notebook 'Adversaria 4' and His Early Training in Chemistry." *Locke Newsletter* 30 (1999): 85–101.

Walton, Izaak. *The Life of Dr Sanderson, late Bishop of Lincoln*. London: Richard Marriott, 1678.

Ward, Seth. *Vindicae Academiarum*. Oxford: Printed by Leonard Lichfield for Thomas Robinson, 1654.

Waterland, Daniel. *Advice to a Young Student with a Method of Study* (1710). 3rd ed. Cambridge, 1760.

Watts, Isaac. *Logic: or, the Right Use of Reason in the Enquiry after Truth* (1725). A new edition, corrected. Edinburgh: Printed by Abernethy and Walker, 1807.

Webb, Beatrice. "The Art of Note-taking" (1926). In Beatrice Webb, *My Apprenticeship*, 2nd ed., 364–70. London: Longmans, Green and Co., 1946.

Weber, Max. "Science as a Vocation." Translated by Michael John. In *Max Weber's Science as a Vocation*, ed. Peter Lassman, and Irving Velody, 3–31. London: Unwin Hyman, 1989. First published as *Wissenschaft als Beruf* (1922).

Webster, Charles. "The Authorship and Significance of Macaria." *Past and Present* 56 (1972): 34–48.

———. "English Medical Reformers of the Puritan Revolution: A Background to the 'Society of Chymical Physitians.'" *Ambix* 14 (1967): 16–41.

———. *The Great Instauration: Science, Medicine, and Reform, 1626–1660*. London: Duckworth, 1975.

———, ed. *Samuel Hartlib and the Advancement of Learning*. Cambridge: Cambridge University Press, 1970.

Weeks, Sophie. "The Role of Mechanics in Francis Bacon's *Great Instauration*." In *Philosophies of Technology: Francis Bacon and his Contemporaries*, 2 vols., ed. C. Zittel, G. Engel, R. Nanni, and N. C. Karafylis, 1:133–95. Leiden: Brill, 2008.

Weinrich, Harald. *On Borrowed Time: The Art and Economy of Living with Deadlines*. Translated by Steven Rendall. Chicago: University of Chicago Press, 2008.

Welch, Mary. "Francis Willoughby, F.R.S. (1635–1672)." *Journal of the Society for the Bibliography of Natural History* 6 (1972): 71–85.

Westfall, Richard S. "Short-Writing and the State of Newton's Conscience, 1662." *Notes and Records of the Royal Society of London* 18 (1963): 10–16.

[Weston, Richard]. *A Discours of Husbandrie used in Brabant and Flanders.* 2nd ed. London: W. Du-Gard, 1652.

Wheare, Degoraeus. *The Method and Order of Reading Both Civil and Ecclesiastical Histories.* Translated by Edmund Bohun. London: Printed by M. Flesher for Charles Brome, 1685.

Whewell, William. *The Philosophy of the Inductive Sciences.* 2nd ed. 2 vols. London: John W. Parker, 1847.

Wierzbicka, Anna. *Experience, Evidence, and Sense: The Hidden Cultural Legacy of English.* Oxford: Oxford University Press, 2010.

Wilding, Nick. "Graphic Technologies." In *Robert Hooke,* ed. Cooper and Hunter, 123–34.

Wilkins, John. *An Essay Towards A Real Character and a Philosophical Language.* London: Samuel Gellibrand and John Martyn, 1668.

Williams, Raymond. *Keywords: A Vocabulary of Culture and Society.* New York: Oxford University Press, 1976.

Willis, John. *The Art of Memory, . . . as it dependeth upon Places and Ideas.* London: W. Jones, 1621.

———. *Mnemonica; or, the Art of Memory.* London: L. Sowersby, 1661.

———. *Mnemonica, sive reminiscendi ars.* London: H. Lownes, 1618.

Wilson, Robert A. "Collective Memory, Group Minds, and the Extended Mind Thesis." *Cognitive Processes* 6 (2005): 227–36.

Wojcik, J. W. *Robert Boyle and the Limits of Reason.* Cambridge: Cambridge University Press, 1997.

Wolf, Christopher, ed. *Casauboniana, sive Isaaci Casauboni varia de scriptoribus librisque judicia . . .* Hamburg: P. Stromeri, 1710.

Wolfe, Charles T., and Ofer Gal, eds. *The Body as Object and Instrument of Knowledge.* Dordrecht: Springer, 2010.

Wood, Anthony. *Athenae Oxonienses: An Exact History of all the Writers and Bishops who have had their Education in the University of Oxford.* A new edition . . . by Philip Bliss. 4 vols. London: Printed for F. C. and J. Rivington, 1813–20. First published in 2 vols, 1691–92.

Wood, Anthony. *The Life and Times of Anthony Wood, Antiquary, of Oxford, 1632–1695, described by Himself.* Edited by Andrew Clark. 5 vols. Oxford: Oxford Historical Society, 1891–1900.

Wood, Paul B. "Methodology and Apologetics: Thomas Sprat's *History of the Royal Society.*" *British Journal for the History of Science* 13 (1980): 1–26.

Woolhouse, Roger S. *Locke: A Biography.* Cambridge: Cambridge University Press, 2007.

Worthington, John. *The Diary and Correspondence of Dr. John Worthington.* Edited by James Crossley. 3 vols. Manchester: Printed for the Chetham Society, 1847–86.

Wotton, William. *Reflections upon Ancient and Modern Learning* (1694). 2nd ed. London: Printed by J. Leake for Peter Buck, 1697.

Woudhuysen, H. R. "Writing-tables and Table-books." *Electronic British Library Journal* (2004), article 3.

Wren, Christopher. *Parentalia: or, Memoirs of the family of the Wrens*. London: Printed for T. Osborn, 1750.

Yale, Elizabeth. "Marginalia, Commonplaces, and Correspondence: Scribal Exchange in Early Modern Science." *Studies in History and Philosophy of Biological and Biomedical Sciences* 42 (2011): 193–202.

———. "With Slips and Scraps: How Early Modern Naturalists Invented the Archive." *Book History* 12 (2009): 1–36.

Yates, Frances A. *The Art of Memory*. London: Routledge and Kegan Paul, 1966.

Yeo, Richard. "Before Memex: Robert Hooke, John Locke, and Vannevar Bush on External Memory." *Science in Context* 20 (2007): 21–47.

———. "Between Memory and Paperbooks: Baconianism and Natural History in Seventeenth-Century England." *History of Science* 45 (2007): 1–46.

———. "Classifying the Sciences." In *Cambridge History of Science*: vol. 4, *Eighteenth-Century Science*, ed. Roy Porter, 241–66. Cambridge: Cambridge University Press, 2003.

———. *Defining Science: William Whewell, Natural Knowledge, and Public Debate in Early Victorian Britain*. Cambridge: Cambridge University Press, 1993.

———. *Encyclopaedic Visions: Scientific Dictionaries and Enlightenment Culture*. Cambridge: Cambridge University Press, 2001.

———. "Ephraim Chambers's *Cyclopaedia* and the Tradition of Commonplaces." *Journal of the History of Ideas* 57 (1996): 157–75.

———. "John Locke on Conversation with Friends and Strangers." *Parergon* 26 (2009): 11–37.

———. "John Locke's 'New Method' of Commonplacing: Managing Memory and Information." *Eighteenth-Century Thought* 2 (2004): 1–38.

———. "John Locke's 'Of Study' (1677): Interpreting an Unpublished Essay." *Locke Studies* 3 (2003): 147–65.

———. "John Locke and Polite Philosophy." In *The Philosopher in Early Modern Europe: The Nature of a Contested Identity*, ed. Conal Condren, Stephen Gaukroger, and Ian Hunter, 254–75. Cambridge: Cambridge University Press, 2006.

———. "Loose Notes and Capacious Memory: Robert Boyle's Note-Taking and Its Rationale." *Intellectual History Review* 20 (2010): 335–54.

———. "Notebooks as Memory Aids: Precepts and Practices in Early Modern England." *Memory Studies* 1 (2008): 115–36.

Young, Robert F., ed. *Comenius in England: The Visit of Jan Amos Komensky . . . to London, 1641–1642*. Oxford: Oxford University Press, 1932.

Zargorin, P. F. *Francis Bacon*. Princeton, NJ: Princeton University Press, 1998.

Zedelmaier, Helmut. "Buch, Exzerpt, Zettelschrank, Zettelkasten." In *Archivprozesse: Die Kommunikation der Aufbewahrung*, ed. H. Pompe and L. Scholz, 38–53. Köln: Dumont, 2002.

———, "De ratione excerpendi: Daniel Georg Morhof und das Exzerpieren." In *Mapping the World of Learning: The Polyhistor of Daniel Georg Morhof*, ed. Françoise Waquet, 75–92. Wiesbaden: Harrassowitz, 2000.

Zedler, Johann Heinrich, ed. *Grosses vollständiges Universal-Lexicon aller Wissenschaften und Künste*. 64 vols. Halle and Leipzig: J. H. Zedler, 1732–50.

Zilsel, Edgar. "The Genesis of the Concept of Scientific Progress." *Journal of the History of Ideas* 6 (1945): 352–49.

Page numbers in boldface indicate detailed discussion of the topic.

27; virtuosi, 57, 60, 63, 65, 70. *See also*
testimony, records of

copying information, xv, 104–5, 116, 156,
186, 216, 255, 279n82, 308n63, 318n137,
323n234. *See also* moving and transferring notes

copyright, 322n216

correspondence: Boyle, Beale and Petty,
xvii, 97, 134–35, 141–43, 145–49, 153–54,
188–89, 209, 211, 255, 299n177, 300n6,
300n13, 304n79, 304n81; collective
note-taking, 221, 225–29, 232–35,
328n66; Hartlib's circle, xvii, 95, 97,
102–4, 106, 108, 112, 123, 128, 148, 150,
255, 298n162, 299n174; Hill, 63; information, xiv–xv, 72, 74; Leibniz, 155;
Locke, 176–77, 183, 188–89, 201, 207–9,
211, 311n7, 312n27; Royal Society, xviii,
90, 220, 235, 259, 299n177; Scientific
Revolution, 219; virtuosi, xiv, 88

Corvinus, Messala, 28, 213

Cosin, John, 59

Coste, Pierre, 175, 214, 324n246

Courten. *See* Charleton (or Courten),
William

Cox, Daniel, 212, 316n82

crafts, 198, 206, 281n117, 293n57. *See also*
trades

Croone, William, 329n94

cross-referencing, xv, 129, 164, 186, 206,
257, 289n115, 313n36

Crusius, Martin, xv, 22, 258

Cudworth, Damaris. *See* Masham family

Cudworth, Ralph, 312n27

Cujas, Jacques, 101

Culpeper, Cheney, 111, 123

Curtius, Sir William, 88

Cutler, Sir John, 226

cyclopedias. *See* encyclopedias and
compendia

daily note-taking. *See* diaries and journals

d'Alembert. *See* d'Alembert, Jean Le Rond

Daniel, Book of, 100

Darwin, Charles, xi, 311n131

Daston, Lorraine, 172, 266n22, 315n74

data, xii, xvi, 71, 105, 127, 134, 172, 234,
242, 245, 247–50, 255, 266n24, 285n12,
334n10. *See also* information

dating. *See* chronological arrangement
and dating

day books. *See* waste books

Daye, John, 47

Dear, Peter, 77

deductive methods: Descartes, 242;
Hartlib's circle, 124–25, 173, 297n150;
Hooke, 250; Locke, 214

Demosthenes, 43

Descartes, René: antibookish rhetoric,
10, 12; Evelyn and, 282n125; information, 5; life's shortness, 290n154;
Locke and, 186, 214; memory, 27, 32,
38–39, 169–70, 214, 242, 323n240;
note-taking, xi, 32, 39; Pell's *Idea*, 125,
128, 299n177, 310n114; philosophical
system, 124, 142, 197, 227, 297n150,
300n7, 333n3

description, 6, 77, 84, 226, 231

desiderata: Boyle, 136, 323n232; collective note-taking, 227; Hartlib's circle,
103–4, 108, 122, 129; Locke, 185; note-
taking, xi; Petty, 299n183; Wren, 190

Devereux, Robert, second Earl of Essex,
280n92, 280n99

diagrams, 94, 169–70, 245–46

diaries and journals: ancients, 271n97;
collective note-taking, 232, 235–37;
Hartlib on, 104, 139; Hay, 189–91;
Hooke, 8, 238, 257; information, xv,
72, 77, 89–90; Locke, 55, 182–83, 187,
189, **195–201**, 202–7, 216, 228, 257,
313n35, 315n80, 316n82, 318n137,
321n198; modern, 273n126; Petty on,
129–30; Scientific Revolution, 220,
258; virtuosi, 1–2, 21 (*see also* Evelyn,
John; Pepys, Samuel). *See also* Boyle,
Robert; Evelyn, John; Hartlib, Samuel:
Ephemerides; notebooks; spiritual
diaries and notes; travel; weather
observations

Dickinson, Edmund, 212

dictionaries, 63, 70, 73–74, 94, 219, 284n4